Stochastic Simulation in Physics

Springer
Singapore
Berlin
Heidelberg
New York
Barcelona
Budapest
Hong Kong
London
Milan
Paris
Santa Clara
Tokyo

Stochastic Simulation in Physics

Springer

P. Kevin MacKeown

P. Kevin MacKeown
Department of Physics
University of Hong Kong
Pokfulam Road
Hong Kong

Library of Congress Cataloging-in-Publication Data

MacKeown, P. K.
Stochastic simulation in physics / P. Kevin MacKeown
p. cm.
Includes bibliographical references and index.
ISBN 9813083263
1. Physics--Data processing. 2. Physics--Mathematical models.
3. Stochastic processes--Data processing. 4. Stochastic processes--Mathematical models. 5. Simulation methods. 6. Monte Carlo method. I. Title.
QC52.M36 1997
530'.01'176--dc21 97-1047
CIP

ISBN 981-3083-26-3

Typesetting: Camera-ready by Author
SPIN 10576338 5 4 3 2 1 0

Contents

List of Figures

Preface

The steadily increasing availability of cheap computer access brought about by the computer age has enormously increased the complexity of problems which can be attacked numerically, whether it be in the field of non-linear classical dynamics or many body quantum systems, and has spawned a whole new discipline of Computational Physics. This branch of physics has established its credentials as a dynamic discipline, complementary to the traditional branches of experimental and theoretical physics. It has broadened the perspective from which many physical problems can be approached, at the same time developing its own methods and concepts. Within the field, perhaps the topic which underwent the most 'inflation' with the rise of the computer is that of stochastic simulation, more colourfully known by its distinguished proponents, Fermi, von Neumann, Metropolis, Ulam and others as Monte Carlo methods. The importance of these methods can be seen from a brief perusal of Physics Abstracts, which reveals how their use has exploded over the recent past.

It is right that these developments be represented in the undergraduate syllabus, and it was in teaching part of a Final Year course on Computational Physics over the past 15 years at The University of Hong Kong that the ideas in this book evolved. The level of knowledge of physics assumed for the most part should correspond to that of a conscientious student at this stage of studies. Apart from the technical interest in the subject as a way of getting quantitative answers to a problem, the subject matter of stochastic phenomena is one which the physics student will benefit from having presented in a variety of contexts, some of which may appear unconventional. Postgraduate students in a number of disciplines should also find at least some of the material of value.

Unlike more mature topics, there is no canon to be followed in the field, topics were chosen because they seem to me to give a feeling for the breadth of applications of Monte Carlo methods in physics. For the same reason there are more references to the research literature than is commonly the case in a textbook. By the very nature of the subject, Monte Carlo methods are invoked in problems involving a high level of complexity. This conflicts with the pedagogical necessity of keeping complications to a minimum in the interests of clarity, and so a compromise has often had to be found. The fact that an example represents a toy model or is treated in just one dimension should not mislead one into underestimating the power of the method in real world problems.

Acknowledgement of permission to reproduce material is due to the Institute of Physics Publishing Limited for Fig. 9.1, excerpted from P. K. MacKeown and D. J. Newman, *Computational Techniques in Physics*, (Bristol: Adam Hilger, 1987), and to Blackwell Publishers and the Royal Statistical Society for Figs 2.3 and 3.4, reproduced from *Applied Statistics*, 1983.

Many students and colleagues over the years have contributed to my education in this field and for this I am very grateful; I would like to mention in particular John Bacon-Shone, Vincent K. W. Cheng, Cheung Kei-Wai, Chu Ki-Cheung, Lee Lap-Chung, Ng Kwai-Kong, Mike Reid and Yeung Hon-Yin for particular enlightenment on some of the topics discussed. Special thanks are due to Chau Hoi-Fung, Vincent K. W. Cheng and Lam Chi-Hang for assistance in the production of the manuscript. It goes without saying that I alone am responsible for such blemishes and confusion which remain.

P. Kevin MacKeown
Hong Kong, March 1997

Chapter 1

INTRODUCTION

> ...a subject that has remained in its infancy for twenty years or more. We are convinced nevertheless that Monte Carlo methods will one day reach an impressive maturity.
>
> Hammersley and Handscomb

1.1 Origins

The epigraph is from Hammersley and Handscomb in the preface to their little gem of a book, *Monte Carlo Methods*, over a quarter of a century ago. In this, as in other things, their judgment was solid; under the classification of 'Monte Carlo Methods' in Physics Abstracts the annual number of articles listed had reached 560 by 1981, increasing over the next decade to over 1800 by 1992. The development of these methods has also grown apace, to the extent that the more general description of *stochastic methods* seems appropriate, although we shall generally refer to them by the traditional title. Starting from tentative and very uncertain beginnings Monte Carlo methods are now established as a technique for solving problems in physics ranging from rapid, brute force estimations of expected performance in the design of experimental apparatus to the probing of very complex questions on the frontiers of quantum field theory.

The solution of problems by numerical methods has always been an important aspect of physics, being required for something as simple as the three body problem in classical mechanics. Numerical methods have become a major rival to traditional analytical techniques in the recent past. As foreseen by von Neumann, they have opened up new avenues for studying linear systems of great complexity, and revolutionised the study of non linear systems. With the advent of massive computing power the computational scientist has established a new role for himself complimentary to the experimentalist and to the theorist, and the techniques and concepts he uses spill over into their thinking. Algorithmic formulation of problems in particular has been raised to the level of respectability of traditional mathematical methods like differential equations. The great increase in computing power brought about by the computer age has enormously increased the complexity of problems which can be attacked numerically, whether it be in the field of non-linear classical dynamics or many body quantum systems, and has spawned a whole new discipline of Computational Physics. Among the techniques developed in this new discipline, stochastic methods have attained considerable importance, as indicated by the statistic quoted. More colourfully known as Monte Carlo methods, they acquired their name — after the famous casino in Monte Carlo — and their modern importance from their use in the Manhattan Project, solving stochastic problems in the development of the atomic bomb. These methods, however, are not restricted to the simulation of physically stochastic systems, but, as we shall see, can also be used to solve problems where no stochastic element is, *a priori*, in evidence. Monte Carlo methods, in fact, form a part of a much wider topic of simulation, e.g. in engineering or operations research — we will confine discussion to problems arising in physics, although many of the solutions will find direct application in others of the physical sciences.

The foundations of the method rely on the fact that behaviour of arbitrary complexity may often be ultimately traceable to very simple rules. It was practiced under the name of 'statistical sampling' — see the book by Shreider — in particular at Los Alamos using slow mechanical or electromechanical devices, much as had been used for years to programme simple manufacturing processes. In fact it had been used by Fermi in Rome in the 1930s to predict results of his experiments on the behaviour of the newly discovered neutron.[1] With the coming into operation of the first electronic computer, the ENIAC at the University of Pennsylvania in 1945, the potential for 'statistical sampling methods' to be rejuvenated was realised by von Neumann and Ulam — for a history of the method by one who was intimately involved in its development, see the account by Metropolis in the book edited by Necia Grant Cooper.

1.2 A Simple Illustration

Stochastic, or Monte Carlo methods can be defined by saying that they employ random numbers to solve a problem, but so concise a definition cannot give the true flavour of the topic, only by way of a concrete example can the nature of the method be grasped. Perhaps the most familiar, and still very illustrative, example of the Monte Carlo method is the 'toy' calculation of the value of π using *Buffon's needle.*[2] This was first described as long ago as 1733 by George Louis Leclerc, later the Comte de Buffon. This illustrates all the most important aspects of Monte Carlo calculations:

- the idea of constructing a stochastic model which has the same answer as the problem one wishes to solve — here the determination of π,
- the concept of estimation and the related aspect of the statistical error, and
- the concept of variance reduction.

If a needle of length l is thrown on to a plane surface ruled with parallel straight lines separated by a distance d the number of lines intersected, ν, can vary between 0 and $1 + \text{int}[L]$, where $L = l/d$. If the throw is totally random, i.e. the centre of the needle is equally likely to fall at any point between the lines and the orientation of the needle is uniformly distributed over the circle, the probability distribution in the quantity $\nu, p(\nu), 0 \leq \nu \leq 1+\text{int}[L]$, can be determined from geometry. The characteristics of this, and other distributions, will be discussed in chapter 2. For the simple case when l is less than the line spacing the only possible values for ν are 0 and 1. In this case, as we shall see,

$$\pi = 2L/\mu,$$

where μ is the mean number of intersections. This mean can be estimated by a Monte Carlo calculation if we have some way of simulating random throws of the needle. This can be done using random numbers and counting the number, n, of intersections which occur in a total of N simulated throws, from which an estimation of μ is n/N. This results in an estimate of π,

$$\overline{\pi}_1 = 2L\frac{N}{n} \pm \frac{\pi}{\sqrt{N}}\sqrt{(\frac{\pi}{2L} - 1)} \tag{1.1}$$

where we have quoted the formal statistical uncertainty, or standard error, on such an estimate, is derived in the next chapter. This latter shows the characteristic $1/\sqrt{N}$ dependence on sample size, which ultimately limits the accuracy with which Monte Carlo estimates can be made. In later chapters we will be concerned with formulating the stochastic models required for the solution of several problems in different branches of physics, woven through which will be a discussion of variance reduction techniques. To set the stage for those topics, we note that our model for estimating π could be improved by superimposing another set of similar lines at right angles. The value of ν can now range from 0 to 2, with a mean $\mu = 4L/\pi$. Repeating the simulation in this case gives the estimate,

$$\overline{\pi}_2 = 4L\frac{N}{n} \pm \frac{\pi}{\sqrt{N}}\sqrt{\frac{(L+2)\pi}{8L} - 1}. \tag{1.2}$$

We note that, for the same sample size N, the statistical uncertainty in this estimate is smaller than in equation 1.1 on the previous page. For the case of $L \to 1$ the improvement in the accuracy is almost a factor of two, with errors of $1.33/\sqrt{N}$ and $2.37/\sqrt{N}$, respectively; this is an illustration of variance reduction, and the importance of exploring different models when applying Monte Carlo methods.

1.3 Scope of the Subject

Our subject naturally breaks down into two parts, foundations (covered in chapters 2, 3 and 5), and applications (covered in the remaining chapters). An integral part of the Monte Carlo solution of any problem is a statistical analysis of the output, which in turn requires a suitable understanding of probability theory. Probability and Statistics form a very large topic. In chapter 2 we highlight concrete results which will be required, time and time again, in later chapters. With this background we are in a position to look into the topic of singular importance in Monte Carlo methods, the provision of 'high quality' random numbers from the uniform distribution. From this requirement, whole new insights into randomness and random numbers have developed. In chapter 3 we look at the several aspects of this, and in chapter 5 describe methods for the selection of random variates from an arbitrary distribution. This chapter also includes a description of Markov chains, a topic of great practical importance in the selection of multivariate configurations, a requirement which occurs in several places throughout the book.

The chapters on applications attempt to show the wide spectrum of problems in physics amenable to Monte Carlo solution, ranging from the almost mathematical (e.g. properties of exotic random walks), to the down to earth practical (such as radiation shielding or reactor criticality calculations). In some cases the methods throw new light on phenomena already well understood from traditional viewpoints, e.g applications of stochastic methods in solving Schrodinger's equation. In other cases stochastic methods have played a vital role in the development of the ideas, as in the phenomenon of self-organised criticality. Because of their generality, they have also facilitated cross-fertilisation between disciplines, between genetics and optimisation in the case of genetic algorithms (chapter 4), and between thermodynamics and optimisation in the case of simulated annealing (chapter 6). Among other topics discussed are percolation (chapter 4), variational integral calculations (chapter 7), polymer configurations (chapter 8), particle transport (chapter 9), statistical mechanics (chapter 10) and the solution of quantum problems (chapter 11). A number of exercises are appended to the chapters which illustrate ideas developed there, and, in some cases, suggest extensions of the topics described.

Monte Carlo methods should only be resorted to for certain problems, when more direct methods are incapable of giving an adequate solution. Perhaps the most powerful advantage of Monte Carlo methods is that almost arbitrary complexity can be introduced into the model under study. For obvious reasons this advantage cannot be illustrated in our examples, which must always be seen as zeroth order attempts to study the system in question. One must not be mislead by the simplicity, in some cases childishness, of the examples described, as this is necessary for a pedagogic presentation. Although they may be unrealistic, the techniques developed are directly transferable to more complex, realistic, problems.

Although in the nature of a textbook, the sources of much of the material is of fairly recent vintage and it cannot be regarded as part of the canon of physics. Rather than interrupt the text, however, with the relevant references, I have left to the end of each chapter a *vade mecum* for the material contained therein. There is a short bibliography of standard texts and reference books, which are referred to throughout by author, at the end of the book. As for equation numbering, equation 2 in section 4 of chapter 8, for example, will be referred to in that chapter as equation 4.2, but elsewhere as equation 8.4.2. With the hope that people of different interests may dip into this book I have tried to make some chapters more or less self-contained; this has meant a small amount of repetition in places, e.g. the all powerful Metropolis algorithm appears in several different contexts.

In practical computations other routines in addition to the purely Monte Carlo algorithms will be necessary. A few of the most intimately related are described (the Hoshen-Kopelman cluster counting algorithm in chapter 4, the Swendsen-Wang algorithm for multi-spin transitions in chapter 6), but most have to be written, or accessed from a library. The compilation by Press *et al.* will often be found invaluable in this context, while the book by Wong is a useful presentation of the underlying numerical methods. All fragments of code quoted are in the FORTRAN language.

1.4 Sources of Errors

Monte Carlo programmes usually are relatively simple, and problem specific, and more likely to be constructed by the user than abstracted from a programme library — there being, of course, exceptions. For example, in the field of particle transport complex code incorporating large banks of data will be found. Such hands-on involvement means that general principles of programming need to be familiar, e.g. the economic evaluation of functions called in great volume. Here we only remind the reader of intrinsic limitations in accuracy of a solution obtained by computing, i.e. sources of error. *Systematic errors*, naturally, can occur if improper estimators, e.g. biassed estimators due to the use of poor quality random numbers, are used. Unlike in most numerical methods, systematic errors in Monte Carlo calculations can often be a cause for concern, especially in calculations which make use of Markov chain methods of variable selection. For this reason anomalies have appeared in the literature which can now be seen to reflect non-ergodic behaviour in the simulations due to insufficient exploration of the state space during sampling. It is very difficult to give blanket advice on this aspect, particular cases will be noted in later chapters. As we have already indicated, *statistical errors* are an inescapable feature of the method, we defer discussion of these until the next chapter. There is, in addition, *computational error*, which can be further analysed in terms of two sources, *round-off error* and *truncation error*, and we briefly describe these here.

1.4.1 Round-off Error

These are a consequence of an approximation in machine evaluation of the result due to the limited accuracy with which numbers are stored and manipulated by the computer. Non-integers are stored in a normalised floating point representation, generally to a base β, in the form,

$$0.d_1d_2....d_k\beta^m \qquad (d_1 \neq 0, d_i < \beta)$$

where k, the word size, is the number of bits devoted to storing a number, and m is an integer. For illustration, if we take the base $\beta = 8$ we have

$$\left(9.15 \right)_{10} = \left(11.114631463146\ldots\ldots. \right)_8$$

$$= 0.11114631463146......8^2.$$

We see that a number exactly represented in one base cannot be so represented in another using the finite number, k, of digits available. Curtailing the representation to k digits yields a number different from that intended. If, in the same spirit, we take $k = 6$ here, the number processed by the programme is

$$0.111146.8^2 \cong (9.1499023438)_{10}.$$

We see that the totality of numbers in a finite interval accessible in the machine is finite. In addition, these numbers are not evenly spaced, or dense; the separation between adjacent numbers is β^{m-k}, so that with fixed k the absolute separation between numbers depends on m, i.e. on the magnitude of the numbers themselves, being smallest between small numbers. Most likely the machine will operate such that during execution the element amongst them which is just less than the number we wish processed will be used. Because of the non-identity of the stored numbers with those of the theory, results will not be exactly equal to those from an ideal computer. For example commutativity of addition will fail in general, i.e.

$$a + b + c \neq c + b + a.$$

For a binary computer with 16- or 32-bit word storage the effect of round-off in any individual operation is not likely to be a serious cause for concern. However the accumulative effect in some extended calculations may lead to erroneous results. If these calculations arose from simulation of a model system the effect might be a manifestation of an irreversibility not inherent in the model itself. Because of the decreasing density of numbers accessible, the computed difference between two large numbers will become less and less distinguishable as the sizes of these numbers increases, ultimately vanishing when the size of the quantised spacing between becomes larger than their difference. Hence taking the difference between large numbers of similar magnitude, i.e. differing only in the least significant digits of their rounded representation, may give rise to a large relative error, and is to be avoided. Just the opposite applies to addition; round-off error will be minimised if priority is given to combining quantities of similar magnitude.

Monte Carlo methods themselves, indeed relaxation methods in general, are quite insensitive to this feature of computation. Round-off can be thought of as adding a little bit more stochasticity to an environment already awash in it. Progressive build-up on earlier operations is not a strong feature, there is often great independence between adjacent steps in the calculation. However, statistical analysis is another matter! A popular way of determining the variance of a sequence of variates $\{x_i\}$ is to take the difference between their mean square and the square of their mean value, i.e.

$$< x^2 > - < x >^2 .$$

In Monte Carlo work these sequences may be very long, and the two sums may suffer from round-off error in their accumulation, and so be poorly determined. Their relatively small difference may then be a hopeless estimate of the variance. In section 2.3.2 we shall discuss the suppression of round-off error in cases like this.

1.4.2 Truncation Error

Truncation error is a consequence of the method used to cast the original problem into one suitable for solution by numerical methods. Examples are the necessary truncation of an infinite series expansion of a function by summing only a finite number of terms, or expressing the derivative of a function at a point in terms of a number of values of the function at discrete points in the neighbourhood of the point. Again truncation error does not very often feature in Monte Carlo methods as such, which usually converge to the solution of the problem posed. Such will be the case, for example, when analogue simulation of the original problem is faithfully carried out. Some techniques may, however, involve truncation error — we have in mind discrete random walks with finite step sizes, although, sophisticated methods may sometimes be employed to overcome this. We note that truncation error may be a feature of the formalism in which the problem was cast before Monte Carlo methods were brought to bear on its solution, e.g. by the use of Finite Difference methods.

Notes

1. Wilson, R. R. (1952), *Monte Carlo Study of Shower Production*, Phys. Rev., **86**, 261-69.

2. Further elaboration on Buffon's Needle, and Laplace's modification thereof, can be found in Mantel, N. (1953), *An extension of the Buffon Needle problem*, Am. Math. Stat., **24**, 674-77, and Perlman M. D. and M. J. Wichura (1975), *Sharpening Buffon's Needle*, Am. Stat., **29**, 157-63.

Chapter 2

STATISTICAL ASPECTS

> Nothing shows a foolish mind more than the attempt to account for any phenomena on the principle of chance or supernatural intervention.
>
> Oscar Wilde

2.1 Random Variables

A *random variable* in physics can be most clearly appreciated by contrasting it with a quantity which is not random, e.g. the charge on the electron. Random variables arise when, for example, repeated attempts under apparently identical conditions to measure a quantity fail to give the same result, within the measurement errors. Examples are the number of hours of sunshine in a day, the distance a 10 MeV neutron will travel in water before it is absorbed, or the reading that will be obtained when a pair of dice are tossed. A possible value for quantities such as these, which we will denote by x, although not unique, is not unconstrained; rather it is distributed in a way which can often be approximated by a simple mathematical function. For physicists, the most natural way to characterise this distribution is in terms of the *probability density function* (or *frequency function*), $p(x)$. This function, the pdf, defines the probability that, when the variable is sampled, a value lying in the range x to $x + dx$ will result; this probability is

$$p(x)dx. \tag{1.1}$$

If the variable is discreet, $p(x_i)$ represents the probability of obtaining the value x_i. As an example, five samplings — x_1, x_2, x_3, x_4, x_5 — of a random variable, might be the output voltage of a scintillation detector assembly produced by successive traversals of it by a charged particle, e.g. cosmic ray muons with the same energy, and might assume the values 0.22, 0.18, 0.21, 0.22, 0.27 volts. Viewing a quantity as a random variable is often, as in the case just quoted, a means of bypassing the complex physical processes underlying the quantity and at the same time setting up a framework for treating the outcome of such processes. Among the random variables most often encountered we have the time of occurrence, or time separation, of *random events.* Normalisation of probability imposes the conditions

$$\int_{-\infty}^{\infty} dx p(x) = 1, \qquad \text{or} \qquad \sum_i p(x_i) = 1.$$

Common practice among physicists, in contrast to mathematicians, is not to distinguish between the name of a variable and the symbol for the value it assumes in a particular case, both denoted by x above. When necessary, e.g. if we are referring to the value of one pdf at a particular value of its variable equal to 3.3, we shall write $p_X(3.3)$, where X is the name of the variable. Mathematical usage also differs in laying emphasis on the *cumulative probability distribution,* which we denote by $P(x)$. This is the probability of a sampling yielding a value less than a given x; it is related to the pdf as

$$P(x) = \int_{-\infty}^{x} dx p(x) \qquad \text{or} \qquad P(x_j) = \sum_{i=1}^{j} p(x_i).$$

As a function of x, $P(x)$ is either constant or monotonically increasing, and in value ranges between zero and one. The mathematician's preference arises because $dP(x) = p(x)dx$, i.e. $dP(x)$ represents the increment of probability and in this form is of more general applicability than equation 1.1 — we will usually be able to get along using the physically more intuitive probability density $p(x)$. When dealing with a sequence of values of a random variable it is important to know whether the individual values were sampled independently from the distribution, or whether the selection mechanism imposed some correlation among the values in the sample.

A specified function of one, or more, random variables, say $y(x)$, is also, obviously, a random variable and will have a pdf associated with it.

A random variable can have more than one component, e.g. $(\sqrt{1/3}, 0, \sqrt{2/3})$ is a single sample of a unit direction vector in space. More generally we can consider a random variable which is a function of a parameter, notably the time, in which case we can talk of a *random function*, $x(t)$ — not to be confused with an ordinary function of a random variable, such as $p(x)$. Our discussion earlier might have referred to the possible values of such a function at one given time. The x-coordinate of a particle undergoing Brownian motion would be an illustration of such a random function over a finite time interval. We defer discussion of such functions until section 8.3.3.

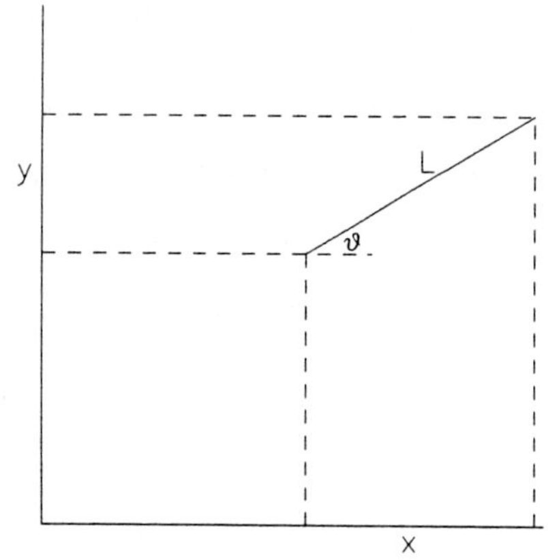

Figure 2.1: Correlated random variables, x and y, defined by a random throw of a needle.

2.2 Distributions

Whatever the origin of a random variable, or vector or function, we consider any one sample of it as being abstracted from a distribution of possible values, e.g. $p(x)$, and imagine that that distribution could be established, e.g. in the form of a histogram, by gathering a sufficiently large number of samples, either sequentially or from an ensemble of identically prepared sources of the variable. The distribution of the variable will depend on one or more parameters. It may also depend on the value of one or more other variables which are themselves random variables and can only be specified in terms of their pdf rather than precise values — we then have to deal with a multivariate distribution.

2.2.1 Simple Distributions

Restricting ourselves first to the case of a single variate, for purposes of illustration we will consider the distribution of a quantity x, namely the projection of a needle of length L on to the x-axis when it is thrown randomly on to a page, as illustrated in Fig. 2.1. Although this is a rather empty quantity in a physical sense, its consideration is of some historical interest, since it is just the quantity considered by the Count du Buffon in his suggestion in 1777 for estimating the value of π by Monte Carlo methods described in the previous chapter. From the figure we see that there is a one to one relationship between x and ϑ lying in one quadrant, i.e. $0 \leq \vartheta \leq \pi/2$, with $x = L\cos\vartheta$. The assumption of random throwing implies an isotropic distribution of ϑ, so we have

$$p(\vartheta)d\vartheta = \frac{2}{\pi}d\vartheta, \qquad 0 \leq \vartheta \leq \frac{\pi}{2}.$$

In terms of x this gives

$$p(x)dx = \frac{2}{\pi}\frac{dx}{\sqrt{L^2 - x^2}}, \qquad 0 \leq x \leq L. \tag{2.1}$$

The corresponding cumulative probability distribution is

$$P(x) = \frac{2}{\pi}\sin^{-1}\frac{x}{L}.$$

These functions are shown in Fig. 2.2 for the case of $L = 1$.

The expectation of a function of a random variable, $z(x)$, which we denote by $\mathbf{E}(z)$, is the weighted value of that function when the variable is drawn from the distribution $p(x)$, i.e.

$$\mathbf{E}[z] = \int dx\, z(x)p(x).$$

The expectation of the variable itself, $\mathbf{E}[\mathrm{x}]$, and of its square, $\mathbf{E}[\mathrm{x}^2]$, also known as the first and second moments of the distribution, are

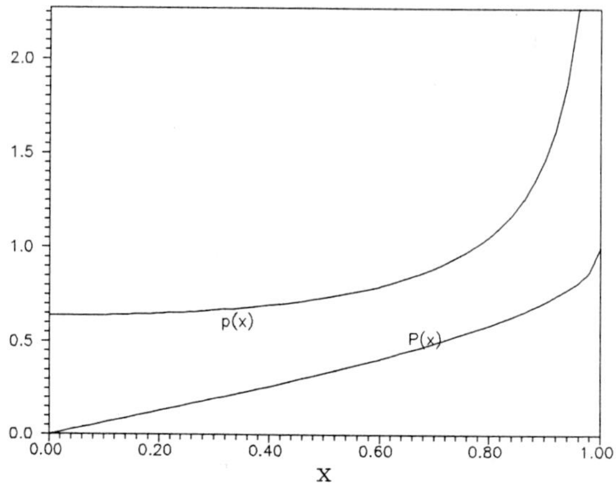

Figure 2.2: The probability distribution $p(x)$ given by equation 2.1, and its associated cumulative distribution $P(x)$.

$$\mathbf{E}[x] = \int dx x p(x), \tag{2.2}$$

and

$$\mathbf{E}[x^2] = \int dx x^2 p(x). \tag{2.3}$$

The first moment is known as the *mean* of the distribution and often denoted by μ, i.e. $\mu \equiv \mathbf{E}(\mathrm{x})$; this parameter may or may not appear explicitly in the form of the distribution. For our example, equation 2.1, we easily find $\mu = 2L/\pi$. The *variance* of the distribution, σ^2, is defined as

$$\sigma^2 = \mathbf{E}[x^2] - \mu^2,$$

and is the averaged squared deviation from the mean, i.e.

$$\sigma^2 = \int dx (x - \mu)^2 p(x). \tag{2.4}$$

The square root of the variance is known as the *standard deviation* of the distribution, σ; it is a characteristic measure of the spread in values of the variable occurring in the distribution. In the case of our example we have

$$\sigma^2 = \frac{L^2}{\pi^2}\left(\frac{\pi^2}{2} - 4\right).$$

We note in passing that it is possible to construct normalised probability distributions for which the mean and variance don't exist or are divergent.

2.2.2 Joint Distributions

More than one random variable may enter into a distribution, in which case we have to deal with the joint distribution of the variables or the distribution of a function of a variable or some combination of variables. If two random variables are involved in a distribution, say x and y, the associated distribution $p(x, y)$ — $p(x, y)dxdy$ being the probability of one variable lying in dx at x *and* the other lying in dy at y — gives rise to means μ_x, μ_y and variances σ_x^2, σ_y^2 in a wholly analogous way with equations 2.2 and 2.4. For example,

$$\sigma_y^2 = \int dxdy(y - \mu_y)^2 p(x, y).$$

Continuing with our example of Buffon's needle, we can consider the projection on to the y-axis as a second random variable, y (see Fig. 2.1); the joint probability in this case can then be written as

$$p(x, y)dxdy = \frac{2}{\pi}\frac{\delta(y - \sqrt{L^2 - x^2})}{\sqrt{L^2 - x^2}}dxdy, \tag{2.5}$$

where we have used the Dirac delta function to impose the unique correlation between the two quantities which exists in this case. A further parameter can be identified in this case, the *covariance*, which is a measure of correlation between the two variables, it is defined by

$$\text{cov}[x, y] = \int dxdy(x - \mu_x)(y - \mu_y)p(x, y) \tag{2.6a}$$

$$= \mathbf{E}[xy] - \mu_x\mu_y. \tag{2.6b}$$

For our example distribution, equation 2.5, using the property that

$$\int dx f(x)\delta(x - x_o) = f(x_o),$$

we find

$$\mathrm{cov}[x, y] = -(4 - \pi)L^2/\pi^2.$$

The *correlation coefficient*, ρ, is defined by

$$\rho = \mathrm{cov}[x, y]/(\sigma_x^2.\sigma_y^2)^{1/2},$$

with values lying in the region $-1 \leq \rho \leq 1$. For the distribution equation 2.5 the correlation coefficient is negative and equals $-(8-2\pi)/(\pi^2-8) = -0.918$, reflecting the very strong negative correlation between the two projections. The covariance vanishes, and with it the correlation coefficient, if the two variables are independent, i.e. if $p(x, y)$ is such that, for example,

$$\int dy(y - \mu_y)p(x, y) = 0.$$

Vanishing covariance among two variables is a necessary condition for them to be independent but is not a guarantee of such independence.

The pdf for a quantity z which is a monotonic function of x, is, from the definition of a pdf, given by the relation

$$p_Z(z) = p_X(x(z))/J, \tag{2.7}$$

where $J = \left| \, dz/dx \, \right|$, or more generally the Jacobian of the transformation. The mean of this distribution, i.e. the expectation of z, is given on page 19; it can also be written as

$$\mathbf{E}[z] = \int dx z(x)p(x) = \int z(x)dP(x).$$

The variance of the variable z is also defined as before, i.e.

$$\begin{aligned}\text{var}[z] &= \mathbf{E}[z^2] - (\mathbf{E}[z])^2 \\ &= \int dz (z - \mathbf{E}[z])^2 p(z) \\ &= \int dx (z(x) - \mathbf{E}[z])^2 p(x).\end{aligned}$$

2.2.3 Distribution of a Combination of Variates

Of frequent interest is the case where z is the sum of two random variables, x and y, which may be from the same or from different distributions, $z = x+y$. In this case $\mathbf{E}[z] = \mu_x + \mu_y$, and, as we show later,

$$\text{var}[z] = \sigma_x^2 + \sigma_y^2 + 2 \text{ cov}[x,y]. \tag{2.8}$$

The actual distribution in z depends of course on the distributions of x and y, and is considered in the following section.

Taking $z = x + y$ where x and y are the two projections of Buffon's needle, equation 2.8 gives for the variance on their sum, using the results quoted earlier, $\sigma_z^2 = [\pi(\pi + 2) - 16]L^2/\pi^2 = 0.153L^2/\pi^2$, which is much less than the variance, $0.935L^2/\pi^2$, on either projection individually. If we are interested in the mean projection, $< z > /2$, it has a variance of 1/4 of the quoted value, i.e. $0.038L^2/\pi^2$, 25 times smaller than on an individual projection. This ability to use negatively correlated quantities to construct a quantity with reduced variance can play an important role in efficient Monte Carlo estimation.

We can extend our discussion to a sum of more than two variables, say N variables, x_i. Of particular interest is the case where all the x_i are from the same distribution, characterised by an expectation $\mathbf{E}[x_i] = \mu$ and variance $\mathbf{E}[x_i^2] - \mu^2 = \sigma^2$. We consider $z/N = N^{-1}\sum_1^N x_i$, and denote it by m. Then

$$\mathbf{E}[m] = N^{-1}\sum_1^N \mathbf{E}[x_i] = \mu.$$

For a given sample size N, particular values of m will occur with some distribution around this mean value and it is the variance on this distribution in m which is our prime interest here. By definition, this variance is

$$\begin{aligned}\sigma_m^2 &= \mathbf{E}[m^2] - \mu^2 \\ &= \mathbf{E}\left[\ \tfrac{1}{N^2}(\textstyle\sum_1^N x_i)^2\ \right] - \mu^2\end{aligned}$$

$$= \mathbf{E}\left[\ \frac{1}{N^2}\sum_1^N x_i^2 + \frac{2}{N^2}\sum_{i=1}^{N-1} x_i \sum_{j=1}^{N-i} x_{i+j}\ \right] - \mu^2$$

$$= \frac{1}{N}\{\mathbf{E}[\ < x^2 >_N + \frac{2}{N}\sum_{i=1}^{N-1} x_i \sum_{j=1}^{N-i} x_{i+j}\ ] - N\mu^2\},$$

where $< x^2 >_N = N^{-1}\sum x_i^2$, and its expectation, which is just the expectation of x^2, is independent of N and equals $\sigma^2 + \mu^2$. So

$$\sigma_m^2 = \frac{1}{N}\left\{\sigma^2 + \mu^2 + \frac{2}{N}\sum_{i=1}^{N-1}\sum_{j=1}^{N-i}\mathbf{E}[x_i x_{i+j}] - N\mu^2\right\}.$$

By switching the dummy indices i and j in the sums, assuming that $\mathbf{E}[x_j x_{j+i}]$ does not depend on j so that $\mathbf{E}[x_j x_{j+i}] = \mathbf{E}[x_i x_{i+j}]$ — the assumption of equilibrium or stationarity — and, via equation 2.6b, that $\mathbf{E}[x_i x_{i+j}] = \text{cov}[x_i, x_{i+j}] + \mu^2$, we have

$$\sigma_m^2 = \frac{1}{N}\left\{\sigma^2 + \frac{2}{N}\sum_{j=1}^{N-1}(N-j)\{\text{cov}[x_i, x_{i+j}] + \mu^2\} - (N-1)\mu^2\right\}$$

$$= \frac{1}{N}\left\{\sigma^2 + 2\sum_{j=1}^{N-1}\left(1 - \frac{j}{N}\right)\ \mathrm{cov}[x_i, x_{i+j}]\right\}.$$

In terms of the correlation coefficients, ρ_j, we have

$$\sigma_m^2 = \frac{\sigma^2}{N}\left\{1 + 2\sum_{j=1}^{N-1}(1 - j/N)\rho_j\right\}, \tag{2.9a}$$

where the j-separation correlation coefficient is defined by,

$$\begin{aligned}\rho_j &= \mathrm{cov}[x_i, x_{i+j}]/\sigma^2 \\ &= (\mathbf{E}[x_i x_{i+j}] - \mu^2)/\sigma^2,\end{aligned} \tag{2.9b}$$

and is independent of i. From the point of view of Monte Carlo work, equation 2.9a is probably the most important formula in statistics; it gives the variance on the distribution of the average, m, of a sample of N variates in terms of the variance of the distribution from which they arise and the correlations between the variates in their sampling.

If we are lucky, the variates in a sequence of interest are independent, in which case we can assume all ρ_j are zero, and we find the familiar result,

$$\sigma_m^2 = \sigma^2/N. \tag{2.10}$$

The pdf for the quantity m, characterised by these values of μ and σ_m^2, depends of course on N, as well as on the distribution from which the x_i derive; in the case that N is not too small, and the x_i are independent, we will see in section 2.3.4, that a universal distribution often applies to m, i.e. one which is independent of the actual distribution of the x values.

In much Monte Carlo work, however, the non-independence of a sequence of variates is not always avoidable, e.g. we shall see in chapter 5 that this is the case when Markov chains are used to select variates, nor even always desirable, as will be discussed in the case of correlated sampling described in chapter 7. In such cases the more general result, equation 2.9a, will be relevant. We will see that it can be written as $\sigma_m^2 = \sigma^2/N'$ where N' is an effective or reduced sample size determined by the correlations.

Equation 2.9 is still very much a statistician's result, all quantities involved refer to expectation values for the population, obtainable in principle only on the basis of infinite samples. It is however the starting point in considering estimation on the basis of finite samples, a topic we consider in detail in section 2.3.

2.2.4 Distribution of Functions of Variates

In the most general case we are interested in the joint distribution of M variables, z_j, each of which, in turn, is a function of n random variables x_i, i.e. $z_j = z_j(x_1, \ldots, x_n), j = 1, \ldots, M$, although in practice the case of $M = 1$ will be of most importance. By introducing the Dirac delta function, something of which statisticians would be wary but which we will get away with, and noting its property that if $f(x)$ is a function with only one root, x_0, in the interval of integration,

$$\delta(f(x)) = \delta(x - x_0) \left| (df/dx)_{x_0} \right|^{-1}, \tag{2.11}$$

we see that equation 2.7 can be written in the alternative form

$$p_z(z) = \int dx p_x(x) \delta(z - z(x)). \tag{2.12}$$

Equation 2.12 is a special case of a more general relation which gives the distribution of functions of a set of random variables in terms of the distributions obeyed by these variables.[1] If there are M such functions, $z_i(x_1, \ldots, x_n)$, of n random variates, x_i, where the variates x_i have a joint probability distribution $p(x_1, \ldots, x_n)$, then the joint distribution of the variables z_i is given by

$$p(z_1, ..., z_m) = \int dx_1...dx_n p(x_1, ..., x_n) \prod_{i=1}^{M} \delta(z_i - z_i(x_1, ..., x_n)). \qquad (2.13)$$

We consider three cases of composition of two variables in this way, with $M = 1$; these give rise to useful results which may be used to select variates from a distribution after a method to select from a simpler distribution has been found, as described later in section 5.2.2. When z is the sum of two variables, x and y, we find its distribution,

$$\begin{aligned} p_Z(z) &= \int dx dy p(x, y) \delta(z - (x + y)) \\ &= \int dx p(x, z - x). \end{aligned}$$

When the two variables, x and y, are independent this just becomes the convolution of the two distributions,

$$p_Z(z) = \int dx p_X(x) p_Y(z - x). \qquad (2.14)$$

For the case of the product of two variables, e.g. $z(x, y) = xy$, we have from equation 2.13

$$\begin{aligned} p_Z(z) &= \int dx dy p(x, y) \delta(z - xy) \\ &= \int dx \int dy p(x, y) \delta(f(y)), \end{aligned}$$

with $f(y) = z - xy$. Using equation 2.11 this becomes

$$p_Z(z) = \int dx \mid x \mid^{-1} p(x, z/x).$$

Again, if x and y are independent we find

$$p_Z(z) = \int dx \mid x \mid^{-1} p_X(x) p_Y(z/x). \tag{2.15}$$

We can, in an analogous way, find the distribution of a quotient, e.g. $z = y/x$ for which the distribution is

$$p_Z(z) = \int dx \mid x \mid p_X(x) p_Y(zx). \tag{2.16}$$

In general the distribution that a physical variable satisfies will not be as obvious as in the case of the thrown needle, but will have to be determined empirically. Often, general principles such as the Central Limit Theorem, to be discussed later, or previous experience, will suggest a parameterised form for the relevant distribution, without however yielding values for its parameters, such as the mean and variance.

2.3 Estimation

2.3.1 Estimating Mean and Variance

We view any sample of the variable as being abstracted from a population conceived of as being characterised by unique values of μ and σ^2, and the nearest we can come to knowing the values of these quantities is to derive estimates of them on the basis of the finite sample available. The available sample may contain $N (N \geq 1)$ values, and we must distinguish between the estimates for the mean, variance and correlation coefficients made on the basis of this sample, which we denote by m, s^2 and r_j, respectively, and the corresponding quantities for the infinite underlying population. Our estimates should be unbiased, i.e. for very large samples the estimates based on the sample of size N should converge to the population parameters. A natural estimator of μ is the sample mean, m,

$$m = < x > = \frac{1}{N}\sum_{1}^{N} x_i; \tag{3.1a}$$

this can be shown to be unbiased. As a natural estimator for σ^2 it is logical, by analogy with equation 2.4, to propose the following expression,

$$\overline{\sigma}^2 = \frac{1}{N}\sum_{1}^{N}(x_i - m)^2 = < x^2 > -m^2, \tag{3.1b}$$

where $< x^2 > = N^{-1}\sum x_i^2$.

This quantity, however, is biased — essentially because the estimate, m, rather than the, unknown, μ is used in its definition. In general such usage gives rise to underestimates of the corresponding parameters of the population, and they must be corrected, as described later, for this bias. In the presence of correlations this correction for the bias is quite complicated. If, however, we can assume no significant correlations in the sample, by adding and subtracting μ in each term in equation 3.1b we have

$$\begin{aligned} \overline{\sigma}^2 &= N^{-1}\sum_{1}^{N}[x_i - \mu - (m - \mu)]^2 \\ &= N^{-1}\sum(x_i - \mu)^2 - 2(m - \mu)N^{-1}\sum(x_i - \mu) + (m - \mu)^2 \\ &= N^{-1}\sum(x_i - \mu)^2 - (m - \mu)^2. \end{aligned}$$

By taking expectation values, i.e. averaging over an infinity of samples of size N we have

$$\begin{aligned} \mathbf{E}[\overline{\sigma}^2] &= \mathbf{E}[N^{-1}\sum(x_i - \mu)^2] - \mathbf{E}[(m - \mu)^2] \\ &= \sigma^2 - \sigma_m^2. \end{aligned}$$

When there are no correlations we have, equation 2.10, $\sigma_m^2 = \sigma^2/N$, so s^2 is an unbiased estimate of σ^2, i.e.

$$\mathbf{E}[s^2] = \sigma^2,$$

where

$$s^2 = \frac{N}{N-1}\overline{\sigma}^2 = \frac{N}{N-1}(< x^2 > -m^2). \tag{3.2a}$$

Another, equivalent, expression for s^2 is

$$s^2 = \sum(x_i - m)^2/(N-1). \tag{3.2b}$$

These last two formulae, although of great importance, presuppose, as has been emphasised, that the individual values have been randomly sampled from the underlying population, i.e. there is no significant correlation among them. This assumption could be investigated by estimating the correlation coefficients ρ_j. If some of these coefficients are significant, in a sense which is discussed in section 2.3.3, s^2 as defined above will not be an unbiased estimate of the variance of the population. The estimation of the correlation coefficients and of the variance in this case which takes account of the finite correlations is described in section 2.3.3.

2.3.2 Practical Considerations

A note on the practical calculation of s^2 in equation 3.2 is necessary. From the point of view of speed, equation 3.2*a* is obviously faster than equation 3.2*b*, since the sums required to calculate the sample mean, m, and mean square, $< x^2 >$, may be carried out in parallel, during the course of a simulation or sampling. However both of these quantities may be numerically very large, depending on the size of the sample and the magnitudes involved.

A round-off error may then exist in one or both of these quantities, due to the limited accuracy with which numbers are stored in the computer resulting in failure to accommodate some of the least significant digits in a number. Just such digits may be significant in the difference between two equally large quantities, as $< x^2 >$ and m^2 may be, so that the relative error in the estimate of their difference may be quite significant. This error will not occur if equation 3.2*b* is used, in which quantities of relatively small, and similar, magnitudes are being subtracted. However in this case two passes will have to be made through the data, the first using equation 3.2*a* to calculate m and then using equation 3.2*b* — this will not only be slower but will require all the data set $\{x_i\}$ to be held in storage. One method of compromising between these two approaches is based on updating; it is not as slow as equation 3.2*a* but is in general more accurate than equation 3.2*b*.[2] As the data becomes available, sums $s_1(k)$ and $s_2(k)$ are formed, which at any stage have the values,

$$s_1(k) = \frac{1}{k}\sum_1^k x_i, \qquad s_2(k) = \sum_1^k (x_i - s_1(k))^2.$$

Starting with $s_1(0) = s_2(1) = 0$, these are updated using datum x_{k+1} as follows

$$\begin{aligned} s_1(k+1) &= s_1(k) + (x_{k+1} - s_1(k))/(k+1), \\ s_2(k+1) &= s_2(k) + (k+1)(x_{k+1} - s_1(k+1))^2/k. \end{aligned}$$

These two quantities tend to m and $(N-1)s^2$, respectively. The subroutine CORREL, listed in the Appendix, processes data according to this formula.

2.3.3 Estimation with Correlated Samples

In general if the ith value in a sequence forming an infinite sample from a distribution $p(x)$ has a specified value x_i, the probability distribution of the $(i+j)$th value is not simply given by $p(x_{i+j})$ as in the case of random sampling, but depends on j in some way. We will assume, however, that it does not depend on i, i.e. at all stages in the sequence the variables are being selected from the same $p(x)$. In time series analysis this is known as the stationarity condition, in our usage in later chapters it will correspond to variables being selected from a distribution which represents configurations of a system in equilibrium. Our primary interest will be in estimating the standard error on the mean of a sample; formally this is just the square root of the variance given by equation 2.9a. However, because the theoretical quantities ρ_j involve a quotient of two quantities each of which needs to be estimated, a covariance and a variance, its estimation is not so straightforward. For that reason we will find it helpful to rewrite equation 2.9a in terms of unnormalised coefficients ρ_j' defined by

$$\rho_j' = \frac{2(N-j)}{N^2} \operatorname{cov}[x_i, x_{i+j}].$$

Equation 2.9 then becomes

$$\sigma_m^2 = \frac{\sigma^2}{N} + \rho, \tag{3.3a}$$

where a full correlation coefficient ρ has been defined,

$$\rho = \sum_{j=1}^{N-1} \rho_j'. \tag{3.3b}$$

Conventionally, equation 3.3a is written in terms of a *correlation length*, τ,

$$\sigma_m^2 = \frac{\sigma^2}{N}\,(\,1 + 2\tau\,)\,, \tag{3.3c}$$

with the quantity τ defined by

$$\tau = N\rho/2\sigma^2. \tag{3.3d}$$

The equivalent number of independent data points, $N' = N/(1+2\tau)$, is less than N in general.

Natural estimators for ρ_j' and ρ suggest themselves, namely

$$\begin{aligned}\overline{\rho}_j' &= \frac{2}{N^2}\sum_{i=1}^{N-j}(x_i - m)(x_{i+j} - m) \\ &= \frac{2(N-j)}{N^2}(< x_i x_{i+j} > -m^2),\end{aligned} \tag{3.4}$$

and

$$\overline{\rho} = \sum_{j=1}^{N-1} \overline{\rho}_j'. \tag{3.5}$$

Subroutine CORREL in the Appendix calculates estimates of the mean and variance, m and s^2, and the correlation coefficients defined by equation 3.4.

As in the case of $\overline{\sigma}^2$, equation 3.1*b*, the use of m instead of μ here causes these estimates of the correlation coefficients to be biased, and in addition there are difficulties with $\overline{\rho}_j'$ as an estimator of ρ_j' due to the finite size of the sample. We will denote unbiased estimators of ρ_j' by r_j. Under idealised circumstances where one is dealing with the whole population the coefficients ρ_j' would all be based on the same, infinite, sized sample, and beyond some $j = J$, characterising the range of the correlations, these coefficients would vanish for a physical variable. In this case we could write

$$\rho = \rho^{(J)} = \sum_{j=1}^{J} \rho_j'.$$

We will denote the unbiased estimate of this quantity by $r^{(J)}$. In the case of a finite sized sample, however, the sample size used as basis for estimating ρ_j' depends on j, it is not N but is shrinking as N-j with increasing j. In the case of any estimated quantity there is an uncertainty, or variance, associated with the estimate and in the case of ρ_j' this uncertainty is likely to increase as j increases because of the decreasing sample size on which it is based. Thus, even if it is true that in the population $\rho_j' \simeq 0$ for all j greater than some value J, this will not in general be so for the calculated quantities $\overline{\rho}_j'$. Even though not individually significant, they may not make a negligible contribution to the sum used to form $\overline{\rho}$. A cut off at some value J, larger than the range of the underlying correlations, but satisfying $J << N$ to exclude the noise contributions, must be made. We thus define the, biased, estimator for $\rho^{(J)}$ by

$$\overline{\rho}^{(J)} = \sum_{j=1}^{J} \overline{\rho}_j'. \tag{3.6}$$

We will discuss below the criterion for selecting a value for J.

It is shown in books on time series analysis, that the bias in equations 3.1b, 3.4 and 3.6 can be corrected for as follows,[3]

$$s^2 = \frac{N}{(N-1)}(\overline{\sigma}^2 + \rho) \tag{3.7}$$

$$r_j = \overline{\rho}_j' + \frac{2(N-j)}{N^2}\left(\rho + \frac{\sigma^2}{N}\right) \tag{3.8a}$$

$$\simeq \overline{\rho}_j' + \frac{2}{N}\left(\rho + \frac{\sigma^2}{N}\right) \quad \text{for} \quad j << N, \tag{3.8b}$$

$$r^{(J)} = \overline{\rho}^{(J)} + \frac{2J}{N}\left(\rho + \frac{\sigma^2}{N}\right). \tag{3.9}$$

We note that for $\rho = 0$, equation 3.7 reduces to the familiar expression 3.2a. A drawback to the above three expressions is that the corrections for bias are expressed in terms of parameters of the population ρ and σ^2. By replacing these parameters by their unbiased estimates, e.g. in $r^{(J)}$ replace σ^2 by s^2 and then ρ by $r^{(J)}$, we find, for $N >> 1$

$$r_j = \frac{N}{N-2J}\,[\,\overline{\rho}_j^{'} + \frac{2}{N^2}\,], \tag{3.10}$$

$$r^{(J)} = \frac{N}{N-2J}\,[\,\overline{\rho}^{(J)} + \frac{2J}{N^2}\overline{\sigma}^2\,], \tag{3.11}$$

$$s^2 = \frac{N}{N-1}\overline{\sigma}^2 + \frac{N}{N-2J}\overline{\rho}^{(J)}. \tag{3.12}$$

It still remains to calculate $\overline{\rho}^{(J)}$, i.e. to decide on the value of J to be used in these expressions. We cannot use its originating definition, i.e. the smallest j for which $\rho_j^{'}$ vanishes in the population because we do not know, *a priori*, the population parameters. We cannot trust the condition $\overline{\rho}_j^{'} = 0$ to determine it reliably, since the $\overline{\rho}_j^{'}$ is a random variable, a single estimate of which, in general, will not coincide with its expectation. Furthermore, as pointed out earlier, the variance on $\overline{\rho}_j^{'}$ will increase with j because of the smaller sample size on which the estimate is based. Several ways have been suggested to overcome this problem, we will just consider one of them.

One plausible criterion under these conditions is to omit terms from equation 3.6 when they are no longer significant at the one standard deviation level, i.e. when the standard deviation on $\overline{\rho}_j^{'}$ is larger than the actual value obtained for $\overline{\rho}_j^{'}$.[4] The variance on the unbiased estimator of $\rho_j^{'}$, i.e. on r_j, is known and the standard deviation can be approximated by

$$\frac{1}{\sqrt{N}}\left\{\frac{4\sigma^4}{N^2} + 2\sum_{k=1}^{\infty}\rho_k^{'}\right\}^{1/2}, \tag{3.13}$$

where it is understood that $j << N$, and $\rho_k^{'} \simeq 0$ for all k greater than some value, K, which also satisfies $K << N$.[5] If we use as estimates of σ^2 and $\rho_j^{'}$

the quantities $\overline{\sigma}^2$ and $\overline{\rho}'_j$, we finally define J as the smallest integer which satisfies

$$\overline{\rho}'^2_J \leq \frac{4\overline{\sigma}^4}{N^3} + \frac{2}{N}\sum_{j=1}^{J}\overline{\rho}'_j, \tag{3.14}$$

where, to exclude noise, the sum includes terms only up to J.

For the validity of the above discussion we must have $J << N$. If this condition is not satisfied, the sample size N, for the degree of correlation existing in it, is too small to enable any meaningful estimate of the population parameters to be based on it.

Finally we are in a position to estimate the standard deviation on the mean of a sample, given by equation 3.3a, which we denote by s_m; inserting the estimates of ρ and σ^2 given by equations 3.11 and 3.12, respectively, we have

$$s_m^2 = \frac{1}{N - 2J - 1}\left[\,\overline{\sigma}^2 + N\overline{\rho}^{(J)}\,\right]. \tag{3.15}$$

In the absence of correlations, $J = 0$ and the result reduces to $s_m^2 = s^2/N$, equation 2.10, illustrating the well known fact that the standard error on the mean only decreases as the inverse square root of the sample size.

In summary, the quantities m, $\overline{\sigma}^2$ and $\overline{\rho}'_j$ ($j = 1, \ldots, N-1$) are calculated via equations 3.1a, 3.1b and 3.4. J is then determined from equation 3.14, $\overline{\rho}^{(J)}$ via equation 3.6 and s_m calculated from equation 3.15.

The overall measure of the correlation in the population, the correlation length, τ, defined by equation 3.3c, can be estimated via equation 3.3d as

$$\tau \simeq \frac{N}{2s^2}r^{(J)}. \tag{3.16}$$

The subroutine CORVAR listed in the Appendix calculates the corrected variance s^2, the standard deviation on the mean and the correlation length in this way.

2.3.4 Distribution of the Mean; The Central Limit Theorem

We have devoted much effort to estimating the variance, or standard deviation σ_m, of the mean of a sequence of variates from a distribution. To be able to say more about the significance of a simulation result, often in the form of an estimated mean, we should like to know what distribution this variance characterises. We can show, in section 2.4.2, that in the case the N variates are from a Gaussian distribution of mean μ and variance σ^2 their sum is also Gaussian distributed with mean $N\mu$ and variance $N\sigma^2$, and so their average is Gaussian distributed with mean μ and variance $\sigma_m^2 = \sigma^2/N$. Fortunately it can be shown that for N *sufficiently large*, if the variates x_i are randomly selected from a distribution (*any* distribution with finite variance) the mean will be Gaussian distributed with variance σ_m^2. This remarkable result is known as the Central Limit Theorem and is fundamental to most interpretation of the results of Monte Carlo simulations. Often $N \geq 10$ will satisfy the large N requirement. The result may, with caution, even be extended to a sum of variates which arise from different distributions.

2.3.5 Confidence Levels

When a parameter of any distribution is estimated the question of its significance arises, how good an estimate is it of the parameter in the underlying population? This is determined from the confidence levels calculated on the basis of the distribution. Even without the knowledge of such quantities some general feel for the significance of a result can be arrived at from the definition of the variance. From its definition, equation 2.4, this quantity is an integral in which all contributions are positive. If we omit a range $\pm\epsilon$ of the integration variable around the mean we have the inequality

$$\sigma^2 > \int_{-\infty}^{\mu-\epsilon} dx(x-\mu)^2 p(x) + \int_{\mu+\epsilon}^{\infty} dx(x-\mu)^2 p(x).$$

Since in these integrals $(x-\mu)^2$ is always larger than ϵ^2, the inequality is strengthened by this substitution, i.e.

$$\sigma^2 > \epsilon^2 \left[\int_{-\infty}^{\mu-\epsilon} dx p(x) + \int_{\mu+\epsilon}^{\infty} dx p(x)\right].$$

But the right-hand side is just

$$\epsilon^2 P(\mid x-\mu \mid > \epsilon),$$

so in this way, we arrive at *Chebyshev's inequality*,

$$P(\mid x-\mu \mid > \epsilon) < \sigma^2/\epsilon^2,$$

which tells us that the probability that the observed value differs in magnitude from its expectation by as much as ϵ is less than σ^2/ϵ^2. For the case where x itself is the mean m based on a sample of N values of the same variate independently sampled from its distribution having variance σ^2, we have, using equation 2.10,

$$\begin{aligned} P(\mid m-\mu \mid > \epsilon) &< \sigma_m^2/\epsilon^2 \\ &< \sigma^2/N\epsilon^2. \end{aligned}$$

The probability can be made arbitrarily small by using a sufficiently large N. This *law of large numbers*, which guarantees convergence to the population mean, is fundamental to all estimation based on sampling.

In many cases, of course, we can be more quantitative in our confidence estimation. For example if the quantity, which we will call m, follows a Gaussian distribution with standard deviation σ_m, as would be the case, by virtue of the Central Limit Theorem, for the mean of a set on N variates from a distribution, then in 68.3% of samples a value lying within $\pm\sigma_m$ of the population mean μ would be found. Conversely there is a 68.3% probability that the interval $m \pm \sigma_m$ contains the population mean. Usually σ_m will not be exactly known and only an estimate, s_m, of it will be available. Often this estimate, s_m, if it is based on not too small a sample, will be accurate enough that it can be used instead of σ_m in the foregoing discussion; if this is not the case the confidence limits of the so called t-distribution must be consulted. In contrast to the case where σ_m is used in the definition, the quantity $t = (m-\mu)/s_m$ is not Gaussian distributed, but follows the Student t-distribution with mean zero, and variance $((N-1)/(N-3))^{1/2}$. One of the earliest applications of Monte Carlo sampling as a research tool was in the determination of the significance levels of this distribution by the statistician W. S. Gosset.[6]

When the quantity of interest follows some other distribution the uncertainty in the estimate of the standard deviation will, in general, be much less important than the fact that the standard deviation will not now determine the interval containing the population mean with the same confidence, and the intervals defined on either side of the estimated mean will not be equally probable. It is usual in these circumstances to consider *central limits* defining symmetric intervals on either side of the estimated mean and encompassing canonical probabilities (e.g. 68.3% or 99.7%, corresponding to one or three standard deviations in the Normal distribution). These limits in the case of the Poisson distribution are described in section 2.4.1.

2.4 Useful Distributions

In this section we summarise the properties of some distributions often encountered in simulations of physical problems. Methods of generating variates from these distributions will be considered in chapter 5. We begin with discrete distributions, in which the variate assumes integral values only. Cumulative distributions in these cases, in principle, can be readily formed from the density functions by summing successive terms. In practice, however, this may be limited by round-off error in the individual terms giving rise to a systematic distortion in such a sum; also the time required to evaluate each term may not be negligible. For both of these reasons it is often much more desirable to use a recursion relation to find successive values of the cumulative distribution.

2.4.1 Case of Discrete Variables

The Binomial Distribution

If the possible outcomes of an experiment can be classified into only two, mutually exclusive, classes — e.g. success or failure, detection or non-detection, etc. — then in a series of N such experiments carried out under identical conditions the number of outcomes falling into one of the classes, e.g. detections, n, will be a random variable. This variable will have a mean value α, the probability of the variate belonging to that class, and will be distributed according to the Binomial distribution, with pdf

$$p_N(n) = \frac{N!}{n!(N-n)!}\alpha^n(1-\alpha)^{N-n}, \qquad 0 \le n \le N. \tag{4.1}$$

The cumulative distribution may be constructed using the recursion formula,

$$p_N(n+1) = \frac{\alpha(N-n)}{(1-\alpha)(n+1)}p_N(n).$$

The mean and variance are given, respectively, by

$$\mu = N\alpha,$$

$$\sigma^2 = \mu(1-\mu/N). \tag{4.1a}$$

Note that N here is a fixed parameter, not a number of independent samplings from the distribution. In general we will have only one sampling, resulting in n successes out of N trials. The estimate of μ is then simply n, and an estimate of σ^2 is $\overline{\sigma}^2 = n(1-n/N)$. This latter quantity is biased, it can be shown that an unbiased estimate, s^2, of the variance is $N/(N-1)$ times this, i.e.

$$s^2 = \frac{N-n}{N-1}n. \tag{4.2}$$

For large values of N the binomial tends to the Gaussian distribution with parameters $\mu = (N\alpha - 1/2)$ and $\sigma = \sqrt{N\alpha(1-\alpha)}$.

The Poisson Distribution

The prototype of a random event is one for which the probability of a single occurrence in a time interval dt at time t is independent of the time t, which we can thus write equal to zero, and simply proportional to dt, dt/τ. We are interested in the probability that a time in the interval t to $t+dt$ from our arbitrary reference time will elapse before the next occurrence of the event, $\pi(t)dt$, i.e. the probability that the next occurrence of the event will fall in the interval dt at t. We will denote by $\overline{\Pi}(t)$ the probability that there will be no occurrence of the event up to the time t. Since $(1 - dt/\tau)$ is the probability that it will not occur in the succeeding interval dt, we have

$$\overline{\Pi}(t + dt) = \overline{\Pi}(t)(1 - dt/\tau).$$

A simple differential equation follows from this which has the solution

$$\overline{\Pi}(t) = e^{-t/\tau}.$$

The cumulative probability is

$$\begin{aligned} \Pi(t) &= 1 - \overline{\Pi}(t) \\ &= 1 - e^{-t/\tau}. \end{aligned}$$

From this we find, by differentiation, the pdf,

$$\pi(t)dt = \tau^{-1}e^{-t/\tau}dt. \tag{4.3}$$

The mean of this distribution, τ, is the mean time between events as is clear from the original definition of τ. The probability of no occurrence of the event within a time t is, by our definition, $\overline{\Pi}(t)$,

$$p(0, t) = e^{-t/\tau}.$$

An important question is, what is the probability $p(n, t)$ of exactly n $(n > 0)$ occurrences within a time t. The probability of n occurring by a time $t + \Delta t$, $p(n, t + \Delta t)$, is the sum of the probability that n will already have occurred by time t times the probability of no occurrence in Δt, plus the probability that n-1 will have occurred by t times the probability of an occurrence in Δt, i.e.

$$p(n, t + \Delta t) = p(n, t)(1 - \Delta t/\tau) + p(n - 1, t)\Delta t/\tau.$$

In the limit $\Delta t \to 0$ this gives the differential equation

$$\frac{dp(n, t)}{dt} + \tau^{-1} p(n, t) = \tau^{-1} p(n - 1, t).$$

Using an integrating factor, the solution is

$$p(n, t) = \tau^{-1} e^{-t/\tau} \int_0^t dt e^{t/\tau} p(n - 1, t).$$

Knowing $p(0, t) = e^{-t/\tau}$, this gives us a way of iteratively determining $p(n, t)$ for any n. Writing the mean number of occurrences, $\mu = t/\tau$ we have

$$p(1, t) = \mu e^{-\mu},$$

$$p(2, t) = \frac{\mu^2}{2} e^{-\mu},$$

and generally we have the Poisson distribution of mean μ with pdf

$$p(n) = \mu^n \frac{e^{-\mu}}{n!}. \tag{4.4}$$

The cumulative distribution can be constructed using the recursion formula

$$p(n + 1) = \frac{\mu}{n + 1} p(n).$$

This is the most important discrete distribution in physics, the conditions determining it being a good approximation to many physical situations, e.g. in the statistics of counting — especially where small numbers of events are involved. It will therefore often arise in simulations of experimental systems and so its properties are of more than usual interest. It can also be thought of as a limiting case of the previous Binomial distribution where $N \to \infty$

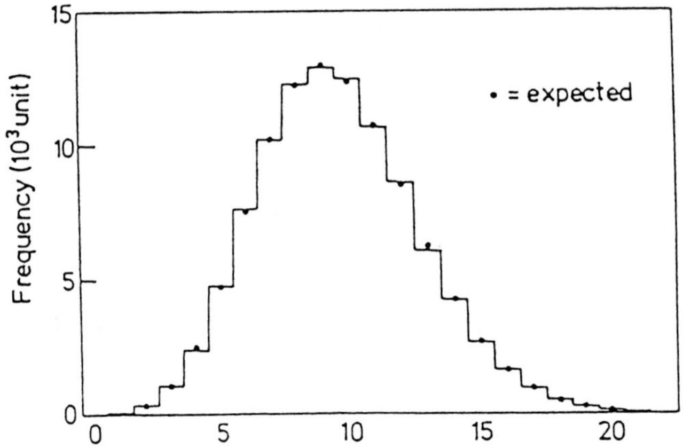

Figure 2.3: A comparison of the observed numbers of γ-rays per 10 ms from a radioactive source where the mean number is 9.54 and the Poisson distribution with the same mean (after Inoue *et al.*, Appl. Statist., **32**, 115-20, 1983).

but the probability of occurrence, $\alpha = n/N$, remains finite. The mean μ is explicitly given and the variance σ^2 is simply equal to μ. In Fig. 2.3 we show the number of γ-rays detected per unit time from a long life radioactive source based on 10^5 time intervals — the histogram — and the numbers predicted by the Poisson distribution with mean $\mu = 9.54$.

The Poisson distribution has the property that if z is a sum of N variates, each of which derives from a Poisson distribution characterised by a mean μ, then z is also Poisson distributed, with parameter $N\mu$. The difference of two Poisson variates, important, for example, in subtracting a noise signal from a total signal,[7] however, is not Poisson distributed. The confidence limits on the mean of a Poisson variable are often of interest. For a sample yielding a value n for the quantity, we know that $\sqrt{n}$ is an estimate of the standard deviation. If the value n, and therefore presumably the mean μ, is fairly large (≥ 20) the Poisson distribution is well approximated by the Gaussian, so the statement that $n \pm \sqrt{n}$ contains the population mean μ with

n	s_n^-	s_n^+		n	s_n^-	s_n^+
0	0.0	1.84		11	3.27	4.42
1	0.83	2.30		12	3.42	4.56
2	1.29	2.64		13	3.57	4.71
3	1.63	2.92		14	3.70	4.83
4	1.92	3.17		15	3.83	4.96
5	2.16	3.39		16	3.96	5.09
6	2.38	3.59		17	4.08	5.21
7	2.58	3.77		18	4.21	5.33
8	2.77	3.95		19	4.32	5.45
9	2.95	4.11		20	4.43	5.56
10	3.11	4.27		50	7.05	8.11

Table 2.1: Fiducial limits for the Poisson distribution.

68.3% probability (or 31.7% probability that it lies outside this interval) is meaningful. If, however, the value of n is 1, such a statement would be obviously wrong, for if one is observed there must be no probability at all that the mean is zero ! What we would like to have are some limits — fiducial limits — which delineate the region within which, with 68.3% probability (or any arbitrary probability β), the mean is likely to lie. Following section 2.3.5, bands with probability equally centred about the observed estimate of the mean are chosen, giving the so called central limits.[8] We denote the width of the lower and upper band, respectively by s_n^- and s_n^+; at large n we have $s_n^- \to \sqrt{n}$ and $s_n^+ \to 1 + \sqrt{n} \cong \sqrt{n}$. Values for $s_n^\pm$ can be found for any significance level, the accompanying table gives the value for $\beta = 0.6826$ corresponding to one standard deviation for the Normal distribution.[9]

The Polya Distribution

A generalisation of the Poisson distribution is the Polya distribution, with pdf given by

$$p(n) = \frac{\mu^n}{n!}(1+b\mu)^{-(n+1/b)} \prod_{k=1}^{n-1}(1+kb), \qquad b \geq 0. \tag{4.5}$$

The mean, μ, is shown explicitly, while the variance is

$$\sigma^2 = \mu(1+b\mu).$$

When the parameter b=0 we have the Poisson distribution, while b=1 results in the Furry distribution for the variable $n+1$.

The Furry Distribution

This distribution, arising in the statistics of cascade processes,[10] has the pdf, $p(0) = 0$ and with $\mu \geq 1$,

$$p(n) = \mu^{-n}(\mu-1)^{n-1}, \qquad n \geq 1. \tag{4.6}$$

Again the mean is explicitly shown, while the variance is $\sigma^2 = \mu(\mu-1)$.

2.4.2 Case of Continuous Variables

The Gaussian Distribution

The Gaussian, or Normal, distribution, is by far the most important distribution encountered in physics. Its pdf has the form

$$p(x) = \frac{1}{\sigma\sqrt{2\pi}} e^{\frac{-(x-\mu)^2}{2\sigma^2}},$$

which explicitly shows the mean and variance. When sampling variates from this distribution it is convenient to work with a standardised distribution centred on mean zero and with unit variance. Denoting this standardised variable $(x-\mu)/\sigma$ by t, its pdf is

$$p(t) = \frac{1}{\sqrt{2\pi}} e^{-t^2/2},$$

and variates Normally distributed with parameters (μ, σ) are related to it by

$$x = \sigma t + \mu.$$

The cumulative distribution function $P(x)$ in this case cannot be written in a simple form, but tabulations of it, commonly denoted as erf(x), are widely available.[11] From such tables we see that a quantity following the Gaussian distribution should lie within $\pm 0.68\sigma$ from its mean in 50% of observations, and within $\pm\sigma, \pm 2\sigma, \pm 3\sigma$ in 68.3%, 95.4% and 99.7% of cases, respectively.

The Gaussian distribution has the property that the sum of any number, N, of variates independently selected from it also follows the same distribution, with mean $N\mu$ and variance $N\sigma^2$. Writing $z = \sum_1^N x_i$ we have the joint probability

$$p(x_1, ..., x_N) = \frac{1}{\sigma^N (2\pi)^{N/2}} \exp\{-\sum_1^N (x_i - \mu)^2/2\sigma^2\},$$

and so, using equation 2.12 we have for the distribution in z,

$$p(z) = \frac{1}{\sigma^N (2\pi)^{N/2}} \int dx_1 ... dx_N \exp\{-\sum_1^N (x_i - \mu)^2/2\sigma^2\} \delta\{z - \sum_1^N x_i\}.$$

This can be rationalised by inserting the following representation of the δ-function, viz.

$$\delta\{z - \sum_1^N x_i\} = \frac{1}{2\pi} \int_{-\infty}^{\infty} dt e^{it(z - \Sigma x_i)}$$

When we write $y_i = (x_i - \mu)/\sigma$ we find

$$p(z) = \frac{1}{(2\pi)^{(N+2)/2}} \int dt e^{it(z-N\mu)} \prod_1^N \int_{-\infty}^{\infty} dy_i \exp -\{\frac{y_i^2}{2} + i\sigma t y_i\}.$$

Using the result that

$$\int_{-\infty}^{\infty} dy e^{-y^2/2\text{-}iay} = \sqrt{2\pi} e^{-a^2/2}, \tag{4.7}$$

we find

$$p(z) = \frac{1}{2\pi} \int dt e^{it(z-N\mu)} e^{-N\sigma^2 t^2/2}.$$

Changing to the variable $u = -\sigma t\sqrt{N}$, this becomes

$$p(z) = \frac{1}{2\pi\sigma\sqrt{N}} \int_{-\infty}^{\infty} du \exp\{-u^2/2 - i\frac{(z-N\mu)}{\sigma\sqrt{N}}u\},$$

which, again using equation 4.7, yields

$$p(z) = \frac{1}{\sigma_N\sqrt{2\pi}} \exp\{-\frac{(z-\mu_N)^2}{2\sigma_N^2}\},$$

i.e. a Gaussian with mean $\mu_N = N\mu$ and variance $\sigma_N^2 = N\sigma^2$. Defining m as z/N, i.e. the mean of N random normal variates, our result gives the mean of m as μ and its variance as σ^2/N, results obtained in a more general context in section 2.2.3. In the more general case m will not be Normally distributed although, when N is large, this will be a good approximation – - see the Central Limit Theorem in section 2.3.4.

The result just derived also holds when the variates are from different Normal distributions, in which case $\mu_N = \sum \mu_i$ and $\sigma_N^2 = \sum \sigma_1^2$.

If x is distributed normally with mean a and variance b^2 the transformed variable $z = e^x$ follows the so called *lognormal* distribution. From equation 2.12 we have its pdf

$$p(z) = \frac{1}{b\sqrt{2\pi}} \int_{-\infty}^{\infty} dx e^{-(x-a)^2/2b^2} \delta(z - e^x),$$

which with the change of variable $t = e^x$ immediately reduces to

$$p(z) = \frac{1}{zb\sqrt{2\pi}} e^{-(\ln(z)-a)^2/2b^2}. \tag{4.8}$$

This distribution could also have been obtained using the chain rule for differentiation, it has a mean of $\exp(a+\frac{b^2}{2})$ and a variance of $(e^{b^2}-1)\exp(a+\frac{b^2}{2})$.

The Exponential Distribution

The interval between successive random events, e.g. the distances between successive collisions of a gas molecule or the time interval between successive decays of a radioactive sample, as we showed in our discussion of the Poisson distribution earlier, is distributed as the exponential distribution, whose pdf is

$$p(x) = \mu^{-1} e^{-x/\mu}, \qquad x \geq 0. \tag{4.9}$$

The mean is explicitly shown while the variance is $\sigma^2 = \mu^2$. The cumulative distribution is simply

$$P(x) = 1 - e^{-x/\mu}.$$

The Gamma Distribution

If we form the sum of n variates x_i, each of which comes from an exponential distribution with mean a, the distribution this quantity follows is the Gamma distribution. With $z = \sum_1^n x_i$, where $p(x_i) = a^{-1} e^{-x_i/a}$, using the results in section 2.2.4 we can show that the pdf for z is [1]

$$p(z) = \frac{a^{-n} z^{n-1}}{(n-1)!} e^{-z/a}. \tag{4.10}$$

The mean of this distribution is $\mu = na$ and its variance $\sigma^2 = na^2$. The distribution can be generalised to n being a non-integer parameter, denoted by b, with the factorial replaced by the Γ-function, $\Gamma(b)$; the mean and variance then equal ba and ba^2, respectively. An important particular case occurs with $a = 2$ and $n = \nu/2$ where ν is an integer, giving the *chi-square distribution*,

$$p_{\chi^2,\nu}(z) = (2^{\nu/2}\Gamma(\nu/2))^{-1} z^{\nu/2-1} e^{-z/2}, \tag{4.11}$$

with mean ν and variance 2ν. Both the gamma distribution and the following beta distribution are found useful in parameterising data.

The Beta Distribution

The pdf of the beta distribution is given by

$$p(x) = \frac{\Gamma(a+b)}{\Gamma(a)\Gamma(b)} x^{a-1}(1-x)^{b-1}, \qquad 0 \le x \le 1 \tag{4.12}$$

with a and b greater than 1. The mean is $\mu = a/(a+b)$, while

$$\sigma^2 = \frac{ab}{(a+b+1)(a+b)^2}.$$

The Uniform Distribution

Perhaps the simplest, and in the context of the present work the most important distribution, is the uniform distribution, with pdf

$$p(x) = \frac{1}{(b-a)}, \qquad a \le x \le b. \tag{4.13}$$

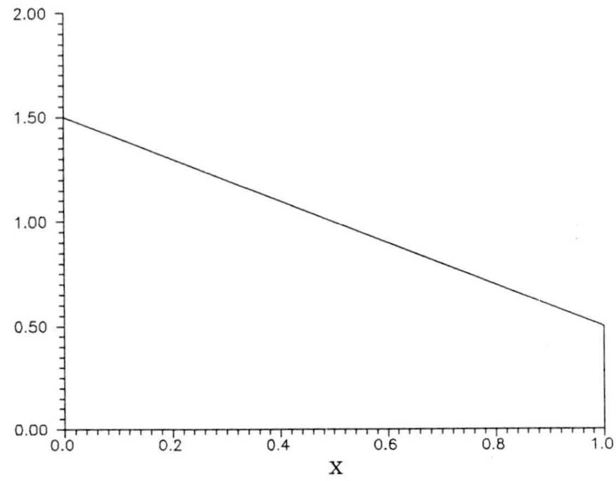

Figure 2.4: The generalisation of the uniform distribution given in equation 4.14, for $\Lambda = -1/2$.

The mean and variance are $\mu = (b-a)/2$ and $\sigma^2 = (b-a)^2/12$. The standardised form — we will denote the variate by ξ — corresponds to $a = 0, b = 1$ and has mean 1/2 and variance 1/12. We note that the related variate $1 - \xi$ has exactly the same distribution as ξ, and their covariance is $\text{cov}(\xi, 1-\xi) = -1/12$.

A generalisation of the uniform distribution is the piecewise-fit distribution, shown in Fig. 2.4, and given by

$$p(t) = 2\Lambda(t - \gamma), \qquad 0 \leq t \leq 1 \tag{4.14}$$

where $\gamma = \frac{1}{2}\text{-}\frac{1}{2\Lambda}$ and $\mid \Lambda \mid \leq 1$. It has mean and variance of $(1 + \Lambda/3)/2$ and $(1 - \Lambda^2/3)/12$, respectively. The corresponding function on an interval (a,b) can be found by replacing Λ by $(b-a)^2\Lambda$.

The Lorentz Distribution

If we define t as the ratio of two variates, x and y, each randomly selected from the same Normal distribution with mean zero and variance σ^2, i.e. both

x and y follow the distribution

$$p_x(x) = \frac{1}{\sigma\sqrt{2\pi}} \exp(-x^2/2\sigma^2)$$

then, using equation 2.16 we have for the distribution in t

$$\begin{aligned} p(t) &= \frac{1}{2\pi\sigma^2} \int_{-\infty}^{\infty} dx \mid x \mid \exp\{-\frac{x^2}{2\sigma^2}(1+t^2)\} \\ &= \frac{1}{\pi\sigma^2} \int_0^{\infty} dx x \exp\{-\frac{x^2}{2\sigma^2}(1+t^2)\}, \end{aligned}$$

i.e.

$$p(t) = \frac{1}{\pi(t^2+1)}. \tag{4.15}$$

This is the standardised form of the Lorentz, or Cauchy, distribution, centred on $z = 0$. The more general form of the distribution involves two parameters, z_o and Γ, and writing $z = z_o + \Gamma t$ takes the form

$$p(z) = \frac{1}{\pi} \frac{\Gamma}{[(z-z_0)^2 + \Gamma^2]}, \tag{4.16}$$

with

$$P(z) = \frac{1}{2} + \frac{1}{\pi} \tan^{-1} \frac{z - z_o}{\Gamma}.$$

Mathematically, the mean of this distribution is not defined, while its second moment diverges so that its variance is infinite; the distribution has no characteristic scale of dispersion. A convenient parameter to use instead is Γ, which is the half width at half maximum. It might at first be thought that there is no place in physics for such a pathology but this is not so. Often known as the Breit-Wigner distribution, it is used to represent the energy dependence of a scattering cross-section, or the mass distribution of a short-lived state in the neighbourhood of a resonance, as well as for describing the intensity profile of spectral lines in many kinds of radiations. 'Long-tailed' probability distributions like this, or the power law distribution (see section 2.5) with no characteristic scale are relevant to some phenomena which, either through non-linearity or quantum aspects, exhibit behaviour on all scales. In terms of spatial behaviour such manifestation of scale invariance is particularly notable in thermal physics — so called critical phenomena. With time as the variable, quantum transitions between states or tunneling through potential barriers can also lead to such distributions.[12] We also note that this distribution is sometimes a useful representation of the Dirac delta function through the relation

$$\delta(z - z_o) = \lim_{\Gamma \to 0} \frac{1}{\pi} \frac{\Gamma}{[(z - z_0)^2 + \Gamma^2]}. \tag{4.17}$$

The Lorentz distribution can be contrasted with the Gaussian in several ways. With $\Gamma = \sigma\sqrt{2\ln 2}$ they have the same full width at half maximum, while taking $\Gamma = \sigma\sqrt{2/\pi}$ results in the same mode for the two distributions. The comparison with the standardised Gaussian for the latter case is shown in Fig. 2.5.

The Power Law Distribution

This distribution is

$$p(x)dx = (\gamma - 1)x_0^{\gamma-1}x^{-\gamma}dx, \qquad \gamma > 1, \qquad x \geq x_0 > 0. \tag{4.18}$$

The mean only exists for values of $\gamma > 2$ and is given by

$$\mu = \frac{(\gamma - 1)}{(\gamma - 2)} x_0,$$

while the variance is infinite unless $\gamma > 3$, in which case it is given by

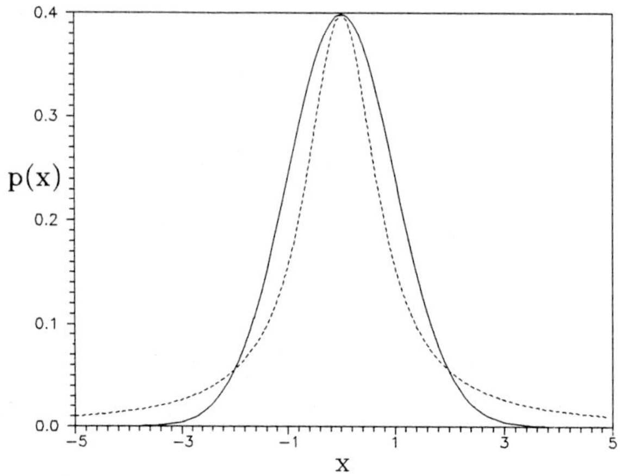

Figure 2.5: A comparison of the Lorentz (dashed curve) and Gaussian (solid curve) distributions taking a value of Γ such that their modes coincide.

$$\sigma^2 = \left(\frac{\gamma - 1}{\gamma - 3} - \left(\frac{\gamma - 1}{\gamma - 2}\right)^2 \right) x_0^2.$$

For values of γ in $1 < \gamma < 3$, it behaves as a Levy distribution.

The Levy Distribution

This distribution has the form,

$$p(x)dx = (\pi x^3)^{-1/2} e^{-1/x} dx, \qquad x \geq 0. \tag{4.19}$$

The mean and the variance of this distribution are infinite.

2.5 Distribution Testing

In simulation work it will sometimes be necessary to test a sequence of variates ($x_i, i = 1, \ldots, N$) to see (i) if they can reasonably be assumed to have been selected from a specified distribution $p(x)$, and (ii) if their selection from that distribution has been at random. We start with the former problem for which we describe two tests.

2.5.1 Tests for Consistency of Distributions

The Chi-square Test

To test a sample of data, $x_i, i = 1, \ldots, N$ against a distribution $p(x)$ by the chi-square test, the N values must be partitioned into k cells to form a histogram, each bin of which will contain n_j values, $j = 1, \ldots, k$. The following statistic (often denoted by χ^2) is then evaluated,

$$T = \sum_{j=1}^{k} \frac{(n_j - N\alpha_j)^2}{N\alpha_j}. \tag{5.1}$$

Here, α_j is the probability of an event occurring in the jth cell based on the distribution $p(x)$, i.e.

$$\alpha_j = \int_{x_j^-}^{x_j^+} dx p(x),$$

and $x_j^{\pm}$ are the limits on the jth bin. The significance of the value for T obtained depends on how much extra information contained in the sample is used, as indicated by ν, the number of degrees of freedom. If no other information in the sample is used, $\nu = k - 1$. If the sample of data is used to determine other parameters, h in number, e.g. estimates of μ and σ^2 to be used in $p(x)$ when evaluating the α_j values, then $\nu = k - h - 1$. Provided some conditions, described below, are satisfied the quantity T is distributed as the chi-square distribution with ν degrees of freedom (equation 4.11)

$$p(T) = p_{\chi^2,\nu}(T),$$

and the significance of the value obtained can be found from tables of the cumulative distribution, $P_{\chi^2,\nu}(T)$.

The conditions referred to are (i) that the number of bins, k, should be large enough to satisfy $\alpha_j << 1$ for all j so that the histogram may mimic the continuous distribution — this is not necessary when testing against a discrete distribution in which case $\alpha_j = p_j$ —, and (ii) the number of events in any one histogram bin should represent a sample from a Normal distribution whose mean is the expected number in that bin, $N\alpha_j$. The binomial distribution, which these numbers actually satisfy, is only a reasonable approximation to a Gaussian when the mean is not too small. Thus the expected number in any one bin should not be less than about five. Subject to these two conditions, sizes of bins should be chosen to encompass equal expected numbers of events based on $p(x)$, with the optimum number of bins being given, approximately, by $k = 3.8(N-1)^{2/5}$. This latter condition optimises the power of the test at the 5% level. Naturally, bins should not be made smaller than the degree of resolution in the data. Perhaps the most useful way of quantifying the significance of T is to ask what is the probability, if the sample really derived from the distribution $p(x)$, of a random sample yielding a value for T equal to or greater than the value calculated; this is given by $1 - P_{\chi^2,\nu}(T)$. The smaller this probability the greater is the suspicion that the sample did not derive from the distribution $p(x)$ — the actual probability decided on is rather subjective, or dependent on the phenomenon in question; for routine comparison purposes a value of 5% is often adopted. A very large value of the probability, i.e. almost coincidence between the observed and expected numbers resulting in a rather small value for T, suggests that the sample being tested was not randomly drawn from

$u\backslash N$	1	2	4	10	20	50
-3.2						99.99
-2.8					99.99	99.98
-2.4					99.98	99.80
-2.0				99.98	99.54	98.88
-1.6				98.49	97.03	95.94
-1.2			96.24	91.43	90.12	89.33
-0.8		81.87	78.39	77.86	77.98	78.22
-0.4	50.99	54.88	58.01	60.82	62.23	63.46
0.0	31.73	36.79	40.60	44.05	45.79	47.33
0.4	21.08	24.66	27.41	29.94	31.25	32.42
0.8	14.43	16.53	18.04	19.31	19.92	20.42
1.2	10.05	11.08	11.65	11.93	11.95	11.88
1.6	7.09	7.43	7.41	7.10	6.79	6.42
2.0	5.04	4.98	4.66	4.10	3.69	3.24
2.4	3.61	3.34	2.91	2.30	1.92	1.53
2.8	2.59	2.24	1.80	1.27	0.96	0.69
3.2	1.87	1.50	1.10	0.68	0.47	0.29
3.6	1.36	1.01	0.67	0.36	0.22	0.12
4.0	0.99	0.67	0.41	0.19	0.10	0.05

Table 2.2: Percentage probability, f, of obtaining for the standardised chi-square statistic in a sample of size N, a value larger than u if the theoretical distribution were being sampled.

the underlying population. Using a standardised variable, u, defined by

$$u = (T - \nu)/\sqrt{2\nu}, \tag{5.2}$$

values of the probability can be interpolated from Table 2.2. There we see that for agreement, in general, u should be less than about 2. The quantity u is a convenient variable both from the point of view of tabulating the distribution, and because the chi-square distribution in terms of this variable tends to the standard Normal distribution when ν becomes very large. Further values of T which for a given ν would be exceeded in $f\%$ of random samples (the $f\%$ confidence level) are tabulated as a function of f in chapter 26 of Abramowitz and Stegun.

The Kolmogorov-Smirnov Test

The utility of the χ^2-test is limited in practice to samples which are not too small by the requirement that at least five events should occur in any binning cell. This will rule out samples which in the physics laboratory would already be considered 'large', say 20, and some other test is desirable. If the sample is being tested against the Normal distribution the Shapiro-Wilks test is a powerful method;[13] for any other distribution (of a continuously distributed variable) a suitable test which is not constrained in this way is the Kolmogorov-Smirnov test.[14] Unlike the χ^2-test, however, this test does not allow for estimation of parameters of the theoretical distribution on the basis of the observed data; in order to use the significance tables these parameters must be given *a priori.*

Implementation of the test is straightforward; arrange the N data values in increasing order, $x_1, \ldots, x_j, x_{j+1}, \ldots, x_N$, where $x_{j+1} \geq x_j$ and form the cumulative distribution of the sample,

$$\overline{P}(x_j) = j/N, \qquad j = 1, ..., N.$$

Compare this in each case with the cumulative distribution which is being tested against, $P(x_j)$, and define the statistic q as $\sqrt{N}$ times the magnitude of the greatest discrepancy between the two, i.e.

$$q = \sqrt{N} max \mid \overline{P}(x_j) - P(x_j) \mid . \tag{5.3}$$

If agreement of the two distributions will be acceptable subject to it being not less likely than $f\%$, (e.g. $f = 5$, corresponding to less than a 5% probability of the theoretical distribution, randomly sampled, giving rise to the recorded data) then that agreement is established if q is less than the corresponding

$q\backslash N$	2	4	7	10	20	30
0.60	79.15	81.41	85.26	86.49	89.47	90.81
0.70	52.01	61.65	63.67	65.53	68.00	69.18
0.80	37.73	43.96	46.05	47.42	49.58	50.61
0.90	26.44	29.30	31.81	32.86	34.70	35.54
1.00	17.16	18.75	20.85	21.84	23.28	23.95
1.10	9.87	11.72	13.08	13.84	14.97	15.48
1.20	4.59	6.74	7.83	8.38	9.23	9.58
1.25	2.70	4.93	5.92	6.41	7.12	7.42
1.30	1.30	3.52	4.42	4.83	5.43	5.68
1.35	(1.314)	2.46	3.24	3.60	4.11	4.28
1.40		1.69	2.35	2.65	3.06	3.30
1.45		(1.468)	1.68	1.92	2.27	2.45
1.50			1.18	1.38	1.66	1.82
			(1.523)	(1.546)	(1.576)	(1.588)

Table 2.3: Percentage probability, f, of obtaining, for the Kolmogorov-Smirnov statistic in a sample of size N, a value larger than q if the theoretical distribution were being sampled. The last, bracketed, figure in each column is the value of q at which the probability has fallen below 1%.

entry for the given N and probability f interpolated from Table 2.3. The actual probability $1 - P_{K-S,N}(q)$ is given in section 3.3.1 of Knuth; for large values of N (in practice $N \geq 30$) it can be approximated by

$$f = 100e^{-2q^2}\{1 - 2q/3\sqrt{N} + O(N^{-1})\}. \tag{5.4}$$

Some further tabulations of the percentage points of the distribution are given in the book of Frodesen *et al.* already cited.

2.5.2 Tests for Correlations

The tests for agreement with a specified distribution, described earlier, do not depend on the order of variates in the sequence. In some Monte Carlo work the order of the variates is not of importance, or certain ordering of the variates may even be desirable, but in the vast majority of cases the assumption of a randomly selected sequence underlies the method and any deviation from randomness would have to be corrected for. Hence, tests for correlations in a sequence, i.e. searches for significant deviations from randomness in the ordering of the variates are of great importance.

There is a great variety of tests, some more powerful than others depending on the kind of correlation being sought. We mention the Global maximum/minimum tests, the runs up/down tests, the maximum/minimum of t tests, the Coupon Collector test, etc. Details of the more important ones are given in the book by Knuth. We will just use a simple one for illustration, Good's serial test, and defer to the end of chapter 3 description of a test using a search for lagged correlations.

Good's Serial Test

This is a test for the uniformity of occurrence of sub-sequences, m-plets, in a sequence of variates.[15] It can be applied to bit sequences; we are mainly interested in applying it to sequences from the uniform distribution, which we assume are reduced to the interval $(0, 1)$. From a sequence of N such variates the $(N - m + 1)$ overlapping m-plets, $(x_1, x_2, \ldots, x_m)$, $(x_2, x_3, \ldots, x_{m+1})$, ... $(x_{N-m+1}, \ldots, x_N)$ defining points in the unit hypercube, are tested for uniformity in their density distribution. The unit cube is divided into k^m equal volume sub-cubes, and the number of points, $n_{i_1 i_2 \ldots i_m}$ falling in each is determined, $i_1, i_2, \ldots, i_m = 1, \ldots, k$. The expected value for these quantities is $e_m = (N - m + 1)/k^m$. The statistics t_m, t_{m-1} and t_{m-2} are formed where $t_m = 0$ for $m < 1$ and

$$t_m = e_m^{-1} \sum_{i_1 \ldots i_m = 1}^{k} \left(n_{i_1 i_2 \ldots i_m} - e_m \right)^2 .$$

Although these quantities t_m are formally like the chi-square statistic, because the overlapping points are not independent the confidence levels for that distribution are not appropriate. However by combining these quantities a statistic, w, defined by

$$w = t_m - 2t_{m-1} + t_{m-2} \tag{5.5}$$

is formed. In the asymptotic limit, i.e. all e_m sufficiently large, the quantity w follows the chi-square distribution with ν degrees of freedom, where $\nu = k(k-2) + 1$.

Notes

1. Gillespie, D. T. (1983), *A theorem for physicists in the theory of random variables*, Am. J. Phys., **51**, 520-523.

2. Chan, T., G. Golub, and R. LeVeque (1983), *Algorithms for computing the sample variance: Analysis and recommendations*, Am. Statist., **37**, 242-47.

3. For example equation 51 of section 8.2 and equation 26 of section 8.3 of Anderson, T. W., *The Statistical Analysis of Time Series*, (New York: John Wiley & Sons, 1971).

4. Daniell, G. J., A. J. G. Hey and J. E. Mandula (1984), *Error analysis for correlated Monte Carlo data*, Phys. Rev., **D29**, 2230-32.

5. Equation 2 of section 8.4.2 of reference 3 above.

6. A tabulation of them can be found in Table 6.10 in Abramowitz and Stegun.

7. Its distribution, and confidence limits on its mean, are discussed in Prosper, H. B. (1985), *The distribution of the difference of two Poisson variates and its application to interval estimation in experiments with low statistics*, Nucl. Instrm. Methods Phys. Res., **A238**, 500-503.

8. Regener, V. H. (1951), *Statistical significance of small samples of cosmic ray counts*, Phys. Rev., **84**, 161-62.

9. The values in the table are based on the 'classical' statistical viewpoint, Eadie *et al.*, section 9.2.

10. See, for example, section 5.23 of Rossi, B., *High-Energy Particles* (Englewood Cliffs, NJ; Prentice-Hall, 1952).

11. For example in Table 7.1 of Abramowitz and Stegun.

12. Physical phenomena with more than one scale occur — a beautiful example is the decay scheme of the K^0 meson, another the propagation of an epidemic among a population by two distinct diffusion mechanisms, e.g. bugs and birds — but the distribution functions for these are not of the pathological type discussed, which describe scale invariant phenomena.

13. Shapiro, S. S. and M. B. Wilks (1965), *An analysis of variance test for normality (complete samples)*, Biometrika, **52**, 591-610.

14. See for example, section 14.4.6 of Frodesen, Skeggestad and Tøfte.

15. Good, I. J. and T. N. Gover (1967), *The Generalized Serial Test and the Binary Expansion of* $\sqrt{2}$, Jour. Roy. Stat. Soc., **A130**, 102-107.

Exercises

2.1 If variables x and y are related by $y = ax$, show that their variances obey $\sigma_y^2 = a^2\sigma_x^2$.

2.2 Show that the variance of the standardised uniform distribution is 1/12.

2.3 The sum of the readings on a pair of dice is distributed as

$$p_n = \begin{cases} \frac{n-1}{36} & 2 \le n \le 7 \\ \frac{13-n}{36} & 8 \le n \le 12. \end{cases}$$

Show that the distribution is normalised, and find its mean and variance.

2.4 Find the mean and variance of the distribution

$$p(x)dx = \frac{1}{2}[\delta(x+h) + \delta(x-h)]dx,$$

where δ denotes the delta function.

2.5 Verify for the distribution in equation 2.2.5 that the covariance is given by

$$\text{cov}[x, y] = -(4-\pi)L^2/\pi^2.$$

2.6 Verify the equivalence of equations 2.3.2a and 2.3.2b as estimates of the variance of a sample.

2.7 If x follows the uniform distribution (equation 2.4.13) find the distribution followed by the quantity y where $y = -\lambda \ln x$.

2.8 Show that if x is sampled from a Gaussian distribution with mean μ and variance σ^2 and y is also sampled from this distribution, the variable $z = x + y$ follows the same distribution with mean 2μ and variance $2\sigma^2$. By induction show that the sum of k variates from the distribution is Gaussian distributed with mean and variance $k\mu$ and $k\sigma^2$, respectively.

2.9 The stretched exponential distribution is

$$p(x)dx = \frac{\gamma}{a\Gamma(\frac{1}{\gamma})} e^{-(\frac{x}{a})^\gamma}.$$

Show that its mean and variance are $\frac{\Gamma(\frac{2}{\gamma})}{\Gamma(\frac{1}{\gamma})}a$ and $a^2\{\frac{\Gamma(\frac{3}{\gamma})}{\Gamma(\frac{1}{\gamma})} - (\frac{\Gamma(\frac{2}{\gamma})}{\Gamma(\frac{1}{\gamma})})^2\}$, respectively, and check the result for the exponential and one sided Gaussian distributions.

Chapter 3

RANDOM NUMBER GENERATION

...random numbers should not be generated by a method chosen at random.

Donald E. Knuth

...strange and unpredictable is not necessarily random.

Park and Miller[1]

3.1 Introduction

The bricks and mortar of any Monte Carlo calculation are numbers x_i chosen to represent a random selection from one or more distributions. A histogram of a sequence of such numbers would approximate the distribution in question, $p(x)$. For example, the number of neutrons emitted when a neutron of specified energy is absorbed by a U^{235} nucleus is a stochastic variable, and its distribution has been experimentally determined; in a mathematical simulation of this process we would generally require that the multiplicities occurring in the calculation mirror this distribution and reflect the same degree of randomness observed in the physical phenomenon. Numbers conforming to a specified distribution, we will find, can readily be produced if we can generate numbers from the uniform distribution; their randomness will also be determined by the extent to which this is also a property of the uniformly distributed variable. The idea of randomness — random numbers really are an abstraction — has been a, generally unfruitful, source of much discussion and argument, and several authors have given paraphrases of their concepts. None of these could be said to crystallise the idea, although several of them do give the impression of getting near the heart of the matter.[2] Focusing on the concept of *non-random* may also be found to be just as helpful. For a physicist carrying out simulations, whose immediate interest in the concept of random numbers is purely technical, a strong intuitive feeling based on experience of phenomena empirically considered to be *random*, like radioactive decay or Brownian motion, leads to satisfaction with a random number sequence if it is 'as random' as these processes. It follows that if a random number sequence can be made to image the randomness of such a physical process it is likely to define the concept of a random sequence adequately for us. Intuitive ideas of how randomness may be achieved, however, cannot always be relied on; witness the remarkable periodicity occurring in a super complicated algorithm, described by Knuth, from which he derives the moral that appears in the quote at the start of this chapter.

Processes such as cosmic ray arrival times, thermionic noise and radioactive decay have been used to generate numbers which meet our expectations and do not contradict any of the mathematical prescriptions for randomness (of which, unfortunately, there are an infinity) applied to them. Tables of such random numbers are available, and we briefly describe their origin in section 3.3. Although there are circumstances where truly random numbers derived in this way are called for, in most Monte Carlo calculations it is only necessary that the sequence of numbers used, whatever its origin, be as far as possible indistinguishable from a true random sequence. So, although, by definition, no mathematical procedure can produce truly random numbers, the predetermined sequence generated by a mathematical formula may be satisfactory if it disguises its origin to this extent. Formulae which generate predetermined sequences which are virtually impossible to distinguish from real random numbers have been found. The numbers produced in this way are properly known as *pseudo-random* numbers; we shall, in general, refer to them simply as random numbers. Random, then, for our purposes means that a sequence of such numbers should have the same probability of passing a statistical test for randomness as true random numbers would have. The philosophy is that if the numbers, when subject to a series of statistical tests, cannot be identified as non-random they are acceptable irrespective of their origin.

Pseudo-random numbers are not to be confused with *quasi-random* numbers; the latter, which are 'less random', are sometimes used in the numerical evaluation of multiple integrals. We shall say a little about them in section 3.2.6.

Basic to all sampling is the availability of numbers uniformly distributed on the interval (0, 1), i.e. $p(x)dx = dx$ for $0 \le x \le 1$, in a random sequence. In the following sections we outline some of the methods used to achieve such sequences — these numbers we will henceforth denote by ξ. The more important generation methods fall into three categories, mid-square, congruential and shift-register methods — we give some account of the latter two.

3.2 Generation of Random Numbers

3.2.1 Congruential Generators

Congruential methods of random number generation start from the relation with m, n and ν non-negative integers,

$$\nu = n \bmod m,$$

which defines ν as being congruent to n modulo m. This is best understood as meaning that ν is the remainder when n is divided by m, or $\nu = n - jm$ where j is the largest integer consistent with ν being non-negative. In particular we note that $m = 2$ gives the bits 0 and 1. Obviously the only possible values that ν can assume are the m integers in the range 0 to m - 1. A sequence of integers $\{\nu_i\}$ can be generated from this relation by taking for n some function of the previous number ν_{i-1}, i.e.

$$\nu_i = f(\nu_{i-1}) \bmod m.$$

Provided an initial seed ν_0 is supplied, a string of distinct values can be generated, of length k, terminating as soon as a value ν_k equal to ν_0 occurs. This could occur on the first exercise, i.e. $k = 1$; we know that the latest it can occur is for $k = m$. The actual value of k will, in general, depend both on the function f and the seed ν_0. The simplest recursion formula, due to Lehmer, is a linear one, i.e. $f(\nu) = a\nu + c$, forming the basis for *multiplicative linear congruential* generators (MLCG), the principal method used in practice. Sometimes a distinction is made between the case c=0 — *multiplicative congruential* method — and $c \neq 0$, the *mixed congruential* method. With suitably chosen values for a, c and m the sequence of integers generated from

$$\nu_i = (a\nu_{i-1} + c) \bmod m \tag{2.1}$$

can have many of the properties of a true random sequence. In particular, it may have the full period, i.e. contain all m distinct integers and thus be independent of the seed ν_0 — although the non-occurrence more than once of any value in the sequence immediately reminds us that it is not simply a random sample of m integers from the uniform distribution on $[0, m-1]$, which would not be constrained in this way. For practical purposes, with large m, the numbers will be sufficiently dense that this should not be a source of complaint. By taking $\xi_i = \nu_i/m$ we find a sequence of numbers distributed on [0,1]. How far, mathematically, such numbers are from an ideal random sequence can be judged by comparing the bit length of the algorithm required to specify equation 2.1, with a, c, m and ν_0 specified, with the criterion for true randomness, viz. that the length of the algorithm for specifying a random sequence requires about the same number of bits of information as contained in the sequence generated.

We would obviously like to have as long a sequence as possible of such numbers before they repeat themselves, in addition they should be uniformly distributed and they should approximate a random selection, i.e. there should be negligible correlations within the sequence. This latter aspect is largely independent of uniformity, for example, if the sequence $\{\xi_i\}$ is uniform and random on [0,1] the sequence $\xi_1, 1-\xi_1, \xi_2, 1-\xi_2, \ldots$ would also pass the test for uniformity but, because of the strong correlations in succeeding pairs, it does not conform to our idea of a random sequence. The number of possible correlations in a sequence is limitless, and at least one occurs by the very definition of the generator as equation 2.1; it will only be possible to test for some of the more suggestive ones. For example, in an ideal random number sequence one would expect the points defined by successive k-tuples, i.e. $\pi_i^{(k)} = (\xi_{k(i-1)+1}, \xi_{k(i-1)+2}, \ldots, \xi_{ki})$ to be uniformly distributed in the k-dimensional unit hypercube. Values of k-tuples like this are often used in practice, in selecting direction vectors in space or selecting variates from a more complicated distribution, for example, we will see in chapter 5 how 12-tuples can be used to approximate selection of a Gaussian distributed variable.

Example 3.1: Properties of a toy generator

We can get a feel for some of these features by considering a few examples of congruential generators — using small values of the modulus m for ease of illustration.

I: with $m = 31, c = 0, a = 7$ and taking $\nu_o = 1$ we find the sequence,

1,7,18,2,14,5,4,28,10,8,25,20,16,19,9,1,. . .

i.e. the sequence repeats after only 15 values. The failure to achieve the full period is an obvious weakness of such a generator, but even more disconcerting is the existence of sub-periods within the sequence. We can notice, for example, $\nu_{12} = 2\nu_9 = 4\nu_6 = 8\nu_3 = 16\nu_0$, or more generally, $\nu_{i+3} = 2\nu_i \bmod 31$.

II: $m = 31, c = 0, a = 3$,
i.e. taking $a = 3$ instead of 7, again starting with $\nu_0 = 1$ we find,

1,3,9,27,19,26,16,17,20,29,25,13,8,24,10,30,28,22,4,12,5,15,14,11,2,6, 18,23,7,21,1,. . .

All the numbers except zero occur (with $c = 0$, zero cannot occur unless it is taken as the seed, in which case it simply reproduces itself — the absence of zero in this way may actually be desirable). This generator may be considered satisfactory as far as its period is concerned, however it also has strong correlations built in, as may be seen from Fig. 3.1(a). Here we have used successive overlapping pairs of the generated numbers, (ξ_i, ξ_{i+1}), $i = 0, 1, \ldots, 28$, to define points

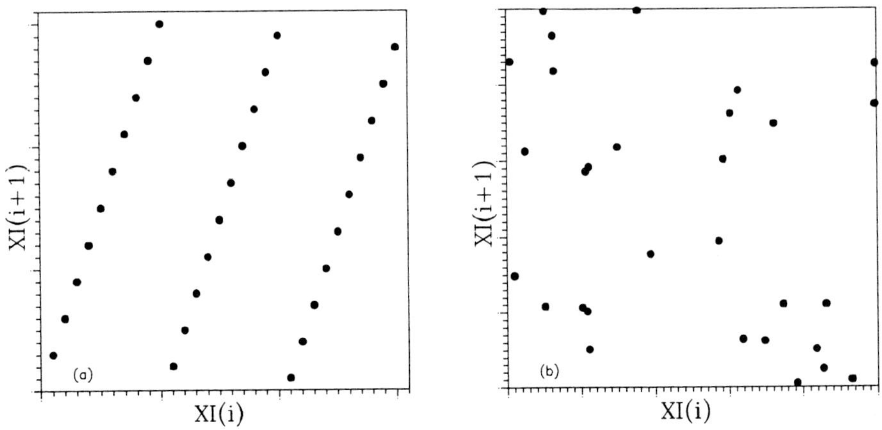

Figure 3.1: (a) Correlation between successive variates from the generator in Example 3.1. (b) Correlations between successive variates generated by a physical process.

in the plane. (Similar behaviour also holds using successive pairs, (ξ_{2i-1}, ξ_{2i}).) The distribution of these points, confined to three parallel planes, is not obviously what one would expect from a sequence of true random numbers, a fact which is emphasised by the analogous plot of 29 points defined by a sequence of physically generated random numbers on [1-30], shown in Fig. 3.1(b), which is based on the first 30 numbers occurring in a random selection (in which repetition of a number does not involve duplication of the sequence). Even more non-uniform 'lattice structure' may be found with such generators when triplets or higher multiplets of numbers are plotted in hyperspace. As a final example we consider a mixed generator with m a power of 2, say $m = 2^5$; the form 2^β, where β is the number of bits used to represent an integer on the computer, is often used since from a practical point of view such a modulus will render the division modulo m rather simple on a binary computer, it will be achieved by default through truncation.

III: $m = 32, c = 1, a = 5$,
we find the sequence which contains all 32 values,

1,6,31,28,13,2,11,24,25,30,23,20,5,26,3,16,17,22,15,12,29,18,27,8,9,14 ,
7,4,21,10,19,0,1,. . .

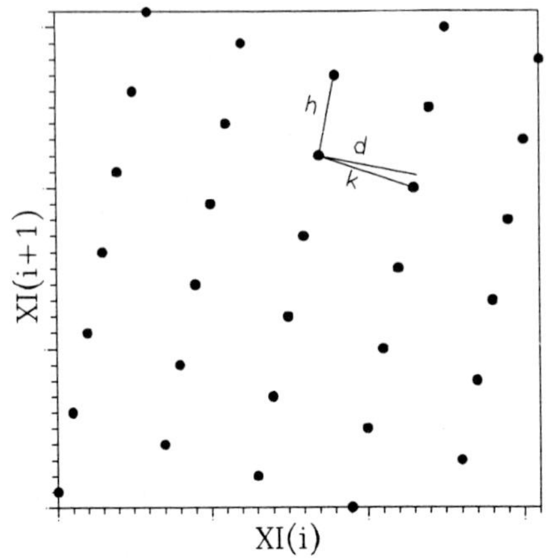

Figure 3.2: The lattice structure of the variates produced by generator III, with the parameters h, k and d used in describing this structure.

> The lattice structure of pairs in the plane is shown in Fig. 3.2; it still does not resemble what one would obtain from a perfect random number generator, however, in some sense it can be seen to be more random than in Fig. 3.1(a).

The necessary properties of the parameters in equation 2.1 to get the full period of a congruential generator are now well understood and the lattice structure of the resulting sequences has been elucidated in recent years. A theorem in Knuth (section 3.2.1.2) shows that in the case $c = 0$ the full period, 1 to (m-1), can be achieved for equation 2.1 by choosing m as a prime number and for a, a primitive element modulo m, i.e. for all prime divisors, p, of (m-1), $a^{(m-1)/p} \bmod m \neq 1$. For the case of $c \neq 0$, Knuth also shows that the full period 0 to (m-1), will be attained for equation 2.1 if the following three conditions are satisfied:

(i) c and m are relatively prime,

(ii) $(a-1)/p$ is an integer for every prime, p, which is a factor of m,

(iii) if m is a multiple of 4 then (a-1) must be divisible by 4.

These conditions, for $c = 0$ and $c \neq 0$, can be seen to be satisfied by our two full period examples, (II) and (III), respectively. Generators with a period of $m/4$ can also be constructed when $c = 0$ and a is not a prime, some of them have good properties and the conditions on them are described by Knuth.

It was long ago realised that the k-tuples of overlapping numbers from a multiplicative generator will always lie on a much smaller number of planes (not more than $(k!m)^{1/k}$ planes) than would those defined by points from a true random number sequence.[3] However, as a comparison of Fig. 3.2 with Fig. 3.1(a) shows, the non-uniformity, or 'granularity', may be much worse in some cases than in others. We can define some measure of the granulation of the points in the plane, or in higher dimensional plots. As an example, for the plots of (ξ_{i+1}, ξ_i) described earlier we can define $r = k/h$, where the spacings h and k are defined in Fig. 3.2. Another possible parameter is the largest spacing between adjacent planes, d in Fig. 3.2, we take ν as its reciprocal, $\nu = 1/d$. For the two generators (II) and (III), (Figs 3.1(a) and 3.2) we find

$$\text{(II)}\ r = 3.11 \qquad \nu = 3.16$$

$$\text{(III)}\ r = 1.24 \qquad \nu = 5.10,$$

small values of r and large values of ν being desirable. Two parameters analogous to these, and their generalisation to higher multiplets, based on the geometry of numbers have been developed to characterise the lattice properties of congruential generators. In Table 3.1 we show some sample values for these two parameters for plots in 2, 3 and 4 dimensions, for various combinations of m, a and c — some of which are used in commercial packages.[4]

			k=2		k=3		k=4	
m	a	c	r	ν	r	ν	r	ν
2^{16}	293	1	1.20	244	1.07	39.5	1.45	14.9
2^{32}	69069	1	1.06	6510	1.29	1440	1.30	230
2^{48}	5^{17}	1	1.87	1.210^7	2.86	4.710^4	1.67	3400
$2^{16}+1$	254	0	1.02	254	22.9	6.5	11.3	6
$2^{16}+1$	75	0	11.7	75	1.59	31.4	3.43	9.2
$2^{32}-1$	630360016	0	1.29	4090	2.92	625	1.64	201

Table 3.1: Parameter values for some multiplicative congruential generators

The second generator in Table 3.1 is a popular one, it is incorporated in the CERN programme library as well as in the package 'Super-duper' described in section 3.2.4, and is seen to have good properties (remember, small r and large ν are desirable). The fifth case is an example of an unsatisfactory generator; however, in general it is not possible to uniquely choose a set of parameters, one generator may be superior to another in respect of two-dimensional correlations but show greater structure in a higher dimension or vice versa.

In the linear congruential generator, equation 2.1, only the immediately previous random variable is used to derive the next one; an obvious generalisation would be to use more of the earlier generated numbers, each associated with a specified multiplier — a compound generator. If the last k generated numbers are used (implying, of course, provision of a k-component seed) we can write

$$\nu_i = (\mathbf{a}.\mathbf{n}^{(i-1)} + c) \bmod m, \tag{2.2}$$

where $\mathbf{n}^{(i-1)} = (\nu_{i-1}, \nu_{i-2}, \ldots, \nu_{i-k})$ and the set of multipliers is $\mathbf{a} = (a_1, a_2, \ldots, a_k)$. By taking a value of just 2 for k and suitable multipliers a_1 and a_2, *a compound generator* producing numbers with a much more acceptable lattice structure may be obtained; a sophisticated versions is described next.

3.2.2 Shift-register Generators

A particular case of interest is the compound generator of the above type, equation 2.2, with $m = 2$, i.e. the generation of bits, known as a shift-register, or Tauseworth, generator. We write these bits as

$$b_i = (a_1 b_{i-1} + a_2 b_{i-2} + ... + a_{p-1} b_{i-p+1} + b_{i-p}) \bmod 2$$

where p indicates the earliest previous bit included in the sequence ($a_p = 1$), and the first p bits are provided; the constant c has been set equal to zero — setting it equal to unity would just give rise to the complimentary bit sequence. Excluding the case where all previous p bits were zeros, it would continue as such indefinitely, there is a maximum of $2^p - 1$ distinct possible iterations before a sequence repeats itself. The conditions on the value of p, and the coefficients $a_i, i = 1, \ldots, p-1$, to obtain this maximum period have been determined from number theory, the details are too abstruse for description here. Of particular interest is the existence of such sequences with only two non-zero coefficients, i.e. b_i is of the form

$$\begin{aligned} b_i &= (b_{i-q} + b_{i-p}) \bmod 2 \\ &= b_{i-q} \oplus b_{i-p}, \end{aligned} \qquad (2.3)$$

where, since the bs are single bits, the addition is equivalent to addition without carrying, or the 'exclusive-or' operation, denoted by $\oplus$. This is defined by $0 \oplus 0 = 1 \oplus 1 = 0, 1 \oplus 0 = 1$, — as an example, in binary representation $53 \oplus 25 = 44$. This exercise is very fast on a binary computer using shift-registers. Suitable values for p and q are tabulated in Knuth. Such a generator can be very fast, but the correlations existing in it and their relationship to the initialising p bits are not very well understood. In addition, although its period is very large, it is still far from the theoretical limit for a generator requiring 250 seed values. The number of distinct ways the k bits in each number of a sequence of p can be combined is 2^{kp}, — 2^k for the simplest generators — enormously greater than the periods described until now. For these reasons they have become to some extent rather eclipsed by the generators described in the following section.

3.2.3 Fibonacci Generators

The prescription described in the last section of constructing a lagged sequence can be done modulo $m > 2$, e.g. taking the pth and qth preceding numbers with, say, $p > q$, and combining them, mod m, according to some operation,

$$\nu_i = (\nu_{i-q} \odot \nu_{i-p}) \bmod m.$$

We can generate a sequence of numbers, given that the first p numbers are provided, e.g. from a MLC generator. The operation denoted by $\odot$ is conventionally the 'exclusive-or' (FORTRAN function IEOR) operation described earlier. It may however be just simple addition or subtraction, and recent studies favour the latter operation, giving rise to a subtracted Fibonacci generator. An especially interesting development is the subtract-with-borrow generator of this type. The algorithm is

$$\nu_i = (\nu_{i-q} - \nu_{i-p} - \beta) \bmod m. \tag{2.4}$$

The 'carry' coefficient, β, which must be set to 0 or 1, arbitrarily, at initialisation, is reassigned in each call as follows; if the quantity in brackets is negative, so that m must be added to it to carry out the modulo m operation, β is set to zero otherwise it assumes the value unity. If m, p, and q satisfy certain conditions, the period of this generator is $m^p - m^q$. An example is described in the next section.[5]

3.2.4 Practical Generators

Most modern generators involve the combination of two or more of the elementary generators described above, an improvement both from the point of enhancing the statistical quality of the numbers and increasing the period of the generator. One such is called Super-duper, originally written at McGill University, Canada, for 32-bit machines, a 16-bit version of which has been used for many of the examples in this book.[6] It combines the numbers produced by a linear congruential generator with one produced by a shift-register generator and has a period in excess of 10^{18}. The former uses equation 2.1 with $m = 2^{32}, a = 69069$ and $c = 0$, with the seed, ν_0, an odd integer to produce a 32-bit integer ν_i. The shift-register generator employs equation 2.3 with $p = 32$ and $q = 17$ to produce bits, b_k, sequences of 32 of which are combined to form an integer μ_i,

$$\mu_i = \sum_{k=1}^{32} b_k 2^{k-1}.$$

These two integer are then combined bit-wise 'exclusive-or' and scaled to give a variate on [0,1] as,

$$\xi = 2^{-32}(\nu_i \oplus \mu_i).$$

The properties of this generator have been exhaustively tested, and it has proved to be one of the most reliable published.

Generators with very long periods have become available in recent years.[7] One such is the generator RANECU[8] which combines two linear congruential generators and has a range of 2×10^{18}. In the Appendix we give a FORTRAN listing of a 16-bit version, which has a range of more than 10^{12}, and which we have used in some of our examples.

If undetected correlations in a familiar generator are suspected a simple modification may improve their quality; the following *Bays-Durham* shuffling algorithm is among the simplest of prescriptions for this purpose — the returned uniform random number is denoted by r.[9]

The method requires a K-1 component vector H and a constant h to be saved between calls; these must be initialised by calling K variates, ξ_i, and setting $H(i) = \xi_i$, $i = 1, \ldots, (K-1)$, and $h = \xi_K$.

At a call set $k = \text{int}[Kh]$, $r = H(k)$, $h = r$, $H(k) = \xi$, where ξ is a new variate, and return r as the required variate. A value of 100 for K would be a reasonable value.

3.2.5 Random Numbers for Parallel Processing

The possibility of carrying out parts or all of Monte Carlo simulations in parallel on distinct processors is now a reality and the furnishing of the basic random numbers for these calculations has to be considered. In particular cases where there is a one-to-one correspondence between the steps of the calculations on the processors, e.g. in multiple simulations of a simple percolation problem, it will be practical to feed sub-strings of variates obtained by subdividing a string of random numbers produced by a central processor generator at regular intervals. In the more usual case where the random number demands of the different calculations or histories are different it would be difficult to arrange an optimum load balance between the processors using this method. Ideally, each processor should have its own autonomous generator, generating sequences which are statistically independent, but when the number of processors becomes large it is not easy to find, for example, the parameters to implement many equally dependable linear congruential generators for this purpose. One of the simplest approaches to the problem is to use a well tested single linear congruential generator, but have each processor independently generate subsequences from it. This can be done by virtue of a simple relation which can be shown to follow from equation 2.1 (Knuth section 3.2.1), viz. that every kth number in a linear congruential sequence can be generated using the relation

$$\nu_{i+k} = (a_k \nu_i + c_k) \bmod m$$

where

$$a_k = a^k \bmod m$$

and

$$c_k = \frac{(a^k - 1)}{(a-1)} c \bmod m = c(a^{k-1} + a^{k-2} + ... + a + 1) \bmod m.$$

This method is known as the *leapfrog* method. If n processors are in use each is equipped with the generator, equation 2.1, with parameters a_k and c_k, $k = 1, \ldots, n$, and provided once from the central processor with a seed — which can be the kth number produced there by equation 2.1 using any seed ν_0. The sequence of numbers generated in the kth processor consists of every nth member of the sequence that would have been produced by the central processor generator if it continued.

3.2.6 Quasi-random Numbers

In some applications of random numbers the lattice properties of the generator, in the sense described in section 3.2.1, may be more important than absence of correlations in a sequence. This is especially true when they are being used to evaluate multidimensional integrals, where uniformity of sampling will give rise to a faster convergence to the answer than randomly distributed points. For this reason so called quasi-random numbers are sometimes used.[10] On the other hand, they will not be suitable, for example, in studying random walks; we will have little use for them but, for completeness, give a brief description of their generation.

Several schemes of quasi-random numbers exist, Sobol' sequences, Korobov sequences, Halton sequences, etc., but we will confine ourselves to the latter type. To a *prime* base p, the van der Corput or Halton number $\eta_i(p)$, $0 \leq \eta \leq 1$, is the number i expressed to the base p and reversed to form a fraction to that base, i.e.

$$\eta_i = 0.p_k p_{k-1} ... p_2 p_1,$$

where, to the base p,

$$i = p_1 p_2 ... p_{k-1} p_k .0.$$

An elegant algorithm due to Halton will produce a sequence of such numbers as follows,

$$\eta_0 = 1/p,$$

$$y = 1/p,$$

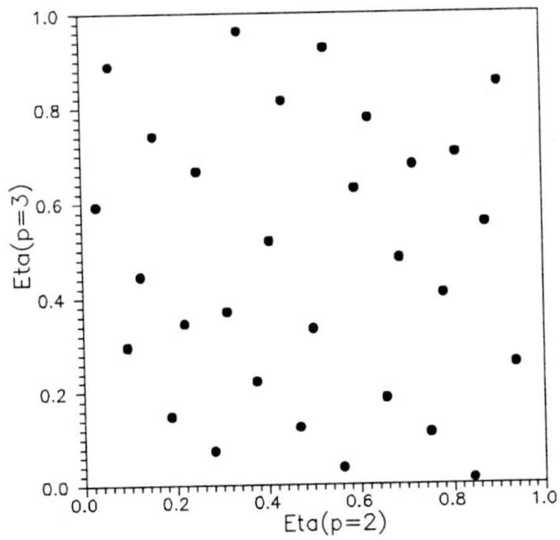

Figure 3.3: Lattice structure of quasi- random numbers in two dimensions.

$$\text{Reassign } y \longrightarrow \frac{1}{p}y, \text{ until } y < 1 - \eta_i \text{ and set}$$

$$\eta_{i+1} = \eta_i - 1 + (p+1)y.$$

Because of the inequality sign occurring, the integrity of the arithmetic must be maintained and the programming of this algorithm is best done in integers; the FORTRAN listing of such a programme is given in the paper by Berblinger and Schlier listed in the notes at the end of the chapter.[10] The coordinates of the ith point in a d-dimensional hypercube can be taken as the Halton numbers corresponding to the first d primes, $\{\eta_i(p_1), \eta_i(p_2), \ldots \eta_i(p_d)\}$. The structure of the first 30 points in two dimensions, i.e. $p_1 = 2$, $p_2 = 3$, is shown in Fig. 3.3. When compared with Fig. 3.1(a) we see a more uniform coverage of the square, the numbers 'strive' to keep away from each other.

3.3 Physically Produced Random Numbers

Monte Carlo techniques were sometimes used before the advent of the electronic age, and the random numbers required were generated using physical rather than arithmetic methods, methods such as throwing a dice — or Buffon's needle! — or spinning a wheel. In this way uniformly distributed numbers could be generated, as well as numbers from other distributions.[11] With the mastery of electronics a greater variety of random physical phenomena could be exploited: thermionic noise, radioactive decay, cosmic ray arrival times. These methods also had the advantage that they could be directly interfaced to a computer, but suffered the disadvantages of being, by today's standards, rather slow, difficult to insure against bias and the fact that a given sequence of generated numbers could not be reproduced. The duplicate of a random number sequence may be required in the debugging of a programme, or in the comparison of two physical models by simulation where it is required to investigate each model in the same 'random environment'. Apart from cases, like a national lottery, where predetermined selection procedures would be unacceptable and physically generated variates imperative, the ever lurking suspicion that pseudo-random generators are not as well understood as we think also means that random number production by physical means is of more than curiosity interest.[12]

Early accounts of the generation of uniform random numbers by physical means, mainly thermionic noise, are given in section 6.4 of Shreider,[13] more recent work includes the use of α-particles,[14] and γ-rays.[15] We will briefly describe the principles underlying these works, first describing how a string of binary bits may be generated. The idea is to image the randomness of a purely random phenomenon, i.e. one whose occurrence in any infinitesimal time interval dt is simply proportional to dt, on to the integers in some interval in order to produce integers at random from that interval. The state of a system triggered randomly by the phenomenon can be interrogated at regular intervals or, alternatively, the state of an already 'flipping' system can be interrogated at times determined by the random process. In the latter case, for example, a scalar records the cumulative count from a stable oscillator of frequency f. The count is interrogated at random times, if it is odd a zero is generated, otherwise we generate a one; the count is then reset. The timing of the interrogation is determined by a signal, in this case the detection of a decay in a radioactive source. Provided the efficiency of detection does not vary and background contributions are negligible this signal can be considered random. Corrections to allow for the slightly different probabilities of even and odd integers can be directly built into the hardware — see the book by Shreider. The same method can be used to produce a string of decimal digits 0-9, by utilising the last digit of the recorded count, with the correction for bias being then correspondingly more important.

In Fig. 3.4 we show a block diagram of an experiment using γ-rays from a radioactive source to gate a scalar recording the cumulative count from a 500 kHz oscillator, and using the last decimal digit of this count.[15] The digits, produced at a rate of about 2 million per hour, were then output to a magnetic tape. All tests conducted on these numbers showed them to be acceptable, and in terms of hyperspace distribution to be superior to numbers generated by some common software generators. Tables of physically generated numbers which have been extensively tested, have been made available on magnetic tape by these authors.

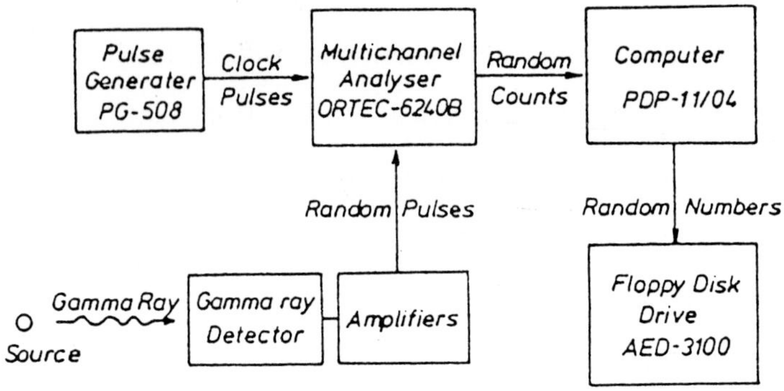

Figure 3.4: Block diagram of the apparatus used to generate uniform random variates from radioactive decay (after Inoue *et al.*, Appl. Statist., **32**, 115–20, 1983).

3.4 Testing Random Number Sequences

Even if the full theory underlying a given generator, whether it be for uniformly distributed numbers or numbers from some other distribution, were understood it would be a foolhardy individual who would bypass the opportunity to test that the generated sequence followed the distribution in question and represented a random selection from that distribution.

Using a congruential generator with full period we know that all the numbers will occur and so the uniformity of the distribution is guaranteed; uniformity in much shorter sequences will, however, be of greater practical importance and must be tested for. This can readily be done, as for any other frequency test, by subjecting a block of numbers to a χ^2-test, described in section 2.5.1. If a block of N numbers is histogrammed into k cells the statistic is

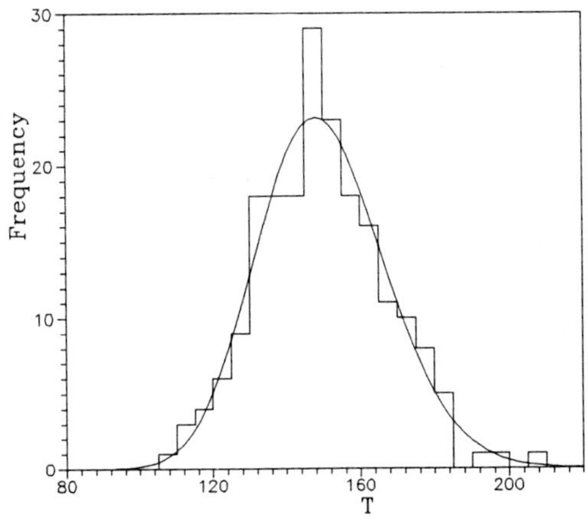

Figure 3.5: The χ^2 variable distribution in testing Super-duper for uniformity, together with the χ^2 distribution.

$$T = \overline{n}^{-1} \sum_{1}^{k} (n_i - \overline{n})^2,$$

where $\overline{n} = N/k$. By testing several successive blocks of numbers in this way the sequence of individual values of T can be tested for agreement with the theoretical χ^2 distribution.

EXAMPLE 3.2: Uniformity of Super-duper

The results of such a uniformity test using strings of $N = 10000$ variates generated by Super-duper, described earlier, are summarised in Fig. 3.5. The numbers were histogrammed into $k = 151$ cells and the statistic T calculated; it should follow the χ^2 distribution with $\nu = 150$ degrees of freedom. This was done for 200 samples, the histogram of these values is compared with $p_{\chi^2,\nu}(T)$ in Fig. 3.5. The cumulative frequency from the histogram was compared with $P_{\chi^2,\nu}(T)$ and the Kolmogorov-Smirnov statistic q, defined by equation 2.5.3, formed. This had a value $q = 0.606$, which, using equation 2.5.4, yields a 46% probability of obtaining a larger value in such a comparison with the predicted χ^2 distribution. From this we conclude that there is no evidence of deviation from uniformity in samples of size 10000 drawn from this generator.

Once uniformity is assured some tests for correlations must be carried out. We illustrate the application of two simple tests, Good's serial test and a test of the correlation coefficients; in serious testing of a generator a full battery of tests would be undertaken,[16] and much larger samples than are taken in these examples used.

EXAMPLE 3.3: Correlations in Super-duper: Good's serial test

For each of 200 samples, each sample containing $N = 10000$ variates, Good's serial test was carried out for overlapping pairs of numbers as described in section 2.5.2. The N-1 points $(\xi_i, \xi_{i+1}), i = 1, \ldots, N-1$, in the unit square were allotted to k^2 sub-squares with $k = 16$, and the following quantities formed,

$$e_1 = (N-1)/k, \qquad e_2 = e_1/k, \qquad c_i = \sum n_{ij}$$

$$t_1 = \sum_{i=1}^{k}(c_i - e_1)^2/e_1$$

and

$$t_2 = \sum_{i=1}^{k}\sum_{j=1}^{k}(n_{ij} - e_2)^2/e_2.$$

The statistic w defined by

$$w = t_2 - 2t_1$$

was calculated. The distribution in the values of w so obtained is shown by the histogram in Fig. 3.6(a). Also shown in that figure is the chi-square distribution $p_{\chi^2,\nu}(w)$ for $\nu = k(k-2)+1 = 225$ degrees of freedom which this distribution should obey to a high accuracy. To compare the consistency of these two distributions we form the corresponding cumulative distributions — they are compared in Fig. 3.6(b) — and again apply the Kolmogorov-Smirnov test. The resulting value

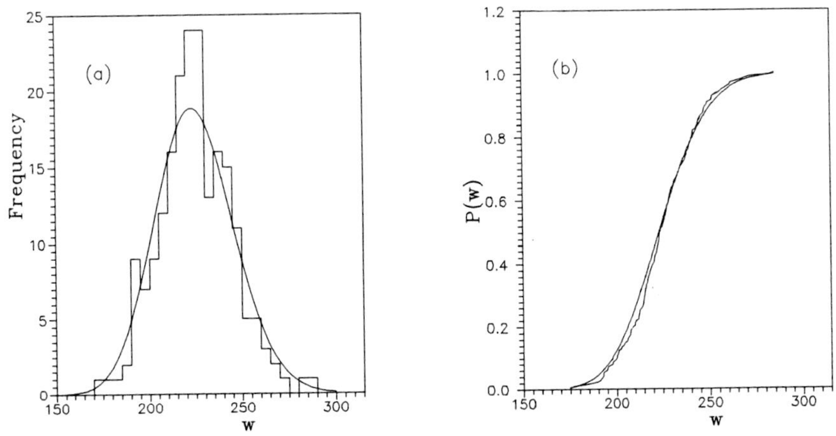

Figure 3.6: (a) The distribution of the statistic w from Good's serial test applied to Super-duper, together with its theoretical curve. (b) The same comparison using the cumulative distributions, the basis for applying the Kolmogorov-Smirnov test.

of the statistic is $q = 0.957$. Using equation 2.5.4 there is a 15.3% probability of getting a value larger than this in a sample of size 200 in such a comparison — not an unacceptable value, but a hint that more extensive testing should be carried out.

EXAMPLE 3.4: Correlations in Super-duper: Correlation coefficients

In this test we calculate the j-separation correlations defined by equation 2.2.9a for $j = 1, 10$ and 100 for a sample of 10000 variates generated by Super-duper. Since the parameters of the parent population are known ($\mu = 1/2, \sigma^2 = 1/12$) in this case estimates of these correlation coefficients are

$$\begin{aligned} r_j &= 12(<\xi_i\xi_{i+j}> -1/4) \\ &= 12(<(\xi_i - 1/2)(\xi_{i+j} - 1/2)>, \end{aligned}$$

where, because of round-off error, the second expression must be used in calculation. The relevant statistic is $u = Nr_j^2$ which, independent of j, can be shown to approximately have the chi-square distribution with one degree of freedom, i.e.

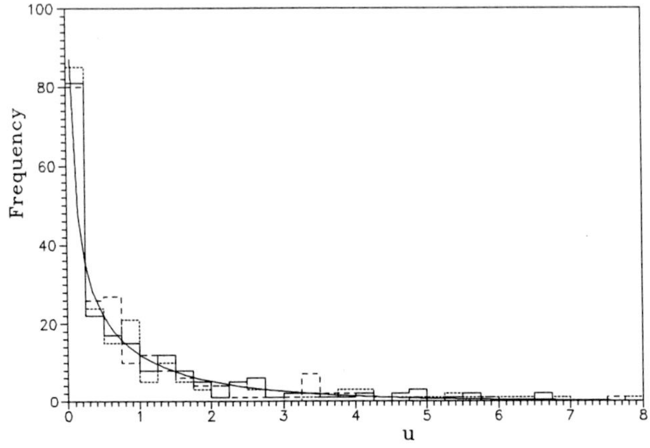

Figure 3.7: The distribution in the statistic u for lagged correlations for Super-duper, compared with the chi-square distribution for one degree of freedom: —— lag $j = 1$, $- - - j = 10$, $\ldots, j = 100$.

$$p_{\chi^2,1}(u) = \frac{1}{\sqrt{2\pi}} u^{-1/2} e^{-u/2}.$$

For each of the three values of j, 200 such coefficients were determined. Their distributions are shown in Fig. 3.7, where they are compared with the above theoretical distribution.

EXAMPLE 3.5: A composite test for correlations

An alternative to a calculation of the correlation of lagged pairs is to combine such pairs, or other multiplets, to form a variate whose distribution, based on independent variates, is theoretically known. For example, using equation 2.2.14 we find that the product two independent uniform random variates, $z = \xi_i \xi_{i+j}$ should be distributed as $p(z) = -\ln z$ for $0 < z \leq 1$. Taking $j = 1$, 10000 non-overlapping pairs from Super-duper were formed; their distribution is shown in Fig. 3.8 and compared there with this theoretical distribution. With 99 degrees of freedom, the chi-square T has a value of 72.7, corresponding to an agreement between the two distributions at the 98% confidence level.

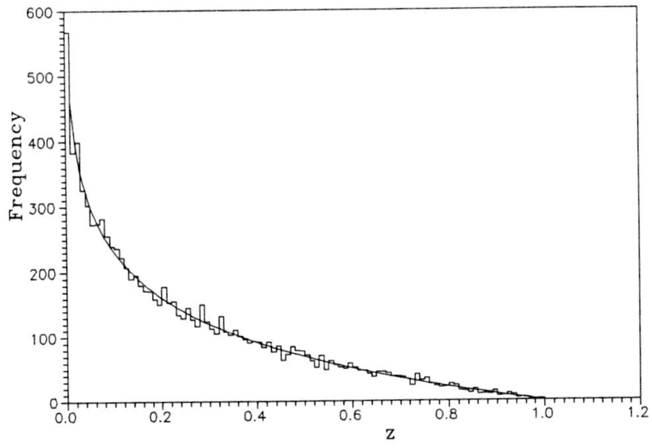

Figure 3.8: Distribution of the variate $z = \xi_i \xi_{i+1}$ generated by Super-duper, together with the theoretical expectation.

EXAMPLE 3.6: A block test for correlations

An important source of information on the quality of random number generators is the results obtained when using the generated numbers to simulate, with high statistical accuracy, a problem whose solution is known. Undetected deficiencies in some generators have been revealed in this way by simulating statistical mechanics problems using so called cluster algorithms — we describe them in section 10.5.3. These algorithms are sensitive to non-randomness, in particular non-uniformity, in successive sequences of a given length. A very simple block test can reveal correlations, and is carried out as follows.[17] A large number, N, of non-overlapping sequences of length m, $(\xi_1, \xi_2, \ldots \xi_m)$, are generated and in each case the mean is found. The number of cases where the mean is less than a half, N_-, and greater than a half, N_+, is tested against expectation using the chi-square statistic, i.e.

$$\begin{aligned} T &= \sum_{i=\mp} \frac{(N_i - N/2)^2}{N/2} \\ &= \frac{(2N_+ - N)^2}{N}. \end{aligned}$$

This quantity should follow the chi-square distribution with one degree of freedom. In Fig. 3.9 we show values obtained for $N = 10^5$

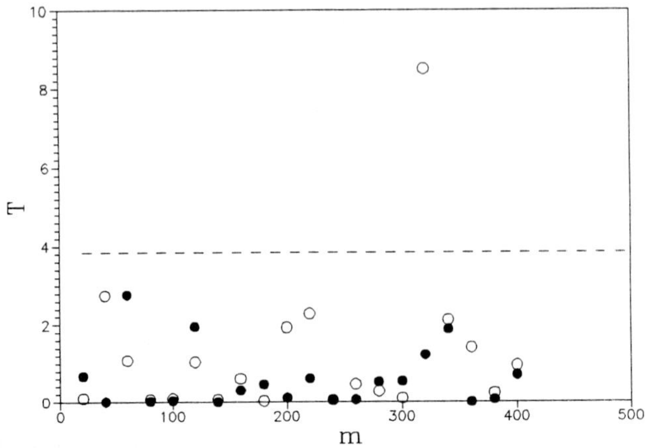

Figure 3.9: Values of the chi-square statistic for different block sizes in the range 20 to 400. ⊙ for Super-duper, • for RANECU. The horizontal line represents the 5% level.

samples of sequences of different length in the range $m = 20$ to 400 from the generator Super-duper (open circles), and from the generator RANECU described in the Appendix (solid circles). The horizontal line at 3.841 represents the 5% level for the chi-square test with one degree of freedom. No systematic divergence from expectation is seen in either case, and one can be fairly confident of the quality of the variates produced by these generators.

Variates from the standard uniform distribution form the basis for the generation of variates from any other distribution, and we will see several ways of employing them for this purpose, but before that we look at a number of phenomena which can be studied using only variates from this distribution in the next chapter.

Notes

1. S. K. Park and K. W. Miller (1988), *Random Number Generators: good ones are hard to find*, Comm. ACM, **31**, 1192-1201.

2. Readable accounts of 'randomness' may be found in, Chaitin,G. J. (1975), *Randomness and Mathematical Proof*, Scientific American, **232**, 47-52, and in Compagner, A. (1991), *Definitions of randomness*, Am. J. Phys., **59**, 700-705.

3. Marsaglia, G. (1968), *Random numbers fall mainly in the planes*, Proc. Nat. Acad. Sci. USA, **61**, 25-28.

4. Some dramatic examples of possible correlations are also shown in Ripley, B. D., *Stochastic Simulation* (New York: John Wiley & Sons., 1987).

5. Marsaglia, G., B. Narasimhan, and A. Zaman (1990), *A random number generator for PC's*, Comput. Phys. Comm., **60**, 345-49.

6. McLeod, A. I. (1981), *An Extremely-portable High-quality FORTRAN Random Number Generator*, Technical report $TR-81-01$, University of Western Ontario Department of Statistical and Actuarial Sciences.

7. The current situation has been reviewed by James, F. (1990), *A review of pseudorandom number generators*, Comput. Phys. Comm., **60**, 329-44; this paper also gives Fortran listings of some practical versions for 32-bit machines.

8. l'Ecuyer, P. (1988), *Efficient and Portable Combination Random Number Generators*, Comm. ACM, **31**, 742-74, where the Pascal code may be found.

9. Bays, C. and S. D. Durham (1976), *Improving a poor random number generator*, ACM Trans. Math. Software, **2**, 59-64.

10. Quasi-random numbers and their uses are described in, James F. (1980), *Monte Carlo theory and practice*, Rep. Prog. Phys., **43**, 1145-89. The programme used here is that given by Berblinger M. and C. Schlier (1991), *Monte Carlo integration with quasi-random numbers: some experience*, Comput. Phys. Comm., **66**, 157-66. Sobol' sequences are described by Press W. H. and S. A. Teukolsky (1989), *Quasi- (that is Sub-) Random Numbers*, Computers in Physics, **3**(6), 76-79.

11. A pioneering study which is still worth reading, physically selecting variates from some important distributions describing electron photon interactions, was done by Wilson R. R. (1952), *Monte Carlo Study of Shower Production*, Phys. Rev., **86**, 261-69.

12. The first draft of this book included the sentence 'a useful pair is $(p = 250, q = 103)$, which gives a sequence of 2^{250} bits or a period of $2^{218} \simeq 410^{65}$ 32-bit numbers before it recycles'. Generators like this, which produced sequences which passed all statistical tests applied, and believed of 'high quality', are reported to give rise to erroneous results to problems with a known solution when employed in a sophisticated algorithm to attack such problems. Subtle, undetected, correlations in the sequences are assumed to be responsible. See Ferrenberg, A. M., D. P. Landau and Y. J. Wong (1992), *Monte Carlo Simulations: Hidden Errors from 'Good' Random Number Generators*, Phys. Rev. Letts., **69**, 3382-84. This all goes to emphasise that one can never take the quality of pseudo-random number generators absolutely for granted.

13. See also Tocher K. D., *The Art of Simulation*, (London; The English Universities Press, 1963).

14. Frigerio N. A. and N. Clark (1975), *A Random Number Set for Monte Carlo Computations*, Trans. Amer. Nucl. Soc., **22**, 283-84. These authors relate how the use of numbers generated in this way resulted in better agreement of experimental results on reactor fluxes with their simulations than obtained when these were carried out with software generators.

15. Inoue H., H. Kumahora, Y. Yoshizawa, M. Ichimura, and O. Miyatake (1983), *Random numbers generated by a physical device*, Appl. Statist., **32**, 115-20.

16. For example, Dudewicz, E. J. and T. G. Rawley, *The Handbook of Random Number Generation and Testing*, (Columbus, OH: American Science Press, 1981).

17. Vattulainen I., T. Ala-Nissila and K. Kankaala (1994), *Physical tests for random numbers in simulations*, Phys. Rev. Lett., **73**, 2513-16.

Exercises

3.1 Use your random number generator to produce a long sequence, say 10000, of integers in the range (0, 9), i.e. write $k = [10\xi]$. Subdivide the sequence into groups of ten digits, and for each group determine how many, n, of its terms are odd. Plot a histogram of these values, $0 \leq n \leq 10$, and compare it, e.g. using the χ^2 test in section 2.5.1, with the binomial distribution, equation 2.4.1, with $N = 10$ and $\alpha = 1/2$.

3.2 Confirming, by simulation, the answer to a problem whose exact solution is already known is helpful in improving one's confidence in a random number generator. Two suitable examples for this purpose are Buffon's needle and the random walk problem (considered in the next exercise). Simulate the random throwing of a needle of unit length on a unit square and count the number of sides intersected in each throw. Uniform random numbers are required to randomly locate the coordinates $(x,\ y)$ of the centre of the needle in the square, and to determine the orientation angle ϑ on $[0, \pi]$. For different sample sizes N use equation 1.1.2 to estimate π and its associated uncertainty, and compare the results with the known value.

3.3 In the simplest random walk (considered in greater detail in chapter 8) a walker starts at the origin and makes successive steps of unit length to the left or right with equal probability. After N steps the probability of being displaced by k steps from the starting point is given by the binomial distribution, equation 2.4.1, for $\alpha = 1/2$ with argument $\frac{N-k}{2}$, i.e.

$$p(k) = p_N(\frac{N-k}{2}) = \frac{N!}{\frac{N-k}{2}!\frac{N+k}{2}!}(\frac{1}{2})^N, k = -N, -N+2, ..., N-2, N.$$

Take $N = 20$ and simulate 1000 walks starting from the origin to build up the probability distribution and compare it, using the χ^2 test in section 2.5.1, with this distribution. As is frequently the case in Monte Carlo calculations, you can use subsets of the data to carry out the test on shorter walks, e.g. $N = 5, 10$. Caution, however, is then necessary in an overall assessment of the results because of a lack of independence between the samples.

3.4 As a search for short range correlations in your random number generator, generate 9999 uniform random variates and use non-overlapping triplets $(\xi_{3i-2}, \xi_{3i-1}, \xi_{3i})$ to generate variates z_i as follows.

Arrange the three in increasing order, i.e. $\xi_1 < \xi_2 < \xi_3$ and set $z = \xi_2$.

If the ξ variates are uncorrelated, the values z should follow the distribution

$$p(z)dz = 6z(1-z)dz.$$

Histogram your z values into 100 cells, superimpose this distribution and calculate the chi-square statistic — there are $\nu = 99$ degrees of freedom.

3.5 Carry out a test of the digits of π (e.g. using successive triplets) as a source of uniform random numbers. A remarkable 'spigot' algorithm which will generate the digits in sequence is given by S. Rabinowitz and S. Wagon, Am. Math. Monthly, **102**, 195-203, (1995).

3.6 Can we use chaos (i.e. a chaotic map) to generate random numbers? The iterated sequence

$$x_{n+1} = 4rx_n(1-x_n)$$

is known as the logistic map.

(a) Take $r = 0.5$ and investigate how x_n is distributed *asymptotically* — present a histogram of the distribution. (Do not take $x_0 = 0$ or 1 or $1 - 1/4r$.)

(b) Repeat the above with $r = 1$. Compare the distribution with

$$p(x)dx = \frac{dx}{\pi\sqrt{x(1-x)}}.$$

Plot the distribution of the variables $y_n = \frac{2}{\pi}\sin^{-1}\sqrt{x_n}$, and $z_n = \ln\left(\frac{x_n}{1-x_n}\right)$, and compare them with the uniform and Gaussian distributions, respectively.

Chapter 4

SIMPLE STOCHASTIC MODELS

> Whatever your views and beliefs on randomness – and they are more likely than not untenable – no great harm will come to you. If the discipline you practice is sufficiently robust...
>
> M. Kac

4.1 Introduction

This short chapter will show how, equipped with a uniform random number generator, we are already in a position to investigate some physical phenomenon of importance, *percolation*, *self-organised criticality* and stochastic optimisation — *genetic algorithms*, and introduce some techniques of importance for later chapters. The idea of percolation arises in considering a fluid contained by a partly porous wall. In such a wall truncated channels of various lengths and directions exist, and if the infiltration, the degree of porosity, is such that a continuous connected path of channels linking the inner and outer surfaces occurs, fluid can pass through the wall — it *percolates*. Such behaviour is of interest, for example, in hydrologic studies and in petroleum engineering. A specimen of material might be tested for its percolating property, or critical porosity, by injecting mercury into it and searching for an electrical conducting path across it. A similar occurrence of electrical current flow will exist in an insulating medium which contains randomly distributed metallic inclusions when the density of such inclusions is high enough. The concept has also been exploited in studying phenomena as diverse as the percolation of charged particles, transported along the tangled magnetic field lines frozen into a plasma and the configuration of galactic clusters and voids in the universe. These ideas can be extended to higher dimensions, and, helpfully, down to two dimensions, which is the case we will use for simplicity of illustration. For the same reason, although practical problems of percolation often occur in a continuous medium,[1] we will only concern ourselves with structured arrays of paths — lattice percolation. There is another way of looking at lattice percolation, in terms of configurations of 'occupied' lattice sites.

4.2 Percolation

4.2.1 Site Percolation

We consider the simple configuration of a square array of $L \times L$ sites each of which is 'occupied' with a probability p and otherwise empty. Nothing more than a uniform random number generator is required to generate configurations; at each site a uniform number ξ_i is compared with p and if $\xi_i < p$ the site is occupied. Examples of individual configurations with $L = 16$, generated taking $p = 0.25$ and 0.6, are shown, respectively, in Figs 4.1(a) and 4.1(b). *Clusters* of occupied sites are readily identified, showing a spectrum of sizes, in the former case ranging from a single site to a cluster of ten sites. In Fig. 4.1(b) there is a cluster of 118 occupied sites. We define a spanning

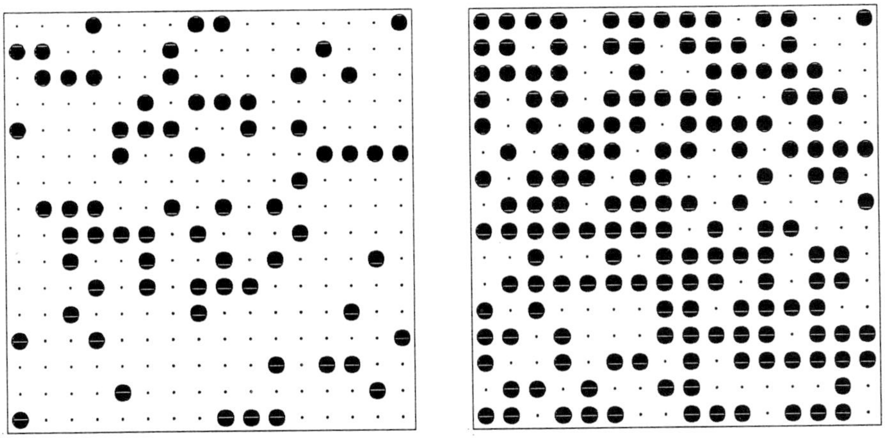

Figure 4.1: Configurations of occupied sites generated with different probabilities, (a) $p = 0.25$ and (b) $p = 0.6$.

cluster as one which extends unbroken from one side of the square to its opposite side; the large cluster in Fig. 4.1(b) is of such a type. If each occupied site is viewed as a pair of orthogonal *bonds*, a conducting path along such bonds between the two sides can be found. Site percolation and bond percolation are not equivalent, however, as can be seen by looking at Fig. 4.2, based on site occupation with probability $p = 0.48$. The corresponding bond configuration is easily visualised. In neither case does a spanning cluster exist. Whereas the pair of bonds associated with a site is required at point b to produce a spanning cluster, or percolating path (north-south), only a single bond is required at a in the figure to produce such a path (east- west). The two types of percolation must be distinguished; the number of sites on an $L \times L$ lattice is obviously L^2 while the number of bonds on the same lattice is $2L(L-1)$. Our discussion here will be for site percolation; analogous, but not identical, considerations hold for bond percolation which we describe in section 4.2.3.

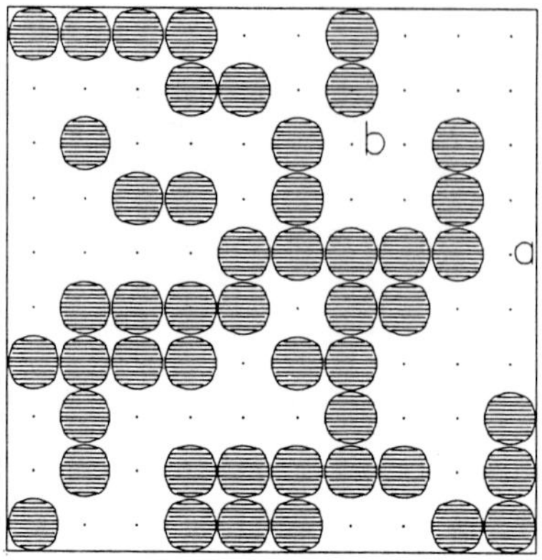

Figure 4.2: The relationship of site percolation to bond percolation.

At a given value of occupation probability p a spanning cluster may or may not exist in an individual configuration of finite size.[2] However, in a large ensemble of configurations generated with that value of p the average probability, π, for such a spanning cluster occurring can be found. This probability will be a function of the probability of occupancy, p, i.e. $\pi = \pi(p)$, and it will clearly be an increasing function of p. It is also plausible that the function will depend on the size of the lattice L, i.e. $\pi = \pi_L(p)$; theoretical interest centres on the $L \to \infty$ limit behaviour of this function. As we will see, a sharp transition in the likelihood of a spanning cluster occurs in the neighbourhood of a critical value of p, p_c, very reminiscent of the spontaneous occurrence of magnetisation in a ferromagnetic material when the temperature passes through the Curie temperature, T_c. In fact, many of the properties of the, geometric, percolation transition have close analogues in thermodynamic phase transitions, and the subject provides a gentle introduction to these phenomena, without having to worry about the complicating 'thermal equilibrium' aspects. It should be added that all of these properties may depend on the structure of the lattice involved, e.g. square, honeycomb, etc. To investigate the percolation probability for finite L at a given value of p by Monte Carlo methods we simply need to generate many configurations of that size and count the fraction displaying a spanning cluster. Visually identifying a cluster and deciding if it is of the spanning type is easy; devising an algorithm to identify clusters, to assign individual sites to clusters, to label the clusters, and finally to decide if a spanning cluster exists, however, is not such a trivial exercise. Although not part of the Monte Carlo aspect of this work, the ability to identify and handle clusters is a valuable technique in sampling configurations in Monte Carlo simulations of multi-particle systems — for example, in the Swendsen-Wang algorithm which we describe in section 10.5.3 — and we will devote a little space to it.

4.2.2 Cluster Counting; The Hoshen-Kopelman Algorithm

We illustrate the concepts involved in cluster determination using the array in Fig. 4.2, which is represented by a 10×10 array $M(i, j)$ with entries 0 and 1 for unoccupied and occupied sites. The Hoshen-Kopelman algorithm requires us to build up two arrays, labels $L(i, j)$ and indices, $N(L_{ij})$.[3] Starting in the top left-hand corner we scan, raster fashion, assigning a number, label, to each occupied site as follows: assign to a site a new integer L_{ij} if neither the site to the north or west of it is occupied and set $N(L_{ij}) = L_{ij}$, while if *only one* of these neighbours is occupied assign its label to the site. For this purpose it is convenient to add a column and a row of unoccupied sites, zeros, on the west and north of the array — not shown in the following illustrations. In this way labels and indices are assigned as follows,

$$\begin{array}{cccccccccclll}
1 & 1 & 1 & 1 & 0 & 0 & 2 & 0 & 0 & 0 & N(1)=1 & N(2)=2 & \\
0 & 0 & 0 & 1 & 1 & 0 & 2 & 0 & 0 & 0 & & & \\
0 & 3 & 0 & 0 & 0 & 4 & 0 & 0 & 5 & 0 & N(3)=3 & N(4)=4 & N(5)=5
\end{array}$$

So far we have assigned five labels, and the array N reads $[1,2,3,4,5]$. Because of the limited overview of the scan a new label 5 has been assigned to a cluster which, visually, we recognise as being the same as that labelled 4. Label 5 is clearly redundant, however we cannot do anything about it at this stage, the purpose of the index array is to enable us to return to purge this label. Continuing we get,

$$\begin{array}{cccccccccccl}
0 & 0 & 6 & 6 & 0 & 4 & 0 & 0 & 5 & 0 & N(6)=6 \\
0 & 0 & 0 & 0 & 7 & ? & & & & & N(7)=7
\end{array}$$

at which stage $N = [1,2,3,4,5,6,7]$.

Now we are at a site where both the west and north site have labels; if they were equal there would be no problem in assigning the label. When, as here, they are unequal at least one of them is redundant. We set our label equal to the smaller one and at the same time set the index of the larger one also equal to this value, i.e. $N(7) = 4$.

We now have for this row

$$\begin{array}{cccccccccll}
0 & 0 & 0 & 7 & 4 & 4 & 4 & 4 & 0 & N(7)=4 & N(5)=4
\end{array}$$

where the same prescription has been applied to the second last site in the row, and $N(5)$ set equal to 4. Continuing in this way we assign labels and indices,

$$\begin{array}{ccccccccccll}
0 & 8 & 8 & 8 & 7 & 0 & 4 & 4 & 0 & 0 & N(8)=7 & \\
9 & 8 & 8 & 8 & 0 & 10 & 4 & 0 & 0 & 0 & N(9)=8 & N(10)=4 \\
0 & 8 & 0 & 0 & 0 & 0 & 4 & 0 & 0 & 11 & N(11)=11 & \\
0 & 8 & 0 & 12 & 12 & 12 & 4 & 4 & 0 & 11 & N(12)=4 & \\
13 & 0 & 0 & 12 & 12 & 12 & 0 & 0 & 14 & 11 & N(13)=13 & N(14)=11
\end{array}$$

Our final array of indices now reads,

i	1	2	3	4	5	6	7	8	9	10	11	12	13	14
Index $N(L(i))$	1	2	3	4	4	6	4	7	8	4	11	4	13	11
Label $L(i)$	1	2	3	4	5	6	7	8	9	10	11	12	13	14

Labels whose index is not the same as the label number are redundant and must now be reassigned to their 'home' cluster as follows:

> If the index of the label is not the same as the label number assign a label number equal to the index and if the index of the label is not.........

After such an exercise each cluster will now have a unique label; for the above case the result is the seven clusters,

```
 1   1  1  1  ·  ·  2  ·  ·   ·
 ·   ·  ·  1  1  ·  2  ·  ·   ·
 ·   3  ·  ·  ·  4  ·  ·  4   ·
 ·   ·  6  6  ·  4  ·  ·  4   ·
 ·   ·  ·  ·  4  4  4  4  4   ·
 ·   4  4  4  4  ·  4  4  ·   ·
 4   4  4  4  ·  4  4  ·  ·   ·
 ·   4  ·  ·  ·  ·  4  ·  ·  11
 ·   4  ·  4  4  4  4  4  ·  11
13   ·  ·  4  4  4  ·  · 11  11
```

We note that the existence of a spanning cluster is established if the same label occurs on opposite faces. A sophisticated algorithm fast, and efficient in use of memory space if only percolating clusters are searched for, is described in the appendix in Stauffer.[1]

EXAMPLE 4.1: Site percolation on a square

We study the percolating properties of an array with $L = 50$; in terms of the asymptotic limit this is a rather small array. Results are based on a sample of 400 configurations generated at each probability p, and the cluster properties are derived using the algorithm described in the previous section. The results are summarised in Figs 4.3(a) – 4.3(c).

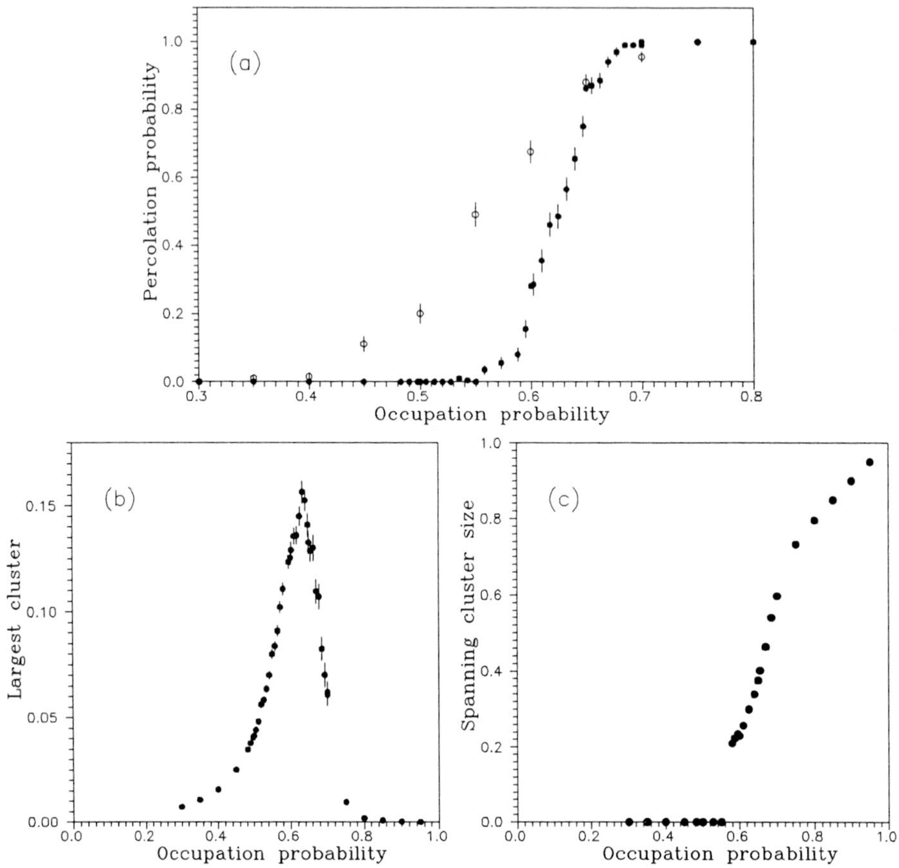

Figure 4.3: (a) The percolation probability, $\pi_L(p)$, for $L = 10$, open circles, and $L = 50$. (b) The fractional size, ξ, of the largest non- spanning cluster for $L = 50$. (c) The fraction of sites, P_∞, which are part of a spanning cluster.

The percolating probability, $\pi_L(p)$, is shown in Fig. 4.3(a). The sharp transition from absence to the presence of a percolating cluster over a narrow range of p values is seen. The results for a system with $L = 10$, shown in the figure by the open circles, illustrate how the sharpness of the transition depends on L; the comparison is suggestive of a true critical value, p_c, in the limit $L \to \infty$. Taking the value of p for which the percolating probability is 50% we find the estimate $p_c \simeq 0.63 \pm .02$ (the most accurately known result in two dimensions is $p_c = 0.5927$). In Fig. 4.3(b) we show the size (the number of sites) of the largest non-spanning cluster normalised to $L^2, \xi(p)$, which is a measure of the coherence length in the system, as a function of p. This plot suggests a singularity in this quantity as $L \to \infty$ at the critical value of p, and the whole behaviour is very reminiscent of the behaviour of, for example, the temperature dependence of the specific heat in the neighbourhood of a second order phase transition. In Fig. 4.3(c) we plot, as a function of p, the fraction of sites in the array which are part of the spanning cluster when present, P_∞. This quantity is a measure of the volume of the system which can contribute to fluid or current flow.

4.2.3 Bond Percolation

The techniques for a Monte Carlo study of percolation have been described earlier in the context of site percolation, mainly because of the greater simplicity of the description in this case. The case of bond percolation is somewhat more important, because of its relevance, for example, to fluid flow or the flow of a current in a random resistor network, but also because it serves as the model for sophisticated algorithms in Monte Carlo studies in statistical mechanics of multi-particle systems, as we shall see in section 10.5. The algorithms in this case are also slightly more complicated.

In a two-dimensional lattice of $L \times L$ sites where a bond is occupied with probability p, the number of bonds can range from 0 to $2L(L-1)$; an example for $L = 6$ with $p = 1/3$ is shown in Fig. 4.4(a). The bond sites are numbered as in Fig. 4.4(b) — here, for programming convenience in writing the labelling algorithm, the array is surrounded on three sides by neutral sites which define additional bonds that will always be unoccupied. A bond array, $B(2L, L+1)$, is defined such that $B_{ij} = 1$ when a bond is occupied, and zero otherwise. The bond array corresponding to the configuration in Fig. 4.4(a) is shown, after labelling and cluster identification, starting with B_{21} in Table 4.1. The labelling and cluster identification method described in section 4.2.2 can be used with some modifications. In assigning the provisional labels,

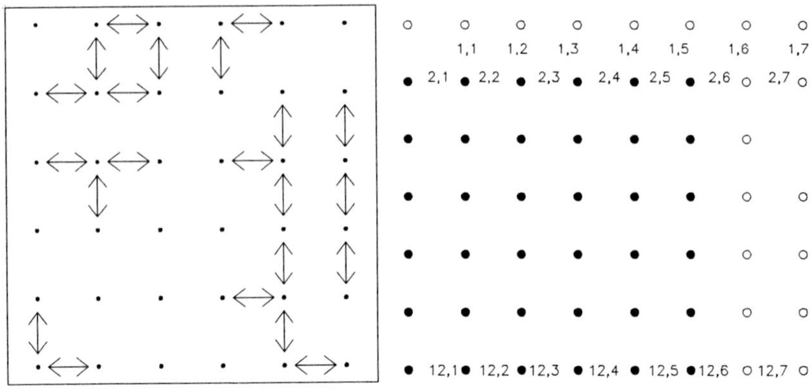

Figure 4.4: (a) An array of bonds, (b) the numbering scheme, B_{ij}.

Table 4.1: Labelled bond array.

0	0	1	0	2	0
0	1	1	2	0	0
0	1	1	0	0	0
0	0	0	0	3	4
0	5	5	0	3	0
0	5	0	0	3	4
0	0	0	0	0	0
0	0	0	0	3	4
0	0	0	0	3	0
6	0	0	0	3	0
0	6	0	0	0	3

'vertical' (or north-south) bonds, B_{ij} where i is odd, and 'horizontal' bonds, where i is even, must be distinguished. At any bond (i,j) in the labelling scan through the array three possible neighbouring bonds must be checked for contact. The neighbours of B_{ij} for vertical bonds are $B_{i-2,j}$, $B_{i-1,j}$ and $B_{i-1,j+1}$, while for horizontal bonds they are $B_{i-1,j-1}$, $B_{i-1,j}$ and $B_{i,j-1}$. A percolating cluster corresponds to a common label in $B_{3,j}$ and $B_{L-1,j}$, $(1 \leq j \leq L)$ or in $B_{i,2}$ and $B_{i,L}$, $(2 \leq i \leq 2L,\ i \text{ even})$.

EXAMPLE 4.2: Bond percolation on a square

The properties of bond percolation for $L = 50$ have been investigated, using the labelling algorithm described. In Fig. 4.5 the percolating

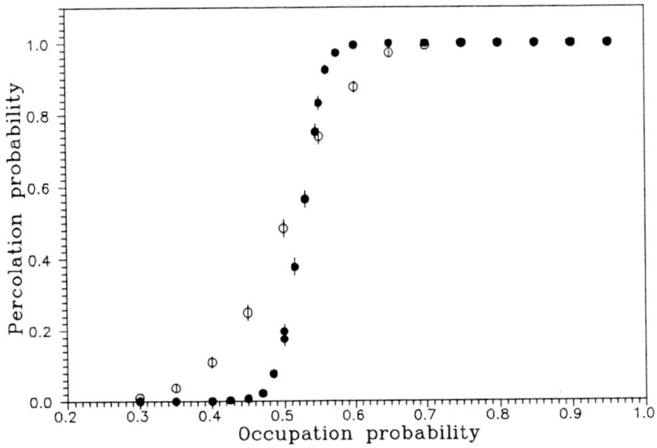

Figure 4.5: The bond percolation probability for $L = 50$, solid circles, and $L = 10$, open circles.

probability is shown, with the results for $L = 10$ superimposed. As expected, the transition extends over a finite range of occupation probability; the best estimate gives $p_c(50) \simeq 0.53$, which is to be compared with the known value $p_c = 1/2$ for an infinite array. In Figs. 4.6(a) and (b) we present the results for ξ and P_∞. We have done this in the neighbourhood of the percolation transition, which we have assumed occurs at $p_c = 0.53$, in the form of log-log plots. In these figures we see that we can approximate the dependence by a power law,

$$\begin{aligned} \xi(p) &\approx |\, p - p_c \,|^{-\nu}, \\ P_\infty(p) &\approx (p - p_c)^\beta. \end{aligned}$$

For $|\ p - p_c\ | << p_c$, the exponents β and ν in these dependences are known as *critical exponents*; in serious studies of the percolation transition, and even more so in the study of phase transitions, these parameters are of great importance in systematising the properties of the transitions, and in extrapolating from actual simulations to the infinite, thermodynamic, limit. For the two-dimensional case considered in the previous example the known values of these exponents are $\beta = 5/36$ and $\nu = 4/3$. Because of our poor estimate of the threshold value $\mathrm{p_c}$ and the small size of our array our estimates are far from these, but the present simple sketch of how they may be estimated is enough for our purposes of illustrating the use of Monte Carlo methods.

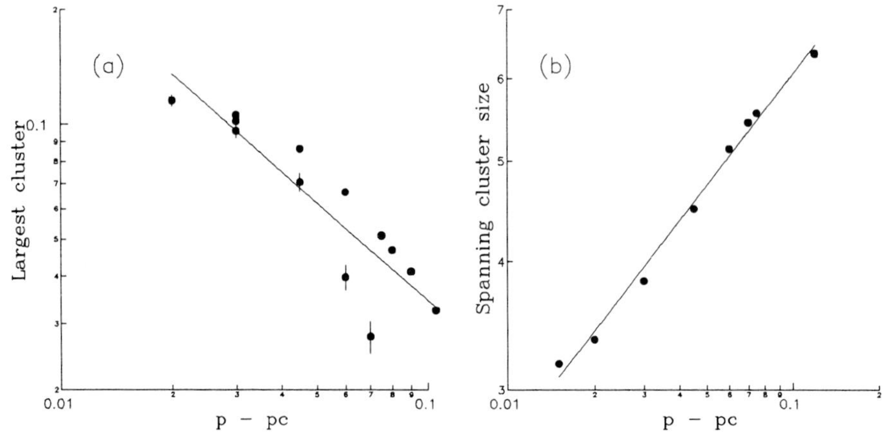

Figure 4.6: (a) The behaviour of $\xi(p)$, and (b) $P_\infty(p)$ in the neighbourhood of the percolating transition.

4.3 Self-organised Criticality

Although computer studies play an important role in arriving at the properties of the phenomenon of percolation, the basic aspects of the subject had been well investigated by conventional methods. In the case of our second topic which can be studied using just uniform random numbers, *self-organised criticality* (SOC) computer methods played a decisive role in its discovery and in clarifying its basic properties.[4] Some complex systems, e.g. an accreting sand pile or the motion of the earth's crust, cannot be readily reduced to the motion of their microscopic components. These systems can attain critical states which are intrinsic to the nature of the system itself, and not the product of external control factors. The behaviour of such systems is known as self-organised. Like percolation, no temperature is involved and the system it describes is in equilibrium, but in this case a dynamic equilibrium which is achieved in a self-consistent way through the interplay of the coupled forces acting within it.

Self-organised critical systems typically are coupled to an energy source which slowly pumps energy into the system. From an arbitrary initial configuration the system evolves by continually reorganising itself to a macroscopic, critical state which characterises it from then on. The stored energy is released thereafter in unpredictable 'avalanches' whose duration is very short compared to the time interval separating them. In terms of size these avalanches can vary from the smallest possible up to arbitrarily large values (limited only by the finite extent of the system). The avalanche sizes follow some distribution, typically an inverse power law, which is a constant property of the self-organised state. Examples of phenomena which are believed to show this behaviour are avalanches in snow and sand piles, earthquakes and tremors, fires in a forest triggered by lightning, outbursts of plasma from the sun (solar flares), 'glitches' in the rotation period of neutron stars, etc. The avalanches occurring in this state are analogous to the fluctuations, e.g. in density, which occur on all size scales in the neighbourhood of a phase transition in a thermal system, and like such critical phenomena, show a *universal* behaviour. This means that the functional dependences of various quantities, e.g. the specific heat, are the same whether the system is magnetic, molecular, superconducting, etc.

The prototype for a critically self-organised system is a sand pile, which, as sand is slowly added, will continually rearrange itself in such a way that its overall morphology remains unchanged in time. As an example we will consider a simpler system, a two-dimensional forest in which trees grow and are destroyed by fire which is occasionally triggered by lightning.[5] This system is not of interest in physics; however, many of its properties are illustrative of more complex physical systems, e.g. the evolution of turbulence in a fluid.

EXAMPLE 4.3: A forest fire model

As in percolation, we consider a two-dimensional array of sites, $L \times L$, at each of which we assume that

(i) trees grow at an empty site with a probability p per unit time,

(ii) occupied sites are set on fire by lightning with a probability f,

(iii) a burning tree propagates the fire to any trees at its immediate nearest neighbour sites and dies.

We are interested in such properties as the nature of the fire fronts which evolve, the spectrum of cluster sizes in the forest, the average over time of the tree density, etc. at times large enough such that all memory of the initial configuration is lost and the critical state has evolved. A site is assigned a value $0, 1, -1$ depending on whether it is empty, occupied or burning. Starting from an arbitrary configuration, without a fire, e.g. completely empty, scans are made through the array to reassign each site value according to the probabilities described. The programming is best done using two arrays of size $(L+2) \times (L+2)$, with the outer layers permanently empty, one array for the previous configuration and one for the updating scan. Three time scales are involved, the longest being that associated with lightning strikes, f^{-1}, the next that associated with tree growth, p^{-1}, and the shortest associated with a tree burning. We take the latter time to be zero, i.e. all trees in a contiguous cluster containing a tree on fire burn down instantaneously. The size of such a cluster can be determined without the complications of the cluster counting algorithm described earlier in the chapter by repeatedly scanning through the array, suppressing growth and lightning as long as there is a tree on fire, and counting the number that burn. With a language which, unlike FORTRAN, allows for recursive calls this can be done more elegantly. It can be argued that the critical state will evolve in the limit where $f/p \to 0$. In this case fire fronts define fractal like surfaces in the forest, an example of an instantaneous fire front in an 80×80 array with $p = 0.01, f = 0.0001$, i.e. $f/p = 0.01$, is shown in Fig. 4.7. In all these results it is assumed that the critical state has set in by 10000 passes through the array. The differential spectrum of tree cluster sizes which burn, $n(S)$, settles down to a power law function, allowing for the finite size of the array. The results for a 100×100 array for the same value of f/p shown, in Fig. 4.8, indicate a dependence of $S^{-\alpha}$ with $\alpha \simeq 0.9$. Such a power law distribution implies that there is no characteristic size of a cluster, the

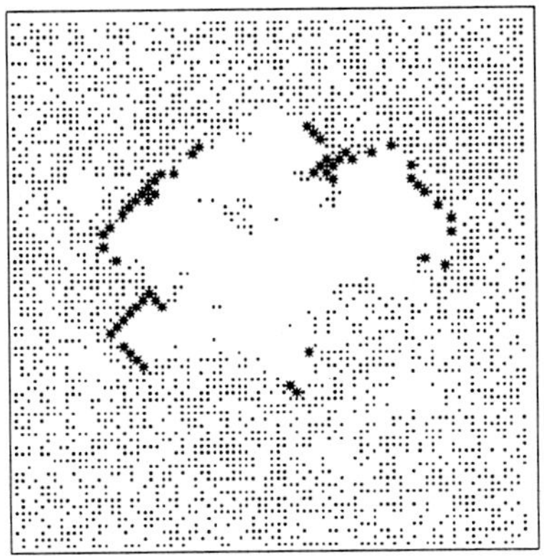

Figure 4.7: The tree population at a given time, stars indicate burning trees.

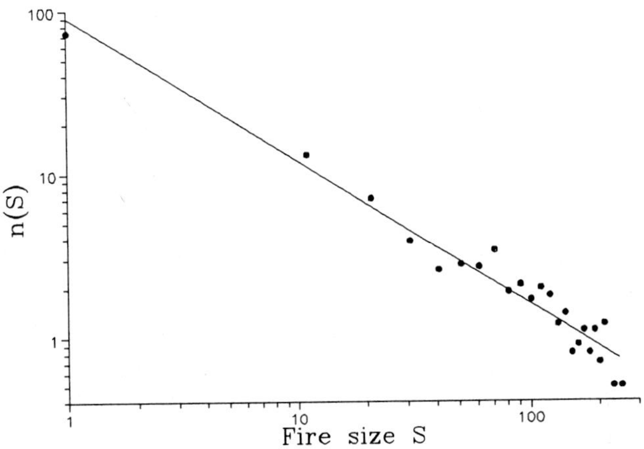

Figure 4.8: The spectrum of fire sizes.

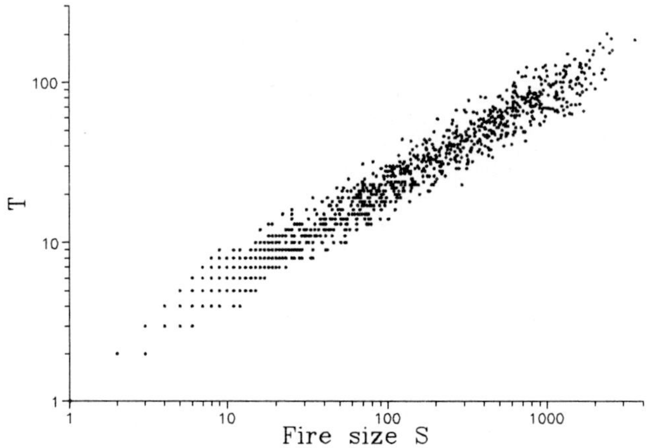

Figure 4.9: The duration of a fire as a function of its maximum size.

mean of the distribution diverges (for a forest of infinite extent), and clusters of all sizes burn. This spectrum is related to the actual spectrum of cluster sizes existing in the forest in the critical state, $D(S)$, as follows,

$$n(S) \propto D(S) \times \text{ probability of a cluster being struck by lightning.}$$

The probability involved is directly proportional to S, and so we find,

$$\begin{aligned} D(S) &\approx n(S)/S \\ &\approx S^{-2}. \end{aligned}$$

Although, relative to the other time scales involved, the time for a fire to burn out has been assumed to be zero, so we can look at the duration of a fire in terms of a time unit which increases by one unit every time a new nearest neighbour set of trees is processed. The time defined in this way will also correspond to the linear dimensions of the fire. This quantity, as a function of fire size is shown in Fig. 4.9 The density of trees in the forest as a function of time for the same value of f/p is shown in Fig. 4.10(a). The mean tree density in the critical state $\bar{\rho}$ as a function of f/p is shown in Fig. 4.10(b), where, because of the strong correlations between configurations in successive passes through the array, the method described in section 2.3.3 has been used

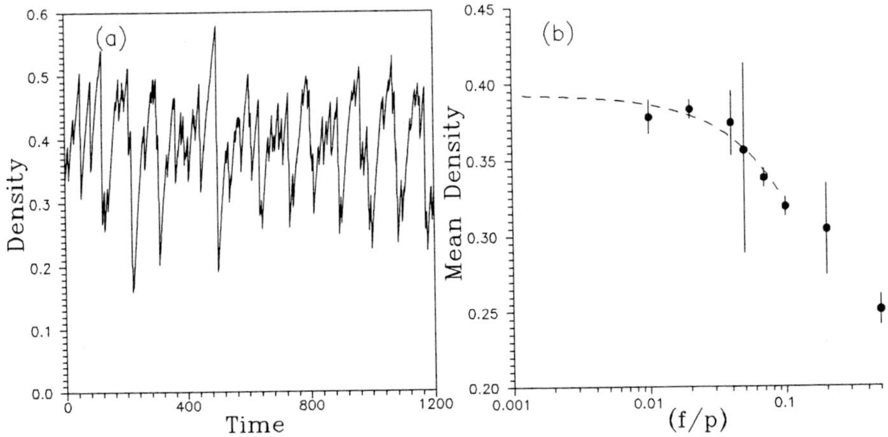

Figure 4.10: (a) The density of trees as a function of time, and (b) the mean tree density as a function of (f/p). The dashed curve in (b) is a linear fit to value of $(f/p) < 0.1$.

in estimating the variance. A linear fit at the smallest values of f/p suggests that as $f/p \to 0$,

$$\overline{\rho} \to 0.396 \pm 0.007.$$

Further study readily suggests itself, e.g. including a natural life for the trees, allowing for lightning to be accompanied by rain such that the probability of a fire propagating from a nearest neighbour is less than unity, inclusion of a wind vector giving rise to a directed fire, etc.

4.4 Stochastic Optimisation

4.4.1 Introduction

Optimisation is the determination of the value of a parameter, or parameters, which will maximise or minimise a given function — we need consider only the latter since the function can, if necessary, be replaced by its negative. Some classical optimisation techniques have invoked random variables, but these have been of very limited usefulness. We will be concerned in this chapter, and in chapter 6, with recent developments in this field, modeled on the behaviour of systems in nature. Systems spontaneously find optimum arrangements in nature, simulation of the methods adopted in such natural behaviour gives us methods for solving general optimisation problems. A slowly cooling system, left to itself, eventually finds the arrangement of atoms which has the lowest energy. This is the behaviour which motivates the method of Optimisation by Simulated Annealing, which we consider in chapter 6. An altogether different optimisation strategy is that which occurs in nature by 'natural selection'. Here new arrangements are continuously being generated and the fittest for survival are retained through competition and go on to produce further specimens some of which may be even more fit, converging on an optimum arrangement. This is the basis for Genetic Algorithms, stochastic algorithms which can be used for determining optimum configurations, e.g. minimum energy configurations, of a multicomponent system, and require only uniform random numbers for their implementation; they compete with other stochastic algorithms, such as Simulated Annealing.

4.4.2 Genetic Algorithms

The nature of the function to be optimised (typically specified by many constraint conditions) and of the parameters involved may be very different from problem to problem. However, if we consider the set of parameters as specifying coordinates in a state space, we seek to locate that point in the space at which the function is a minimum. The philosophy behind the method is based on mimicry of the natural evolution of chromosomes in a biological system which gives rise to a 'fittest' organism as determined by natural selection. To mimic this our algorithm requires three components:

(i) a chromosome-like specification of the parameters involved,

(ii) a prescription for mutation and/or reproduction of chromosomes, and

(iii) intervention, which models natural selection, to select chromosomes for preferential reproduction, on the basis of their 'fitness', e.g. their effectiveness in optimising a cost function.

Genetic algorithms may be very complex,[6] they may even entail recursive evolution of the computer programme which is doing the calculation, but as usual we will focus on a simple scheme by way of introduction. The easiest way to describe the method is in the context of a simple example. Consider a square array of $L \times L$ sites populated by $N = L^2$ 'molecules', a fraction f of type A and (1-f) of type B, the numbers of each type remaining fixed. An energy ϵ is associated with each bond between similar molecules and an energy zero with each bond between unlike pairs. What is the minimum energy for such a system?

Taking $f = 1/2$ — this is known as a binary alloy — we can represent an arbitrary configuration of a 4×4 array as shown, or, using binary notation, as in the table below. Reading by rows, we can also represent it by a chromosome (1000010011111010) containing two gene types. (We could equally write it as a natural chromosome chain, i.e. reading rows 2 and 4 from right to left, but our choice is simpler for programming.)

ABBB	*1 0 0 0*
BABB	*0 1 0 0*
AAAA	*1 1 1 1*
ABAB	*1 0 1 0*

The energy of this configuration is 11ϵ; the general expression for the energy is $\frac{\epsilon}{2}\sum_i \sum_{j(i)} \delta_{ij}$, where for each i, the sum over j runs over all the immediate nearest neighbours of i, and δ_{ij} is the Kronecker symbol.

A 'farm' of M such chromosomes, i.e. arbitrary configurations randomly generated, is initially chosen, for illustration we will just take $M = 4$, e.g. we might have

$$
\begin{array}{lll}
4 & (1000010011111010) & E = 11\epsilon \\
3 & (1110010011001010) & E = 12\epsilon \\
2 & (1001100011111000) & E = 13\epsilon \\
1 & (1111111010000000) & E = 18\epsilon
\end{array}
$$

each with equal numbers of the two gene types, and their fitness, e.g. their associated energy E, calculated. Here a chromosome represents an order configuration, in general application it may be the binary representation of a parameter or coordinate or series of parameters, in which case there is no requirement of conservation of gene species. A weight w_i can be assigned to each. There is a lot of flexibility in prescribing a reproduction/survival schedule, we concentrate on a relatively simple one for clarity of presentation. As a chromosome's weight we could take its rank, $w = M$ for the one with minimum energy, down to $w = 1$ for the most energetic as shown above. A new generation of configurations is produced as follows. A first, dominant, parent is chosen, heavily weighted by its fitness and mated with another chosen at random to produce an offspring, which may then undergo mutation. For example, we take No. 4, the fittest, and mate it with No. 1.

A prescription for the offspring is as follows. Its basic pattern is that of the dominant parent. The sites where the genes of the two parents are different are marked, indicated by $\wedge$ in the figure, the numbers of each

1	1	1	1	1	1	1	1	0	1	0	0	0	0	0	0	0	
4	1	0	0	0	0	1	0	0	1	1	1	1	1	0	1	0	dominant
		$\wedge$	$\wedge$	$\wedge$	$\wedge$		$\wedge$			$\wedge$	$\wedge$	$\wedge$	$\wedge$		$\wedge$		parent
			†	†			†			*	*	*					

type in the sets from the dominant parent being n_0 and n_1. From whichever group is the smaller, say n_0 (here both n_0 and n_1 are 5 so either can be chosen), a random number $k, 1 \leq k \leq n_0$ of their sites is chosen (a case with $k = 3$ is marked with † in the figure) and the genes switched with a similar number of sites, randomly chosen, from the set where they are of type 1, indicated by * in the above illustration. This gives the basic structure of the offspring:

$$1011011010001010 \tag{4.1}$$

This structure can then suffer mutation with probability p_m. These mutations can be of different types; we describe three, each with their own probability.

Reptation, with probability p_r: detach a segment of k genes from the head of the chromosome and append them to the tail; k can be a random number in some range $1-k_{\max}$, for example, taking $k = 3$ yields the offspring (1011010001010101).

Inversion, with probability p_i: two points along the chain are selected at random and the intervening sequence reversed. For example choosing sites 6 and 10 in equation 4.1 results in the offspring (1011001011001010).

Scramble, with probability p_s: as in inversion, choose two sites along the chain at random and reassign the intervening genes randomly — in the present case being sure to conserve the number of each type. For example reassigning sites 5 to 10 in equation 4.1 might produce (1011001011001010).

Clearly such mutations can be combined, and others devised. The mutation schedule, i.e. the set of values of p_m, p_i, p_s, p_r and $k_{\max}$ may need to be arrived at by trial and error, a possible set might be (1/4, 1/3, 1/3, 1/3, N/5).

A complete new generation of M offspring can be produced in this way to replace the previous generation, or the M fittest from among the $2M$ configurations may be retained as parents of the next generation. Alternatively, only m, $1 \leq m < M$ offspring are produced after which their fitness is tested before proceeding further, and they are used to replace earlier chromosomes of lesser fitness; m may be as small as one or two — this is known as steady state reproduction. The strategy to be used will depend on the nature of the problem.

The method can very rapidly converge on the minimum energy arrangement of the molecules — a purely homogeneous mixing of the two types with zero energy if ϵ is positive, and a complete segregation of the two types when ϵ is negative. We now look at an example involving numerical parameters rather than an order sequence as in the binary system.

EXAMPLE 4.4: Charged particles on a line

We consider N similar particles which interact among themselves with a repulsive Coulomb-like force in one dimension and which are also subject to a confining potential of the form $V(x) = 1536x^4 - 3200x^3 + 2362x^2 - 720x + 81$. We will take the potential produced by a charge at a distance x to be $\alpha / \mid x \mid$. How will the charges distribute themselves at equilibrium? The shape of the confining potential is shown in Fig. 4.11. In the absence of the repulsion, $\alpha = 0$, all the particles will congregate in the deepest well to yield the configuration of minimum energy, zero. In the general case the charges will take up the set of positions $\{x\}$ which minimises their total energy $U(\{x\})$. Taking each charge equal to unity, the expression for this is

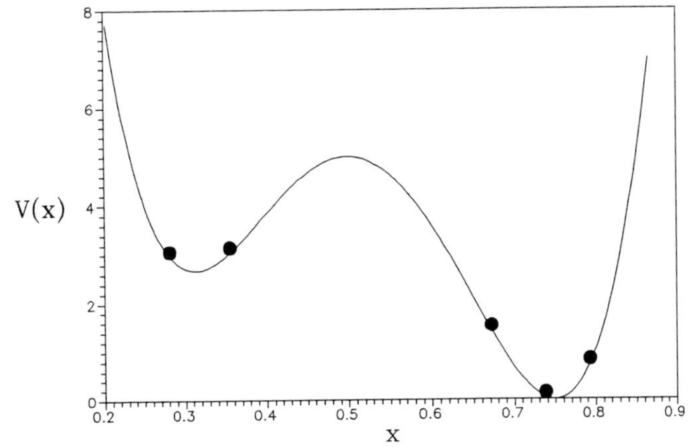

Figure 4.11: The curve shows the potential $V(x)$ in Example 4.4. The final disposition of the charges for the case of $\alpha = 0.1$ is indicated, tied to the curve for purposes of illustration.

$$U(\{x\}) = \sum_{1}^{N} V(x_i) + \sum_{i=1}^{N} \sum_{j>i} \frac{\alpha}{\mid x_i - x_j \mid}.$$

We consider the case of five particles, so we seek the point in a five-dimensional space which minimises U. Each of the coordinates is represented in binary format; in the present case, for reasonable values of α, all values will lie in $0 < x < 1$, and we use an 8-bit representation, in the general case the range of the parameters will have to be scaled to achieve this. This will limit our location resolution to one part in 2^8. A chromosome can be generated from 40 randomly generated bits to represent an initially random configuration of the five particles, each successive block of eight bits representing the coordinates of a particle. The chromosomes are best represented by logical arrays, e.g. in a FORTRAN programme a logical array CHROME(NFARM,40) can represent NFARM chromosomes, and bits can be assigned by the statement CHROME(I, J) = (XI.LE..5) where XI is a uniform random variate. Conversion to numeric data is done when the energy needs to be evaluated. A farm of ten zeroth generation chromosomes was generated and a rank $1-10$ assigned to them in order of decreasing energy. A weight proportional to the square of its rank was assigned to each. The first generation was produced from the zeroth generation by selecting dominant parents on the basis of their weight and mating them with

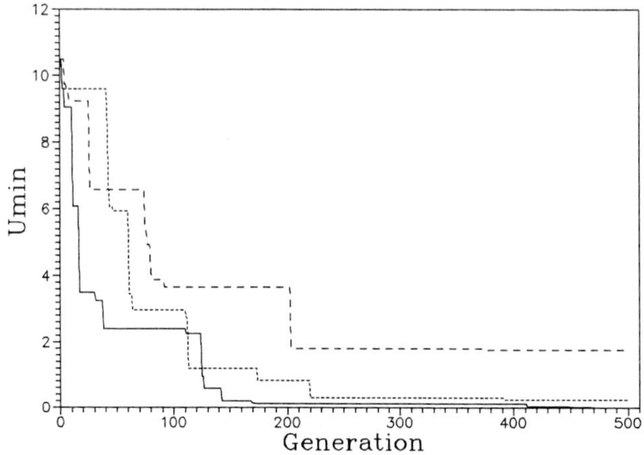

Figure 4.12: The lowest encountered energy as a function of generation number with an exchange probability $p_e = 0.1$ for different reptation and inversion probabilities. Short dashes: $(p_r = 0.1, p_i = 0)$, long dashes: $(p_r = 0, p_i = 0.1)$, solid line: $(p_r = 0.1, p_i = 0.1)$.

another chosen randomly from the farm. The method used is a variant of that described in the previous section, genes in the dominant parent which differed from those in the other partner were replaced with a probability of p_e. The resulting offspring was mutated with a probability p_i by inversion, or with a probability p_r by reptation, where up to k_{max} ($k_{\text{max}} = 10$) genes were transferred from the head to the tail. The ten least energetic among the 20 resulting chromosomes were ranked, and taken as the parents for the next generation, and the procedure iterated.

For $\alpha = 0$, where we know the answer, we show in Fig. 4.12 the minimum encountered energy as a function of generation number for some reproduction schedules having $p_e = 0.1$. In all cases the result converges to zero. Although the determination of an optimum schedule is rather dependent on the actual problem some degree of mutation is usually necessary to prevent freezing into some localised minimum in the parameter space. In Fig. 4.13 we show the results for $\alpha = 0.1$, for which the asymptotic energy is found to be 15.210ϵ. Both the lowest energy to date, solid curve, and the mean for the ten particles in the current generation, dashed curve, are shown. Regions where these curves run together indicate temporary freezing into a local minimum from which later mutation enables escape. The asymptotic value is

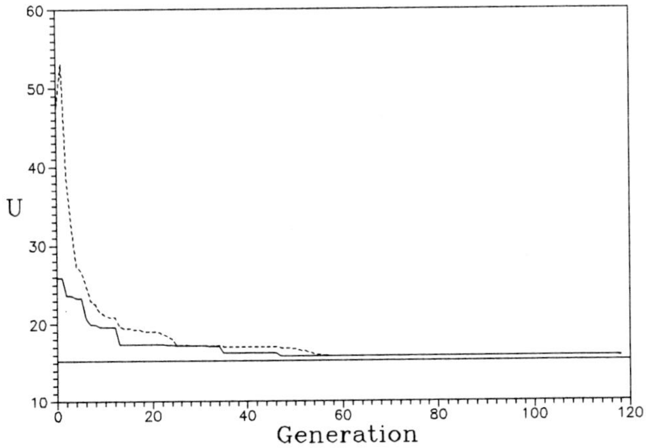

Figure 4.13: The minimum encountered energy, solid line, and the average energy of the ten chromosomes, dashed line, as a function of generation number compared with the asymptotic energy.

also shown by the horizontal line. Finally in Fig. 4.11 we show the disposition of the charges at equilibrium, where they have been tied to the curve for illustration purposes. The solution of this problem using simulated annealing is considered in exercise 6.1 in chapter 6.

There are other problems, e.g. involving only simple random walks, where random numbers from a uniform distribution are all that is required, more usually, however, random numbers from other distributions are necessary for solving Monte Carlo problems. In the next chapter we describe the generation of such random variables.

Notes

1. Stauffer, D., *Introduction to Percolation Theory*, (London: Taylor & Francis, 1985). The ideas can be extended to other fields, e.g bond percolation on a finite lattice in three dimensions has been applied to the fragmentation properties of colliding nuclei at high energies; W. Bauer, D. R. Dean, U. Mosel and U. Post (1985), *New Approach to Fragmentation Reactions: the Nuclear Lattice Model*, Phys. Lett., **150B**, 53-56.

2. Clearly, if the occupied sites form a percolating cluster the unoccupied do not in the transverse direction, and vice versa. This is not so in three dimensions where both types of site can percolate simultaneously in any directions.

3. Hoshen, J. and R. Kopelman (1976), *Percolation and cluster distribution. I. Cluster multiple labeling technique and critical concentration algorithm*, Phys. Rev. $B\mathbf{14}$, 3438-45.

4. The concept of SOC is described in Bak, P., C. Tang and K. Wiesenfeld (1987), *Self-Organized Criticality: An Explanation of* $1/f$ *Noise*, Phys. Rev. Lett., **59**, 381-4, and elaborated on by the same authors, *Self-organized criticality*, Phys. Rev. **A38**, $364-74$, (1988). Its relevance to earthquakes is discussed by Chen, K., P. Bak and S. P. Obukhov (1991), *Self-organized criticality in a crack-propagation model of earthquakes*, Phys. Rev. **A**, 625-30, and by Olami Z., H. J. S. Feder and K. Christensen (1992), *Self-Organized Criticality in a Continuous, Non-conservative Cellular Automaton Modeling Earthquakes*, Phys. Rev. Lett. **68**, 1244-47

5. Drossel B. and F. Schwabl (1992), *Self-Organized Critical Forest-Fire Model*, Phys. Rev. Lett., **69**, 1629-32.

6. See, for example, L. Davis, *A Genetic Algorithms Tutorial* in L. Davis (Ed.) *Handbook of Genetic Algorithms*, (New York: Van Nostrand Reinhold, 1991), and Smith, R. W. (1992), *Energy mimimization in binary alloy models via genetic algorithms*, Comp. Phys. Comm., **71**, 134-46.

Exercises

4.1 An interesting 'anti-percolating' calculation serves as a test of the quality of uniform random numbers. Consider a cubic array of $N \times N \times N$ sites, initially all occupied. In one unit of time a total of N^3 sites (i, j, k) are successively selected at random, e.g. $i = [N\xi + 1]$ etc., and the site emptied if it is not already empty. The number of sites remaining occupied after a time t should decay as e^{-t} if the selection in the cube is truly at random. Test your random number generator, taking $N = 20$, by comparing your results with this prediction. See D. Stauffer, 'Random Number Generation' in K. H. Hoffmann and M. Schreiber, (Eds.), *Computational Physics*, (Heidelberg: Springer-Verlag, 1996).

4.2 In the forest fire model of SOC, show that the size spectrum of tree clusters in the forest in the critical state $D(S)$ is related to the spectrum of fire sizes which occur, $n(S)$, by $D(S) = S^{-1}n(S)$.

4.3 For S assumed a continuous variable, $S \geq 1$, show that the following distributions have the same mean,

$$p(S)dS = \frac{e^{1/2}}{2} e^{-S/2} dS \qquad S \geq 1,$$

$$p(S)dS = \frac{3}{2} S^{-5/2} dS \qquad S \geq 1.$$

Show that the variance of the former is $\sigma^2 = 4$, and it thus has a scale of 2 units associated with it, whereas the variance of the latter is infinite, reflecting the lack of any characteristic scale.

4.4 Develop a programme to study the evolution of a lattice forest as described in Exercise 4.3. Include also the effects of 'rain', i.e. introduce a probability $p* \leq 1$ that a tree on fire will set a nearest neighbour on fire. Investigate the properties of the fires which occur, e.g. the distribution of burning clusters, the mean tree density, etc. for different values of p^*.

4.5 Consider an array of M magnetic dipoles arranged in an $L \times L$ square array, with $L = 8$, where each dipole can assume two values (projections) which we denote by $\sigma = -1$ or $\sigma = +1$ The energy of any configuration of the dipoles is

$$U = \frac{1}{2} \sum_{i=1}^{M} \sigma_i \sum_{j(i)} J_{ij} \sigma_j .$$

Here, for each i, the sum over j runs over all the immediate nearest neighbours of i. To make allowance for edge effects supply *periodic boundary conditions*, i.e. surround the array by layers on each side which are images of the furthermost row (column) in the array — see section 3.2 of chapter 10. Every dipole in the array proper now has four nearest neighbours and interactions with these boundary spins are included in the definition of the array's energy. Effectively the array is extended to an $(L+2) \times (L+2)$ array with the corner elements absent.

Use the method of Genetic Algorithms to find the spin configuration of lowest energy. Consider first the case of $J = -1$.

Start with a random spin configuration, i.e. assign the spin σ at each of the M points equal to +1 or -1 at random. Assign the periodic boundary spins and calculate the energy for this configuration. Convert the configuration into a chromosome, i.e. a string of M binary bits (or bipolar bits ± 1, it doesn't matter here). Repeat all this nine more times to generate a starting farm of $N = 10$ chromosomes, each characterised by its energy. Rank the chromosomes in order of decreasing energy, and assign a weight equal to its rank to each. Calculate also the mean energy.

Generate a new family of ten chromosomes as follows (the prescription is somewhat different, and simpler, than that described earlier). From among the ten in the preceding generation select one with a probability proportional to its weight, and one of the remaining nine at random to act as parents. From them produce two children by *crossing*. To do this, choose a random gene $k(> 1)$ on a chain and interchange all the genes from k to M between the two chains. For example if the parent chains were

1100110101110

and

0001110010101

and the choice of k was 7, the new chromosomes would be

1100110010101

and

$$0001110101110$$

Repeat this four more times, each time selecting two parents from the old generation as above, (one or both of them may have been used before) to produce a total of ten chromosomes of the new generation. Destroy all the members of the old generation. Before applying the fitness test to the new configurations they should be subjected to mutation with a probability of p_r per bit. The most efficient (if not entirely rigorous) way of doing this is to select a chromosome at random, select a gene in it at random and change it (a spin flip), and repeat this two-fold selection until NMp_r bits have been changed. Take $p_r = 0.02$.

Transfer each new configuration to the matrix array, i.e. update the boundary spins appropriately, and calculate its energy. Calculate also the mean energy of the ten, and archive the configuration with lowest energy to date if present. Present plots of the lowest energy encountered to date, the lowest energy in a family, and the mean family energy versus generation number.

Terminate the process when the minimum energy configuration has been identified, or after 200 generations.

The solution to the above case, $U/M = -2J$, is trivial and does not require genetic algorithms. You should now do the spin glass case, which is far from trivial. In this case the nearest neighbour individual values of J_{ij} assume values ± 1 at random. When the coupling constants are randomly ± 1 it is not at all clear what configuration of the spins will result in the lowest energy. Generate a spin glass, i.e. generate a set of nearest neighbour $J_{ij} = J_{ji}$ randomly having values $+1$ or -1, $(2L(L+1)$ distinct values in total when the periodic boundary is allowed for) and use these to define the energy. Start with an arbitrary farm of chromosomes as above and repeat the process to find the ground state configuration of your system.

Different spin glasses will have different ground state energies, but it is believed that the average energy over the ensemble of all such systems tends to a limiting value of -1.40 per spin as the system size goes to infinity. For more on this project see section 10.4, and P. Sutton and S. Boyden (1994), *Genetic algorithms: A general search approach*, Am. J. Phys., **62**, 549-52.

Chapter 5

SAMPLING FROM A DISTRIBUTION

> Anyone who considers arithmetical methods of producing random numbers is, of course, in a state of sin.
>
> John von Neumann

5.1 Introduction

Numbers from a uniform distribution can be very useful in many situations, whether it be in selecting a random permutation of objects, or, as in the previous chapter, selecting occupation probabilities. However, most applications of Monte Carlo methods require sequences of a variable selected from a distribution other than the uniform one. In this chapter we describe techniques for selecting variates, both discrete and continuous, from an arbitrary distribution, including several examples of importance in physics.[1]

5.2 Standard Methods

Henceforth we denote variates drawn from the uniform random distribution

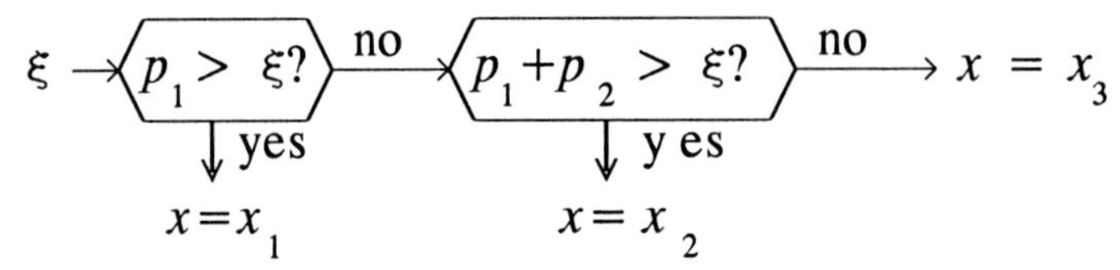

Figure 5.1: Using a uniform random number to select from among three possibilities.

on [0,1] by ξ; we also note that the quantity $(2\xi - 1)$ is uniformly distributed on [-1,1]. Such variates are useful for simulating a choice between some events with known probabilities of occurrence, or in selecting samples of a variate which is uniformly distributed, e.g. the polar angle in the azimuthally symmetric scattering of a particle, in which case we can write $\phi_i = 2\pi\xi_i$. More usually, however, we want instances of a variable which follows some other distribution, $p(x)$ say, where the quantity x may represent a discrete or continuous variable.[2]

If x can assume two discrete values, x_1 and x_2, with probabilities p_1 and p_2 $(= 1 - p_1)$, respectively, we can select a value by calling a uniform random variate ξ and setting $x = x_1$ if $\xi < p_1$ and $x = x_2$ otherwise. By extension, if the variable can assume three discrete values with probabilities p_1, p_2 and p_3 the algorithm in Fig. 5.1 will obviously select from it.

EXAMPLE 5.1

If in the case of a neutron interacting with a nucleus in a medium, σ_s, σ_a, and σ_f represent the cross-sections for scattering, absorption and fission, the probabilities for the processes are $p_s = \sigma_s/\sigma_T$, $p_a = \sigma_a/\sigma_T$ and $p_f = 1 - p_s - p_a$ where the total cross-section is $\sigma_t = \sigma_s + \sigma_a + \sigma_f$. In simulating neutron transport a decision can be made on the neutron's fate on the basis of a single uniform variate and this prescription.

By induction we arrive at the prescription for sampling variates from any distribution where the variable assumes discrete values x_i with frequencies $p(x_i)$, namely choose the value x_k where k is the smallest integer such that

$$P(x_k) \equiv \sum_{j=1}^{k} p(x_j) \geq \xi, \tag{2.1}$$

in other words, the smallest value of x for which the cumulative probability

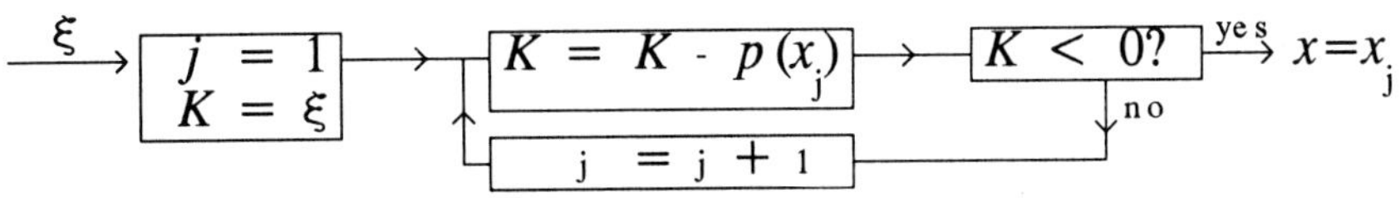

Figure 5.2: The general algorithm to select from a discrete distribution.

equals or exceed ξ. A concise version of the algorithm is shown in Fig. 5.2. The important Poisson distribution, equation 2.4.4 can be selected from in this way, although as we will see in example 5.3 a faster method can be devised in this case.

5.2.1 Method of Inversion

Equation 2.1 can be generalised to sample from a continuously distributed variable on the range $a \leq x \leq b$, with distribution $p(x)$, by writing

$$P(x) = \xi \tag{2.2}$$

and inverting the relation to solve for x, where

$$P(x) = \int_a^x p(x)dx, \qquad 0 \leq P \leq 1.$$

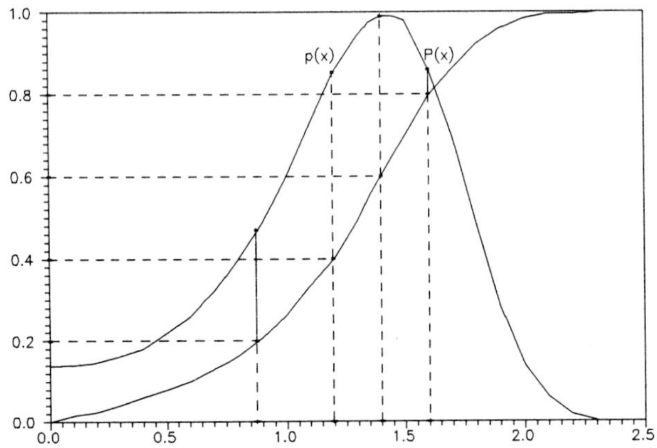

Figure 5.3: Illustrating how uniform selection of values of the cumulative probability results in the variates having the distribution $p(x)$.

It is instructive to see the origins of this prescription graphically. In Fig. 5.3 we show a distribution $p(x)$ together with its cumulative distribution $P(x)$. We see that by selecting intervals on $P(x)$ at random directly proportional to their size, we are selecting, at random, intervals dx proportional to the area supported by $p(x)$ on that interval.

EXAMPLE 5.2

For the exponential distribution, equation 2.4.9, we have $P(x) = 1 - e^{-x/\mu}$, and thus from equation 2.2

$$x = -\mu \ln(1 - \xi).$$

We can equally write this as

$$x = -\mu \ln \xi, \tag{2.3}$$

since if ξ is uniformly distributed on [0,1] so also is $1-\xi$, and we shall generally write this important result in this way.

EXAMPLE 5.3

The result of the previous example together with the physical origin of the Poisson distribution, equation 2.4.4, leads to a faster method of selection for this important distribution. We know that the Poisson distribution for the number of events occurring in a time t when the mean number is μ, is a consequence of the distribution of the time interval between random events being exponentially distributed, characterised by a mean rate of occurrence $1/\tau$, where $\mu = t/\tau$, (section 2.4.1). Taking $t = 1$, i.e. $\tau = 1/\mu$, if we select time intervals t_i from $p(t) = \frac{1}{\tau}e^{-t/\tau}$,

$$t_i = -\tau \ln \xi_i,$$

until the accumulated time $\sum_1^k t_i$ exceeds the unit interval, we know we have selected one more than the number of events falling in the interval t, i.e we should set $n = (k-1)$. Exponentiating this relation gives a faster algorithm, we should find the smallest k for which

$$\prod_{i=1}^{k} \xi_i < e^{-1/\tau},$$

and set $n = (k-1)$.
When μ is so large that the distribution can be approximated by a Gaussian, we can set $n = \text{int}[\mu + \eta\sqrt{\mu} - 0.5]$, or zero if this quantity is negative, where η is a variate from the standardised Gaussian.

EXAMPLE 5.4

For the power law distribution, equation 2.4.18, we have

$$P(x) = 1 - (\frac{x}{x_0})^{1-\gamma}, \qquad x \geq x_0$$

from which equation 2.2 yields the prescription

$$x = x_0 \xi^{-1/(\gamma-1)}.$$

EXAMPLE 5.5

As another example of the inversion method we consider the piecewise-fit distribution, equation 2.4.14, given by

$$p(t) = 2\Lambda(t - \gamma), \qquad 0 \le t \le 1.$$

Using equation 2.2, variates from it can be selected using the prescription

$$t = \gamma + \operatorname{sgn}(\Lambda)\sqrt{\gamma^2 + \xi/\Lambda}$$

where $\operatorname{sgn}(\Lambda)$ arises from the requirement $p(t) \ge 0$. Using the appropriate value of Λ to select t, variates from the distribution on the interval (a,b) can be selected as $x = a + (b - a)t$.

Many other examples of the use of this method can be found, i.e. where equation 2.2 can be explicitly solved for x. However, even when this is possible, as in examples 5.1 – 5.4, it may involve the evaluation of higher level functions and thus be relatively slow. It may be preferable in such cases to adopt a faster, less direct method. As a typical time for the generation of a uniform random variate is 10 μsec, it is desirable that the selection time for a variate from another distribution be of the same order of magnitude.

Many examples of distributions can be found for which equation 2.2 cannot be explicitly solved — the case of the Gaussian distribution is perhaps the most obvious. Other methods of sampling must be devised. Methods particular to a specific distribution can sometimes be devised; a comprehensive account is not suitable here.[3] We describe a few important ones later in the chapter. We now look at some other general methods.

5.2.2 Method of Composition

By combining the convolution relations (equations 2.2.14 –2.2.16) for pairs of probability distributions with inversion, prescriptions can be built up for many common distributions. We illustrate the principles of the method by two such distributions, the Gamma and the Lorentz distributions.

EXAMPLE 5.6

As an example of the use of equations 2.2.14–2.2.16, we note that the variate $u = -\ln \xi$ is distributed as e^{-u}. If w is defined as the sum of two variables u and v each of which follows this distribution, then the variate $w = -\ln \xi_1 - \ln \xi_2$, using equation 2.2.14 can easily be shown to follow the distribution $p(w)dw = we^{-w}dw$, $w \geq 0$. Again by adding another variable x, which has the same exponential distribution, i.e. $y = w + x = u + v + x$, we find its distribution,

$$p(y)dy = \frac{y^2}{2}e^{-y}dy, \qquad y \geq 0.$$

It can be readily concluded that the sum of n standard exponential variates,

$$z = -\sum_{i=1}^{n} \ln \xi_i = -\ln \prod_{i=1}^{n} \xi_i,$$

follows the standardised Gamma distribution (equation 2.4.10, with $a = 1$) for integer n, i.e.

$$p(z)dz = \frac{z^{n-1}}{(n-1)!}e^{-z}dz, \qquad z \geq 0.$$

EXAMPLE 5.7

As we showed in section 2.4, the ratio of two variables, each uniformly distributed on [0,1] is a quantity which is distributed as the standardised Lorentz distribution, i.e. $z = \xi_1/\xi_2$ is distributed as,

$$p(z)dz = \frac{dz}{\pi(z^2+1)}.$$

5.2.3 von Neumann's Method

This *Acceptance-Rejection* method is of much wider applicability than methods described so far. We illustrate it for the case of the variable x restricted to an interval, $a \leq x \leq b$, — it need not be finite. We write the distribution of interest in the form,

$$p(x)dx = CQ(x)\pi(x)dx,$$

where $\pi(x)$ is a distribution from which we are able to sample using one of the methods described elsewhere; it is known as the importance function. The domain on which the distribution $\pi(x)$ is defined, and normalised, must contain the interval (a,b) but need not coincide with it. $Q(x)$ is simply the ratio of the distributions p and π, normalised so that its maximum value is unity — for which reason the constant C is factored out — i.e.

$$CQ(x) \equiv p(x)/\pi(x) \qquad a \leq x \leq b$$

$$\equiv 0 \text{ otherwise.}$$

The method is as follows:

(i) select a variate x' from the distribution $\pi(x)$, and

(ii) evaluate $Q(x')$ and compare it with a variate ξ selected from the uniform distribution. If $Q(x') > \xi$ accept x' as a sample variate from the distribution, otherwise reject it and return to step (i).

The method can be simply justified as follows: the probability of a value being selected in an interval dx' at x' in a single pass of this prescription is $\pi(x')dx'$ times the probability of acceptance of x'. But the probability of acceptance of x' in step (ii) is just $Q(x')$, so we have the probability

$$= \pi(x')\frac{p(x')}{C\pi(x')}dx' = \frac{1}{C}p(x')dx',$$

i.e. it is directly proportional to $p(x')$. The quantity C^{-1} is seen to be an *efficiency* factor; C represents the average number of passes which must be made with the algorithm per variate selected. Further arguments for the plausibility of the method, as well as questions of efficiency, are best investigated in the context of some examples.

EXAMPLE 5.8

When the range of the variable is finite the method can always be employed by taking the uniform distribution as the importance function for selection. For the distribution,

$$p(x)dx = \frac{2x}{(2-x)^3}e^{-x/(2-x)}dx, \qquad 0 \le x \le 2$$

we simply take $\pi(x)dx = \frac{1}{2}dx$, $0 \le x \le 2$. The function $p(x)$ has a maximum value $p_0 = 1.2585$ at $x = \sqrt{2}$ so we can write

$$Q(x) = p(x)/p_0, \qquad C = 2p_0 = 2.517,$$

which immediately tells us that the efficiency, C^{-1}, is only about 40%. Reference to Fig. 5.4 clarifies the origin of this inefficiency. Selecting from $\pi(x)$ using $x' = 2\xi_1$, the pairs of values (x', ξ_2) define points uniformly distributed over the rectangle of unit height. From step (ii) we accept only those values of x' from the subset of points which lie below the curve $Q(x)$. The more the function $Q(x)$ lies close to its maximum value, unity, throughout the interval the greater will be this probability of acceptance.

A uniform importance distribution can always be used in such cases but, as we have seen in example 5.7, for greatest efficiency the rejection function $Q(x)$ should be as near as possible to a constant in the region where variates from the distribution $\pi(x)$ are likely to be selected. This can be achieved in a systematic way by using an importance function which involves a parameter $\beta, \pi(x, \beta)$. The efficiency will then be a function of this parameter, $C^{-1}(\beta)$,

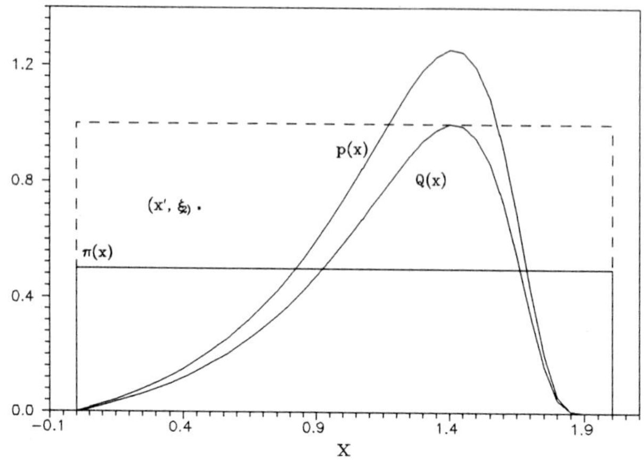

Figure 5.4: Decomposition of the probability distribution in example 5.8 using the uniform distribution as the importance function. Points (x', ξ_2) not falling under the curve $Q(x)$ are rejected.

and the particular value of β can be chosen which will maximise the efficiency. The following example illustrates these considerations.

EXAMPLE 5.9

We consider the standardised gamma distribution with non-integer parameter b, i.e.

$$p(x)dx = \frac{x^{b-1}}{\Gamma(b)} e^{-x} dx, \qquad x \geq 0.$$

The function is defined for values of $b>0$ but the method of selection described here only works for $b \geq 1$. We take for our importance function the exponential,

$$\pi(x, \beta)dx = \beta e^{-\beta x} dx.$$

The rejection function is

$$CQ(x) = \frac{x^{b-1}}{\beta\Gamma(b)} e^{-x(1-\beta)}.$$

It has a maximum at $x = (b-1)/(1-\beta)$, at which point its value gives C,

$$C = \frac{(b-1)^{(b-1)}}{\Gamma(b)} e^{1-b} \frac{1}{\beta(1-\beta)^{b-1}}.$$

This can be minimised with respect to β, yielding a maximum efficiency for $\beta = 1/b$. Thus we have,

$$\pi(x)dx = e^{-x/b}dx/b, \qquad Q(x) = (\frac{x}{b})^{b-1} e^{(b-1)[1-x/b]}.$$

The efficiency $C^{-1} = b^{-b}\Gamma(b)e^{(b-1)}$ decreases with b, it has a value of 100% for $b = 1$ but has fallen off to 29% for $b = 10$.

EXAMPLE 5.10

Variates from the Fermi-Dirac energy distribution can be obtained by writing $E = x/\beta$, where x is distributed as

$$p(x)dx = B(\omega)\frac{\sqrt{x}}{e^{x-\omega}+1}dx, \qquad x \geq 0.$$

with $\beta = 1/kT$, $\omega = \beta\mu$ where μ is the chemical potential and $B(\omega)$ is the normalising constant.[4] We take as an importance function

$$\pi(x)dx = \frac{2}{\sqrt{\pi}}\sqrt{x}e^{-x}dx.$$

By writing $x = u^2$ this takes the form

$$\pi(u)du = \frac{4}{\sqrt{\pi}}u^2 e^{-u^2}du, \qquad u \geq 0$$

the standardised Maxwell-Boltzmann distribution. We have

$$Q(x) = \frac{1}{e^{\omega - x}+1},$$

and an efficiency factor

$$C^{-1} = \frac{2e^{-\omega}}{B(\omega)\sqrt{\pi}}.$$

This efficiency is good at small values of ω, at $\omega = -2$ it equals 95%, but falls off rapidly for $\omega > 2$, where it has a value of 38%, and where an alternative method will be called for.

5.2.4 Method of Superposition

If we can write the probability distribution as a superposition of probability distributions $p_k(x)$, each of which we are able separately to sample from, we can use this method.[5] Write

$$p(x)dx = dx \sum_{k=1}^{N} w_k p_k(x), \qquad a \le x \le b$$

where all the functions $p_k(x)$ are normalised on the interval, all $w_k > 0$, and, by normalisation, $\Sigma w_k = 1$. The method is:

(i) select with probability w_k an index k, and

(ii) select a variate from the distribution $p_k(x)$.

The total probability of selecting a value of x in dx is the probability of 'using' the distribution p_1 times the probability of selecting it in dx from $p_1(x)$ plus the probability of using p_2 times the probability of selecting it in dx from $p_2(x)$ plus ..., i.e. $w_1 p_1(x)dx + w_2 p_2(x)dx + \ldots = p(x)dx$. Any of the other methods described here can be used to sample from the individual distributions. In the case where von Neumann's method is applied to the individual distributions we have

$$p(x)dx = dx \sum_{k=1}^{N} a_k Q_k(x) \pi_k(x),$$

and the average number of passes required to select a sample variate is Σa_k.

EXAMPLE 5.11

As a simple example of the use of this method, consider the selection of a scattering angle from the angular distribution for Thompson scattering of unpolarised radiation, which has the form

$$\frac{1}{\sigma_T}\frac{d\sigma}{d\Omega} = \frac{3}{16\pi}(1+\cos^2\vartheta),$$

where σ_T is the total cross-section. Writing $\cos\vartheta \equiv x$, $d\Omega = 2\pi dx$, we have

$$\begin{aligned} p(x)dx &= \frac{3}{8}(1+x^2)dx, \qquad -1 \le x \le 1, \\ &= \sum_{k=1}^{2} w_k p_k(x)dx \end{aligned}$$

where

$$w_1 = 3/4 \qquad p_1(x)dx = \frac{1}{2}dx, \quad -1 \le x \le 1, P_1(x) = \frac{1}{2}(1+x)$$

$$w_2 = 1/4 \qquad p_2(x)dx = \frac{3}{2}x^2dx, \quad -1 \le x \le 1, P_2(x) = \frac{1}{2}(1+x^3).$$

Thus, with probability 3/4 we select from the uniform distribution on [-1,1], i.e. set $x = 2\xi - 1$, and with probability 1/4 from the second distribution, set $x = (2\xi - 1)^{1/3}$.

EXAMPLE 5.12

As another illustration of this method we take the Planck frequency spectrum for black-body radiation, (the Bose-Einstein distribution). Writing $x \equiv \hbar\omega/kT$ the distribution takes the form,

$$p(x)dx = \frac{15}{\pi^4}\frac{x^3}{e^x - 1}dx, \qquad x \ge 0.$$

By expanding the denominator we can write it in the form,

$$p(x)dx = \frac{15}{\pi^4}\sum_{k=1}^{\infty} x^3 e^{-kx} dx$$

$$= \sum_{k=1}^{\infty} \frac{90}{\pi^4} \frac{1}{k^4} \frac{(kx)^3}{3!} e^{-kx} d(kx)$$

$$= \sum_{k=1}^{\infty} w_k p_k(x) dx,$$

where $p_k(x)$ is the Gamma distribution in kx with parameter $n = 4$, and the coefficients $w_k = 90/\pi^4 k^4$. Thus we have the prescription: call five uniform random variates, select a value of k, i.e. find the smallest K such that

$$\frac{90}{\pi^4} \sum_{k=1}^{K} \frac{1}{k^4} \geq \xi_1,$$

and (via example 5.6) write

$$x = -K^{-1} \ln(\xi_2 \xi_3 \xi_4 \xi_5).$$

5.2.5 Tabular Interpolation

An obvious basic method which can be used for a discrete or continuously distributed variable is to consult a table of frequencies tabulated in advance. In the case if a discrete variable the method follows the description at the beginning of the chapter; for a continuously distributed variable its range is divided into N equiprobable intervals, i.e the relation $P(x_i) = i/N$ is solved for $x_i, i = 0, 1, \ldots N$ once and the array of $(N+1)$ values of x_i stored in memory. Linear interpolation can be used on the table as follows: use a standard uniform variate ξ to form the quantity $N\xi$, and write it as the sum of its integer and fractional parts, $N\xi = I + r$. Use I to select the cell to be selected from the table and write

$$x = x_I + r(x_{I+1} - x_I).$$

Tabular interpolation is most likely to be of use when the distribution in question is an empirical one without an analytical representation, e.g. based on some set of measurements.

5.3 Special Methods

Many distributions give rise to singular and ingenious prescriptions.[3] We describe a few examples, with some emphasis placed on the important case of the Gaussian distribution.

EXAMPLE 5.13

If three uniform deviates are called and arranged in ascending order, $\xi_1 < \xi_2 < \xi_3$, and we set $x = \xi_2$ the probability that x lies in dx at x is the product of the probability that ξ_2 lies in dx at x and that ξ_1 is less than x and that ξ_3 is greater than $x + dx$, i.e.

$$\begin{aligned} p(x)dx &\propto dx.P(\xi_1 < x).[1 - P(\xi_3 < x + dx)] \\ &\propto dx.x.[1 - (x + dx)] \\ &\propto x(1 - x)dx + O(dx^2). \end{aligned}$$

The distribution selected from is $p(x)dx = 6x(1 - x)dx$, $0 \leq x \leq 1$.

EXAMPLE 5.14

If we set $x = \xi$, where $\xi = \max(\xi_1, \xi_2, \ldots, \xi_N)$, the probability distribution for x can be deduced as follows: the probability that x lies in dx at x is

$$\textstyle\sum_{i=1}^{N}\{(\text{probability of } \xi_i \text{ in } dx \text{ at } x)(\text{probability all } \xi_{j\neq i} < x)\}.$$

The first term in the sum is just dx, and, since the ξ's are independent, the second term is $\prod_{j\neq i}(\text{probability of } \xi_j < x)$, i.e. an $(N\text{-}1)$-fold product of the cumulative distribution for the uniform variate, $\prod_{j\neq i} P_{\xi_j}(x) = x^{N-1}$. Thus, each term in the sum is independent of i, and the distribution of x is

$$p(x)dx = Nx^{N-1}dx, \qquad 0 \leq x \leq 1.$$

The mean of this distribution is $\frac{N}{N+1}$ and its variance is $\frac{N}{(N+1)^2(N+2)}$.

EXAMPLE 5.15

Here we consider the standardised Gaussian distribution. The most direct way to sample from this distribution, known as the Box Muller method depends on the trick of considering two separately independent variables, t and u, each obeying the distribution, i.e.

$$p(t)dt = (2\pi)^{-1/2}e^{-t^2/2}\,dt$$

and

$$p(u)du = (2\pi)^{-1/2}e^{-u^2/2}\,du.$$

By making the transformations to new independent variables r and ϑ, $t = r\cos\vartheta$ and $u = r\sin\vartheta$, we can transform the joint probability

$$p(t)p(u)\ dtdu = (2\pi)^{-1}e^{-(t^2+u^2)/2}\,dtdu$$

into the form

$$p(r)p(\vartheta)\ drd\vartheta = (2\pi)^{-1}re^{-r^2/2}\,drd\vartheta, \qquad r \geq 0.$$

This is in the form of two separable distributions, each of which can be sampled from by the method of inversion,

$$r = [2\ln(1/\xi_1)]^{1/2},$$

and

$$\vartheta = 2\pi\xi_2.$$

Transforming back then leaves us with *two* standard Gaussian variables,

$$\eta_1 = [2\ln(1/\xi_1)]^{1/2}\cos(2\pi\xi_2),$$

$$\eta_2 = [2\ln(1/\xi_1)]^{1/2}\sin(2\pi\xi_2),$$

which are independent.

Although exact, there are clearly drawbacks to this method. It will be relatively slow because of the need to call the square-root, logarithmic and geometric functions. In addition the independence of the two variates will critically depend on the lack of correlations between the uniform variates which means that a high quality uniform random number generator needs to be employed.[6] Faster methods, mostly based on finding efficient importance functions using the von Neumann method, but tedious to spell out, can be found in the literature — in example 5.6 we describe a non-exact but interesting, and relatively fast method using the Central Limit Theorem. We note that the sum of squares of two such random variables is $2\ln(1/\xi)$, and the sum of squares of N (even) of such variables is $-2\ln\prod_{i=1}^{N/2}\xi_i$.

EXAMPLE 5.16

The variance on a quantity which is a sum of N independent variates from a distribution is just the sum of their variances, so that the variance on the sum of N variates from the uniform distribution is $N\sigma_\xi^2 = N/12$, using the fact that $\sigma_\xi^2 = 1/12$. The mean of such a sum is $N/2$. The quantity can be scaled to one of zero mean and unit variance by writing

$$\eta = \sqrt{12N}\,\Big(\frac{1}{N}\sum_1^N \xi_i - 0.5\Big).$$

As noted in section 2.3.4 for sufficiently large N by the Central Limit Theorem, such a sum is Gaussian distributed. A conveniently 'large' value for N is 12, so that the quantity

$$\eta = \sum_1^{12}\xi_i - 6$$

has mean zero, unit variance and is, approximately, Gaussian distributed. Although it requires calling 12 uniform variates it will be much faster than the method in the previous example — this will be especially true in an environment of parallel processing where the numbers can be simultaneously generated — since no high level functions are required. In most cases it will be as good an approximation to a Gaussian variate as is warranted by the application of this distribution to the phenomenon in question.

EXAMPLE 5.17

A variate v from the Maxwell-Boltzmann velocity distribution can be written as $v = u/\sqrt{\alpha}$, where $\alpha = m/2k_BT$, and u is a variate from the standardised distribution

$$\begin{aligned} p(u)d\mathbf{u} &= \pi^{-3/2}e^{-u^2}d^3\mathbf{u} \\ &= \frac{4}{\sqrt{\pi}}u^2e^{-u^2}du, \qquad u > 0. \end{aligned}$$

The easiest way to select from this distribution is to note that the first form in Cartesian coordinates can be written as the product of the probability for each of the three components, u_i, i.e.

$$\prod_{i=1}^{3}\{\frac{1}{\sqrt{\pi}}e^{-u_i^2}du_i\}, \qquad -\infty < u_i < \infty$$

each of which is Gaussian distributed with mean 0 and variance 1/2. Thus each component can be sampled as $u_i = \eta_i/\sqrt{2}$, and the total velocity as $u = \sqrt{(\eta_1^2 + \eta_2^2 + \eta_3^2)/2}$, where the η_i are standard Gaussian variates. Noting example 5.15, where the sum of squares of two standardised Gaussian variables equals $-2\ln\xi$, we have

$$u = \sqrt{\eta^2/2 - \ln\xi},$$

where η and ξ are from the standardised Gaussian and uniform distributions, respectively.

EXAMPLE 5.18

An important physical distribution which can serve as a 'laboratory' for selection methods is the *Watt* distribution, which characterises the energy distribution of neutrons emerging from neutron fission in uranium. It involves two parameters, and is written as,

$$p(E)\ dE = Ce^{-E/T} \sinh \sqrt{bE}\ dE, \qquad E \geq 0,$$

with $C = \frac{2}{\sqrt{\pi b}} T^{-3/2} e^{-bT/4}$. Approximate values are $T = 0.965$ MeV and $b = 2.29$ MeV^{-1}. The three components of the velocity of a molecule according to kinetic theory are Gaussian distributed,

$$p(\mathbf{v}) dv_x dv_y dv_z = (\frac{\alpha}{\pi})^{3/2} e^{-\alpha v^2} dv_x dv_y dv_z.$$

In a frame moving with constant, non-relativistic, velocity $\mathbf{V}$ relative to the gas the distribution of the velocity, $\mathbf{v}' = \mathbf{v} - \mathbf{V}$, $(d^3\mathbf{v}' = d^3\mathbf{v})$, is readily shown to be,

$$p(v')dv' = 2\sqrt{\frac{\alpha}{\pi}} \frac{e^{-\alpha V^2}}{V} e^{-\alpha v'^2} \sinh(2\alpha V v') v' dv'.$$

In terms of the kinetic energy, $E = mv'^2/2$, we find the energy distribution — the Watt spectrum,

$$p(E)\ dE = Ce^{-E/T} \sinh \sqrt{bE}\ dE, \qquad E \geq 0,$$

with

$$C = 2\sqrt{\frac{\alpha}{\pi}} \frac{e^{-\alpha V^2}}{mV}, \qquad T = m/2\alpha, \qquad b = 8\alpha^2 V^2/m.$$

We thus have a direct method to sample from this energy spectrum; as in the previous example, select three components of velocity, v_x, v_y, v_z, each from a Gaussian with mean 0 and variance $1/2\alpha$, and set $E = mv'^2/2$ where $v'_x = v_x, v'_y = v_y, v'_z = v_z - V$ and $v' = \sqrt{v'^2_y + v'^2_y + v'^2_z}$.

Using the methods described in this chapter so far, independent sequences of a variate can be generated by repeated application of the appropriate algorithm. In some applications of random variates in simulation it is desired to have the sequence of numbers generated satisfy some overall constraint, as might follow, for example, from a requirement of energy conservation. This results in a modification of the distribution being selected from, effectively changing it into a multivariate distribution. We consider such constrained variate sampling in the next section.

5.4 Multivariate Sampling

Generating samples from multivariate distributions is a frequent requirement of Monte Carlo applications in physics and can call for much ingenuity as there are often correlations imposed on the sample. Examples would be a uniform distribution of points on a sphere or the multivariate Gaussian distribution. Another important case is the Boltzmann distribution of particle coordinates in a system at thermal equilibrium, given in terms of the interaction energy Φ by

$$p(\mathbf{x})d\mathbf{x} = Ke^{-\beta\Phi(\mathbf{x})}d\mathbf{x}$$

with $K = \int dx_1 \ldots dx_N e^{-\beta\Phi(\mathbf{x})}$.

EXAMPLE 5.19

Points uniformly distributed on the surface of a sphere in N dimensions follow the distribution

$$p(\Omega)d\Omega = \frac{\Gamma(\frac{N}{2})}{(2\pi)^{N/2}}d\Omega \tag{4.1}$$

where $d\Omega$ is an element of area on the unit sphere in N dimensions. These points can be specified by their direction cosines, $\Omega = (\omega_1, \omega_2, \ldots, \omega_N)$, so we need to choose the components of an isotropically distributed vector in the N dimensional space. We note that in two dimensions, a point defined by two independent variates, x_1 and x_2, from a Gaussian of zero mean will be distributed in the plane with radial symmetry, since their distribution

$$p(x_1, x_2)dx_1dx_2 \propto e^{-(x_1^2+x_2^2)/2\sigma^2}dx_1dx_2$$
$$\propto e^{-r^2/2\sigma^2}\, dA,$$

depends only on the radial variable. Independent of the value of σ, any such pair of points could be used to represent an isotropically distributed direction. The direction cosines, $x_1/\sqrt{(x_1^2+x_2^2)}$, $x_2/\sqrt{(x_1^2+x_2^2)}$, would give coordinates of a point uniformly distributed on the unit circle. The analogous result in three dimensions is also easy to see, and the result generalises to a space of higher dimensions. Thus the algorithm is: select N standard Gaussian variables, η_i, and set $\omega_i = \eta_i/\sqrt{\sum \eta_i^2}$.

Perhaps the most common, and, fortunately, most tractable problem in multivariate selection is that from the Gaussian distribution. We consider generating a vector of N variates from the general Gaussian distribution

$$p(\mathbf{x})d\mathbf{x} = \frac{(\det \underline{\underline{\alpha}})^{1/2}}{(2\pi)^{N/2}} e^{-\frac{1}{2}(\mathbf{x}-\underline{\mu})^T \underline{\underline{\alpha}}(\mathbf{x}-\underline{\mu})}\, d\mathbf{x}, \tag{4.2}$$

where $\underline{\underline{\alpha}}$ is an $N \times N$ symmetric matrix, the inverse of the variance- covariance matrix $\underline{\underline{\alpha}}^{-1}$. A vector $\mathbf{y}$ with means zero is generated $(\mathbf{x} = \mathbf{y} + \underline{\mu})$ from the distribution

$$p(\mathbf{y})d\mathbf{y} = \frac{(\det \underline{\underline{\alpha}})^{1/2}}{(2\pi)^{N/2}} e^{-\frac{1}{2}\mathbf{y}^T \underline{\underline{\alpha}}\mathbf{y}}\, d\mathbf{y}.$$

A linear, orthogonal, transformation of the components of $\mathbf{y}$ can be found, $\mathbf{z} = \mathbf{U}^{-1}\mathbf{y}$, which transforms the exponent in the exponential into a sum of squares, i.e. the distribution in the new variable $\mathbf{z}$ becomes a product of independent Gaussian distributions of its components. This takes the form

$$\frac{\prod_k \Delta_{kk}^{1/2}}{(2\pi)^{N/2}} e^{-\sum_k \Delta_{kk} z_k^2/2},$$

where $\underline{\underline{\Delta}} = \mathbf{U}^T \underline{\underline{\alpha}} \mathbf{U}$ is a diagonal matrix, and $\det(\underline{\underline{\Delta}}) = \det(\underline{\underline{\alpha}})$. The N values of z_i are then chosen, using standard Gaussian variates η_i, as

$$z_i = \eta_i / \sqrt{\Delta_{\rm ii}} \tag{4.3}$$

and the inverse transformation, $\mathbf{y} = \mathbf{U}\mathbf{z}$, carried out.

We will not consider here the general problem finding the matrix $\mathbf{U}$.

EXAMPLE 5.20

In this example we look at the simple case of two variables having the same variance and with a correlation coefficient ρ,

$$p(\mathbf{y})d\mathbf{y} = \{2\pi\sigma^2\sqrt{1-\rho^2}\}^{-1} \exp -\{\frac{y_1^2 - 2\rho y_1 y_2 + y_2^2}{2(1-\rho^2)\sigma^2}\} d\mathbf{y}. \tag{4.4}$$

A diagonalising transformation is

$$\mathbf{U} = \frac{1}{\sqrt{2}} \begin{bmatrix} 1 & 1 \\ -1 & 1 \end{bmatrix},$$

yielding the diagonal form

$$\underline{\underline{\Delta}} = \mathbf{U}^T \underline{\underline{\alpha}} \mathbf{U} = \frac{1}{\sigma^2} \begin{bmatrix} \frac{1}{1-\rho} & 0 \\ 0 & \frac{1}{1+\rho} \end{bmatrix}.$$

If η_1, η_2 are standard Gaussian variates we have

$$z_1 = \sigma\eta_1\sqrt{1-\rho}\ , \qquad z_2 = \sigma\eta_2\sqrt{1+\rho},$$

and, transforming back, $\mathbf{y} = \mathbf{U}\mathbf{z}$ finally gives

$$y_1 = \frac{\sigma}{\sqrt{2}}\sqrt{1-\rho}\,\eta_1 + \frac{\sigma}{\sqrt{2}}\sqrt{1+\rho}\,\eta_2, \tag{4.5a}$$

$$y_2 = -\frac{\sigma}{\sqrt{2}}\sqrt{1-\rho}\,\eta_1 + \frac{\sigma}{\sqrt{2}}\sqrt{1+\rho}\,\eta_2. \tag{4.5b}$$

The methods described in the previous two examples yield independent multivariate samples with the required correlations existing in each sample, but many cases will not be amenable to such explicit treatment. There exists a method of great power and generality, practical even for very large samples, which can be applied, but it suffers from the fact that the sequences generated are not independent. This is the method of *Markov chains* which we describe in the next section.

5.5 Markov Chain Methods

5.5.1 Introduction

There exists another, more abstract, method of generating variates from a probability distribution which makes use of the phenomenon of a stationary distribution associated with a *Markov chain.* The method is especially suitable in selecting multidimensional variates — configurations — and is of great practical importance in computational statistical mechanics and quantum field theory. When the variate has many components, e.g. the coordinates of a point in phase space, normalisation of the relevant probability density function, e.g. the square of a quantum mechanical wavefunction or the Boltzmann probability function, becomes far from a trivial task. In other words we have the distribution in the form

$$p(\mathbf{x})d\mathbf{x} = Cf(\mathbf{x})d\mathbf{x}$$

where the explicit form of the function f is known but the normalising constant $C = \{\int f(\mathbf{x})d\mathbf{x}\}^{-1}$, an integral over the whole multidimensional space, is not known and not easily determined. The Markov chain method of selection enables variates to be selected without a knowledge of this constant.

5.5.2 Markov Processes

Physically a Markov process is an evolution of a sequence of states of a system, where at every stage of the evolution the transition to the succeeding state depends only on the current state of the system, including, most generally, the time, without any memory of its earlier behaviour. We will restrict ourselves to processes with no explicit time dependence — *homogeneous Markov processes.* Such processes may be terminal or they may perpetually evolve. When the possible states are discrete — we will assume them finite in number for simplicity — we can characterise the direct transition from any one state, $\mathbf{x}_i$, to another, $\mathbf{x}_j$, by the relative probability of its occurrence, denoted by W_{ij}. Confining ourselves to perpetually evolving processes the properties of this quantity are: (i) W_{ij} is non-negative — by definition, (ii) for at least one value of $j \neq i$ it must be non-zero, otherwise the evolution will terminate, and (iii) by conservation of probability, W_{ij} must satisfy,

$$\sum_j W_{ij} = 1 \text{ for all } i.$$

The array of such elements defines a *stochastic matrix* $\mathbf{W}$. Starting with an arbitrary state the system will evolve through a stochastic sequence of states, a Markov chain. How the properties of Markov chains can be used to generate variates from a distribution is best introduced by way of an example.

EXAMPLE 5.21

We consider the stochastic matrix $\mathbf{W}$, where

$$\mathbf{W} = \begin{bmatrix} \frac{1}{3} & \frac{1}{3} & 0 & 0 & 0 & \frac{1}{3} \\ \frac{1}{30} & \frac{19}{30} & \frac{1}{3} & 0 & 0 & 0 \\ 0 & \frac{1}{12} & \frac{7}{12} & \frac{1}{3} & 0 & 0 \\ 0 & 0 & \frac{1}{6} & \frac{1}{2} & \frac{1}{3} & 0 \\ 0 & 0 & 0 & \frac{1}{3} & \frac{8}{15} & \frac{2}{15} \\ \frac{1}{96} & 0 & 0 & 0 & \frac{1}{3} & \frac{21}{32} \end{bmatrix},$$

and use it to execute a random walk among six states, denoted by

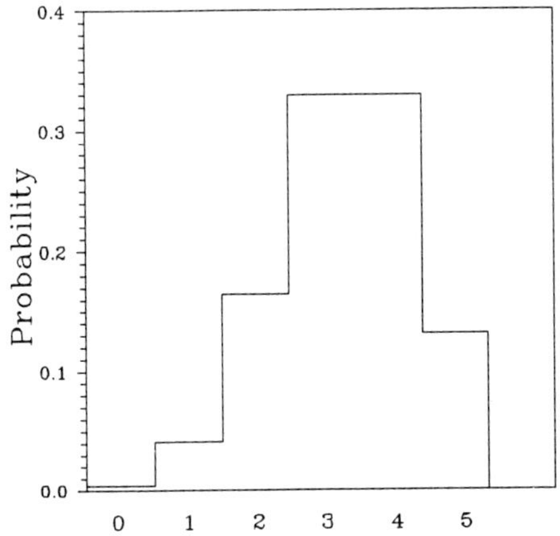

Figure 5.5: The distribution selected from using the Markov chain prescribed by the matrix **W** in Example 5.21.

$x_1, x_2, x_3, \ldots, x_6$. This is done as follows: if, for example, at some stage in the walk the current value of the variable is x_3, the probability that the next value will be x_1 is $W_{31} = 0$, that it will be x_2 is $W_{32} = 1/12$, etc. In particular the next value may also be x_3 again, with probability 7/12. By calling a single uniform random deviate, ξ, the decision can be made in the familiar way; in favour of x_1 if $W_{31} > \xi$, else x_2 if $W_{31} + W_{32} > \xi$, else ... If the choice should fall, say, to x_4 a new value for ξ is called and the walk continued on the basis of the probabilities in row four of the matrix. It is clear that an initial value for x will have to be specified to get the process going. This matrix will certainly generate an arbitrarily large sample of values in this way; what is not so transparent is whether strings of these variables can be taken as representative samples from any probability distribution, or, if this is the case, to what extent such samples depend on the initial starting value. It can be shown that for this matrix (the necessary conditions are described in the next section) the relative frequency of occurrence of the six values of x will stabilise to definite ratios. In fact they will be distributed such that on average, x_1 occurs with probability 1/243, x_2 with probability 10/243, x_3 with probability 40/243, etc. In other words the matrix generates the distribution $p(x)$ shown in Fig. 5.5; with x representing the integers 0 to 5, it is just the Binomial distribution, equation 2.4.1, for $N = 5$ with success parameter $\alpha = 2/3$.

In the theory of Markov chains, the random walk dictated by the matrix **W** in the above example is *ergodicergodic behaviour*, i.e. all values of the variable are accessible from any particular one in a finite number of steps, and the distribution $p(x)$ is the *stationary distribution* of the walk whose one-step probability matrix is **W**. Not every stochastic matrix gives rise to a stationary distribution, i.e. where memory of the initial state is lost. A central problem in the theory of Markov chains is to decide whether a given matrix does give rise to such a distribution, and how to find it. Our general problem is the reverse of this, given a probability distribution $p(\mathbf{x})$ how do we find a stochastic matrix, $\mathbf{W}(p)$, which will generate a chain of variables which mimic a sample selected from the distribution p.

5.5.3 Markov Chain Algorithms

As suggested by the discussion in section 5.5.2, transition operators can be found which, independent of the starting point, will, after 'sufficient' steps, generate variates which are distributed as $p(\mathbf{x})$. The formal requirements on the stochastic matrix **W** are:[7]

(i) $W_{ij} \geq 0$, and $\sum_j W_{ij} = 1$ for all i — the stochastic condition,

(ii) state j can be reached from state i with probability W_{ij} so the probability of being in state j after a step is the sum of the transition probabilities weighted by the probability of the system already being in state i, i.e. $\sum_i p_i W_{ij} = p_j$ for all j — the convergence condition, and

(iii) for some finite m, $(W^m)_{ij} > 0$, i.e. state j must be accessible from state i with finite application — the ergodic condition.

Under these conditions, the theory of Markov chains shows that

$$\lim_{n\to\infty}(W^n)_{ij} = p_j,$$

independent of i. In other words, whatever the starting point, after sufficient steps the probability of occurrence of any particular state in the chain tends to its absolute probability as determined by the distribution function p. This distribution is said to be the *stationary distribution* of the algorithm. We can say that the system of linear equations in (ii) for a specified set of W_{ij} uniquely determines the values p_j. This uniqueness is not reciprocal, the conditions allow great variety in the possible matrices $\mathbf{W}$ which will generate the distribution p. In practice, a stronger, sufficient, condition is used in the construction, viz.

$$p_i W_{ij} = p_j W_{ji}.$$

This is known as the principle of *detailed balance*; in physical situations it will often correspond to microscopic reversibility. It is not essential for selection algorithms, e.g. it does not hold for the Heat Bath algorithm described in section 10.2.3. Summing over i, we then find condition (ii) automatically satisfied, since for any j,

$$\sum_i p_i W_{ij} = \sum_i p_j W_{ji} = p_j \sum_i W_{ji} = p_j,$$

where condition (i) has been used in the last step. Condition (iii) may pose a problem in some applications, enormously large values of m corresponding to 'frozen in' regions of the state variables, so called quasi-ergodic regions.

To find a form for the matrix $\mathbf{W}$ we write W_{ij} as a product of a targeting probability for the state j and an acceptance probability of that state. For the former it is simplest to take any symmetric stochastic matrix **G**, then in terms of $\eta_{ij} \equiv p_j/p_i$, $(\eta_{ji} = 1/\eta_{ij})$, suitable choices for the elements of **W** are, for $i \neq j$,

$$W_{ij} = \eta_{ij} G_{ij} \quad \text{if } \eta_{ij} < 1, \tag{5.1a}$$

$$W_{ij} = G_{ij} \quad \text{if } \eta_{ij} > 1. \tag{5.1b}$$

The element W_{ii} is determined by

$$W_{ii} = 1 - \sum_j W_{ij} \tag{5.2}$$

$$= 1 - (\overset{(>)}{\sum} W_{ij} + \overset{(<)}{\sum} \eta_{ij} W_{ij}) > 0,$$

where the two terms in the brackets correspond to sums over values of j for which $\eta_{ij} > 1$ and $\eta_{ij} < 1$, respectively. It is easily seen that the condition of detailed balance, $W_{ij} = \eta_{ij} W_{ji}$, is satisfied for such a choice. More formally we can write,

$$W_{ij} = G_{ij}.\min[1, \eta_{ij}].$$

The major advantage of using Markov chains is that selection depends on η_{ij}, a ratio of probabilities, so that the overall normalisation of these probabilities, which can be very difficult when the dimensions of the variate in question is large, is not required. The method also suffers from decided weaknesses. Principal among these is the non-independence of successive variates sampled; the probability of a value x_j following a value x_i is completely specified by the algorithm — in our illustration above, for example, x_4 never succeeds x_1. In independent sampling obviously no such constraint can exist. For the same reason any sequence of samples depends to some extent on a preceding sample, and, ultimately, on the starting value. To select characteristic samples not biased by this arbitrary value a number of the initial variates in the chain should be discarded from consideration. The problem of the correlation in sequences can be allowed for by statistical methods, discussed in section 2.3.3. How rapidly the sequence of variates generated from an arbitrary starting value converges to a representative sample from the distribution $p(x)$ depends on the choice for $\mathbf{W}$ and on the starting value. The number of early samples which should be discarded prior to this convergence will, in general, have to be determined empirically, e.g. on the basis of the expectation of some parameters of the distribution.

5.5.4 Practical Algorithms

If $\mathbf{G}$ is a symmetric matrix each row of which contains $\nu + 1$ non-zero elements, each equal to $1/(\nu+1)$ with $\nu \geq 1$, and the remaining elements equal to zero we have the *Metropolis algorithm* for a discrete variable,[8]

$$W_{ij} = \left(\frac{1}{\nu+1}\right)\frac{p_j}{p_i} \qquad \text{if } p_j > p_i, \tag{5.3a}$$

$$= \left(\frac{1}{\nu+1}\right) \qquad \text{if } p_j < p_i, \tag{5.3b}$$

and

$$W_{\text{ii}} = 1 - \sum_{j \neq i} W_{ij}. \tag{5.4}$$

This corresponds to a possibility of only ν distinct values being accessible directly from any one particular point, (although, of course, all points are ultimately accessible).

Another choice for $\mathbf{G}$ sometimes used[9] has elements

$$G_{ij} \propto \frac{\min[1, \eta_{ij}]}{1 + \eta_{ij}}.$$

It is then easy to show that the elements of $\mathbf{W}$ are

$$W_{ij} \propto \frac{\eta_{ij}}{1+\eta_{ij}}, \tag{5.5}$$

with W_{ii} again given by equation 5.2. This is known as *Barker's algorithm.*

The merits of rival algorithms is decided principally by (i) how fast the influence of the starting point is lost, (ii) how little correlation exits in the generated sequences, and (iii) how well they avoid prolonged entrapment in isolated regions of the space of the variable — quasi-ergodic behaviour. The role of these factors may also depend on the use being made of the variates, and the determination of an optimum prescription is something of a black art. More by default and convenience than empirically determined superiority, the Metropolis algorithm is very commonly adopted and we shall encounter it frequently in subsequent chapters.

For simplicity, our discussion of Markov Chain methods has been in the context of discrete states, but the greatest merit of the method is seen when it is extended to a variable, x, continuously distributed as $p(x)$. A target point can be written as $x' = x + \delta$, where δ is chosen from a distribution $g(\delta)$. Care must be taken that the points considered as destinations in making a transition from a point x are symmetrically distributed about this point, i.e. δ must be selected from a distribution which is symmetric about zero, e.g. $\delta = \pm\Delta$ with equal probability, or δ selected from a uniform distribution on $\pm\Delta$ about the point. This latter corresponds to a targeting matrix

$$\begin{aligned} G(x \to x') &= g(\delta) = \frac{1}{2\Delta}, & -\Delta \leq \delta \leq \Delta \\ &= 0, & |\,\delta\,| > \Delta. \end{aligned}$$

In this case prescriptions like equations 5.1 and 5.2 can be used for the transition functions, with the definition $\eta = p(x')/p(x)$, i.e

$$W(x \to x') = g(\delta) \,.\, \min[1, \eta]. \tag{5.6}$$

Strictly speaking, $G(x \to x')$ need not be symmetric in its arguments; if it is not, equation 5.6 should be replaced by

$$W(x \to x') = G(x \to x') \min\left[\, 1, \frac{G(x' \longrightarrow x)}{G(x \longrightarrow x')}\eta \,\right]. \tag{5.7}$$

The Markov Chain method is a very powerful technique for sampling from multivariate distributions, but one needs to remain alert to its limitations, notably questions of efficiency, ergodicity, i.e. exhaustive exploration of the space of the variables, and correlations. By efficiency we mean the fraction of tentative points which lead to a new coordinate value to be accepted. In each dimension such tentative values will typically be selected at random from a region of $\pm\Delta$ about the current value, the smaller Δ the more likely the tentative value is to be selected.

EXAMPLE 5.22

For illustration only (in practice the method would never be used for selecting a one-dimensional variate) we look at some results of selecting from the one-dimensional standardised Gaussian distribution,

$$p(x) \propto e^{-x^2/2},$$

in every case taking $x_0 = -1$. Fig. 5.6(a) shows how the acceptance falls off with Δ as expected — the first 20 variates at every value of Δ having been ignored in order to wipe out the influence of the starting point. Small values of Δ, however, slow down the exploration of the space of the distribution and lead to strong correlations in a sample; in Fig. 5.6(b) we see how estimates of the standard deviation of the distribution, based on samples of 50 variates, are thereby underestimated, only tending towards, but not reaching, the value of the population for large values of Δ. The correlations at large values of Δ arise from the low acceptance there; partial 'freezing in' in high

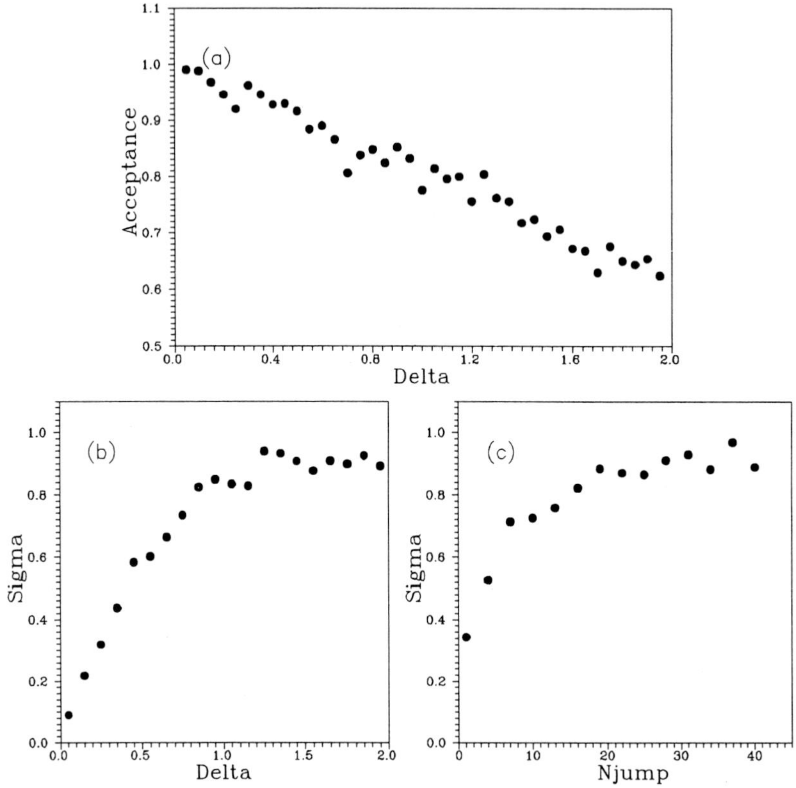

Figure 5.6: Markov chain selection from the Gaussian distribution, (a): the acceptance probability for a new value as a function of the step size Δ, (b): the standard deviation of samples of size 50 as a function of Δ, and (c): the standard deviation of samples of size 50 ($\Delta = 0.2$), when only every kth variate is included, as a function of k.

probability regions of the distribution, in this case around the mode, can occur. In either case, because successive values sampled are not being randomly chosen from the distribution, they are correlated. One way to suppress correlations in a sample is instead of accepting every generated value of the variate, to limit the sample to every kth value. In Fig. 5.6(c) we show, for a fixed value of $\Delta = 0.2$, how the estimate of the standard deviation, based on samples of 50, tends to that of the Gaussian population as k (= NJUMP) increases.

The potential pitfalls of the method are already clear from a straightforward example, indicating the caution necessary when selecting in multidimensional space from distributions whose structure is often not transparent, e.g. there are real hazards of being 'frozen in' regions of a multimodal distribution. We look at other examples in chapter 10.

Other Markov chain based algorithms include one for the selection of self-avoiding walk configurations, described in chapter 8, one based on the Fokker-Planck equation, and the *Demon* algorithm, useful in selecting thermodynamic configurations, which is described in section 10.2.3.

Notes

1. Early work on random number selection is described in chapter 3 of Hammersley and Handscomb.

2. Our discussion here is basically for illustration and takes little account of questions like the relative efficiency of different methods, or the influence of round-off errors with a finite word size, etc. In developing a major Monte Carlo project a professionally constructed library, especially of common generators like the Gaussian, is advisable; detailed considerations of such generators can be found, for example, in chapter 9 of Fishman (1978).

3. A very comprehensive summary of of selection methods for distributions of importance in physics is the report, Everett, C. J. and E. D. Cashwell, *A Third Monte Carlo Sampler*, Los Alamos Report $LA - 9721 - MS$, (Oak Ridge, TENN: U. S. Department of Energy Office of Scientific and Technical Information, 1983).

4. Tables of $1/B(\omega)$ can be found in Appendix C of Kittel, C., *Thermal Physics* (New York: John Wiley, 1969).

5. This method is known as the method of Composition in some of the literature.

6. The effects are only observed for very ill chosen generators of pseudo random numbers; see, for example, Golder, E. R. and J. G. Settle (1976), *The Box-Muller Method for Generating Pseudo-random Normal Deviates*, Appl. Statist., **25**, 12-20.

7. Hammersley and Handscomb (1964), chapter 9.

8. Metropolis N., A. W. Rosenbluth, M. N. Rosenbluth, A. H. Teller and E. Teller (1953), *Equation of state calculations for fast computing machines*, J. Chem. Phys., **21**, 1087-92.

9. Barker, A. A. (1965), *Monte Carlo calculations of the radial distribution function for a proton-electron plasma*, Aust. J. Phys., **18**, 119-33.

Exercises

5.1 Use pairs of uniform random variates to generate 1000 variates x, according to the following algorithm,

IF $2\xi_2 \leq \xi_1(3-\xi_1)$ THEN set $x = 1-\xi_1$, ELSE $x = (\sqrt{9-8\xi_2}-1)/2$.

Use the χ^2 statistic to test the hypothesis that these variates are from the distribution

$$p(x)dx = \frac{3}{2}(1-x^2)dx, \qquad 0 \leq x \leq 1.$$

5.2 A triangular distribution is defined by

$$\begin{aligned} p(x)dx &= x dx \qquad 0 \leq x \leq 1 \\ &= (2-x)dx \qquad 1 < x \leq 2. \end{aligned}$$

Show that selection can be made as follows,

IF $\xi \leq 1/2$ THEN $x = \sqrt{2\xi}$ ELSE $x = 2 - \sqrt{2(1-\xi)}$.

Show that $z = x + y$, where x and y are variates from the uniform distribution on [0,1] obey this distribution, and thus that a much faster method of selecting is to set $x = \xi_1 + \xi_2$.

5.3 What distributions would be sampled from by the prescriptions

(i) $x = \sin^2 \xi\pi/2$

(ii) $x = \frac{1}{\alpha}\ln[\xi/(1-\xi)]$?

5.4 Explain how values of a standard Gaussian variable, apart from a sign, $|\eta|$, can be selected using von Neumann's method with an importance function $\pi(\eta) = e^{-\eta}$, and show that the fraction of successful choices is $\sqrt{\pi/2e} \simeq 0.76$.

5.5 Remembering that $\sin^2 + \cos^2 = 1$, explain why, if the numbers ξ are truly random, the two Normal variates selected in example 5.15 are independent. For the situation when the ξ values are from a multiplicative congruential generator, see section 6.7.3 of Bratley, Fox and Schrage.

5.6 Show that the logistic distribution with zero mean,

$$p(x)dx = \frac{\alpha e^{-\alpha x}}{(1+e^{-\alpha x})^2}dx, \qquad -\infty < x < \infty,$$

can be sampled from by inversion using a single random number ξ as

$$x = \alpha^{-1}\ln\left(\frac{\xi}{1-\xi}\right).$$

Use this distribution as an importance function in von Neumann's method to select variates from the standardised Gaussian

$$p(\eta)d\eta = \frac{1}{\sqrt{2\pi}}e^{-\eta^2/2}\,d\eta.$$

Show that we must have $\alpha > \sqrt{2}$, and that the sampling efficiency is greatest using this value.

5.7 The distribution in the previous question can be modified to yield a distribution on $[-\infty, u]$

$$p(x)dx = \frac{\alpha(1+e^{-\alpha u})e^{-\alpha x}}{(1+e^{-\alpha x})^2}dx, \qquad -\infty < x < u,$$

show that selection can be made from

$$x = \alpha^{-1}\ln\left(\frac{1-\xi}{e^{-\alpha u}+\xi}\right).$$

5.8 The distribution in example 5.11 can be generalised to an asymmetric distribution of the form

$$\frac{1}{\sigma_T}\frac{d\sigma}{d\Omega} = \frac{3c}{4\pi(c+3)}(1 + b\cos\vartheta + c\cos^2\vartheta).$$

Show that this is a suitably normalised probability distribution provided $b^2 < 4c$. Show that, even when b is negative, the method of superposition can be used to sample from the distribution. In particular, for the case $b = -3/2, c = 3/2$ show that

$$\begin{aligned}\cos\vartheta &= \frac{1}{2} + \operatorname{sgn}(\frac{28\xi-27}{8})(|\frac{28\xi-27}{8}|)^{1/3} \quad \text{with probability } 7/12\\ &= 2\xi - 1 \quad \text{with probability } 5/12.\end{aligned}$$

5.9 In the multivariate Gaussian distribution, equation 5.4.2, the variance-covariance matrix $\underline{\underline{\alpha}}^{-1}$ is a symmetric, positive definite matrix. Such a matrix can be factored as a product of a lower triangular matrix with its transpose — a Cholesky factorisation — $\underline{\underline{\alpha}}^{-1} = \mathbf{L}\mathbf{L}^T$. Here the elements of $\mathbf{L}$, $l_{ij} = 0$ for $j > i$. If $\eta_1, \eta_2, \ldots \eta_N$ are N standard Gaussian variates, it can be shown that correlated variates from the distribution, equation 5.4.2, can be written as,

$$x_i = \mu_i + \sum_{j=1}^{i} l_{ij}\eta_j.$$

Since $l_{ij} = 0$ for $j > i$ this can be written as,

$$x_i = \mu_i + \sum_{j=1}^{N} l_{ij}\eta_j,$$

or

$$\mathbf{x} = \mu + \mathbf{L}\eta.$$

Comparing this with the case of a single Gaussian, $x = \mu + \sigma\eta$, we see that the matrix $\mathbf{L}$ plays the role of an effective variance which correctly takes the covariances into account.

Use this method to show that for the distribution in example 5.20, i.e.

$$p(\mathbf{y})d\mathbf{y} = \{2\pi\sigma^2\sqrt{1-\rho^2}\}^{-1}\exp-\{\frac{y_1^2 - 2\rho y_1 y_2 + y_2^2}{2(1-\rho^2)\sigma^2}\}d\mathbf{y},$$

$\underline{\underline{\alpha}}^{-1} = \mathbf{L}\mathbf{L}^T$, where

$$\mathbf{L} = \begin{bmatrix} \sigma & 0 \\ \sigma\rho & \sqrt{1-\rho^2} \end{bmatrix},$$

with the result

$$x_1 = \sigma\eta_1,$$

$$x_2 = \sigma\rho\eta_1 + \sigma\sqrt{1-\rho^2}\eta_2.$$

Show that these variates have the same mean, variance and covariance as those derived in example 5.2.

5.10 A discrete variate may assume any one of six values with probabilities $1/32, 5/32, 5/16, 5/16, 5/32, 1/32$. Show that if only the two values adjacent (cyclic) to any value are targeted, each with probability $1/2$, the Metropolis algorithm gives rise to a transition matrix $\mathbf{W}$, where

$$\mathbf{W} = \begin{bmatrix} 0 & \frac{1}{2} & 0 & 0 & 0 & \frac{1}{2} \\ \frac{1}{10} & \frac{2}{5} & \frac{1}{5} & 0 & 0 & 0 \\ 0 & \frac{1}{4} & \frac{1}{4} & \frac{1}{2} & 0 & 0 \\ 0 & 0 & \frac{1}{2} & \frac{1}{4} & \frac{1}{4} & 0 \\ 0 & 0 & 0 & \frac{1}{2} & \frac{2}{5} & \frac{1}{10} \\ \frac{1}{2} & 0 & 0 & 0 & \frac{1}{2} & 0 \end{bmatrix}$$

5.11 Labelling the variables in the previous example $n_0 = 0$ up to $n_5 = 5$ start with n_0 and use the matrix to generate 100 samples of the variate. Compare their distribution with the Binomial distribution

$$p_N(n) = \frac{N!}{n!(N-n)!}\alpha^n(1-\alpha)^{N-n}, \qquad 0 \le n \le N,$$

with parameters $N = 5$ and $\alpha = 1/2$.

Calculate the variance of the sample, (a) by taking $\frac{100}{99}(< n^2 >$-$< n >^2)$, and (b) by batching as follows: subdivide the sample into 10 successive groups of 10, calculating for each group its mean $< n >_{10}$. Take the mean of these means, $\overline{< n >_{10}}$ and the mean of their squares $\overline{< n >_{10}^2}$, and take the estimate $\frac{10\sqrt{10}}{9}\{\overline{< n >_{10}^2} - \overline{< n >_{10}}^2\}$.

Compare these values with the variance of the population $\sigma^2 = N\alpha(1-\alpha)$.

5.12 Select 2000 variates from the distribution,

$$\pi(x)dx = Ke^{-\beta(x^2 - a^2)^2}\, dx, \qquad x \geq 0,$$

$\beta = 1/2, a = \sqrt{2}$, using the Metropolis algorithm as follows (ξ always means a *new* uniform random number):

set $\Delta = 0.1, x_0 = 3/5$,

$x' = x_{i-1} + \delta$, where $\delta = (2\xi - 1)\Delta$,

$x_i = x'$ if $\pi(x')/\pi(x_{i-1}) > \xi$,

(note that this is automatic if $\pi(x') > \pi(x_{i-1}), \xi$ need not be called)

otherwise $x_i = x_{i-1}$.

Draw a histogram of the numbers and superimpose the function on it ($K = 1.0023$).

Chapter 6

MONTE CARLO OPTIMISATION

> Ah lad, upon the road of life
> 'Tis best to dance with Chance's wife
> And let the rains that come in time
> Erase the footprints of the crime.
>
> Patrick Kavanagh

6.1 Introduction

We have already described, in section 4.4, the essential aspects of optimisation, and seen how the simulation of a natural stochastic process, natural selection, can be used in solving general optimisation problems. We will be concerned in this chapter with a recent development in this field, the method of *Simulated Annealing* which is based on Monte Carlo sampling. Considering the set of parameters of the problem as specifying coordinates in a state space, we seek to locate that point in the space at which the function is a minimum. By drawing on the analogy between such a parameter space and the phase space associated with a system in statistical mechanics, a very fruitful parallel may be established between the optimisation problem and the determination of the equilibrium state of a thermodynamic system. Examples of physics problems where optimisation is required would include the layout of components in a micro-circuit design or the stable geometry for a molecular structure, as well as the fitting of parameterised functions to experimental data, often attempted using least squares minimisation techniques.

6.2 Annealing

6.2.1 Physical Annealing

As an introduction to the Monte Carlo technique of optimisation by simulated annealing (OSA), we consider the thermal behaviour of a physical system as it cools down. The configuration of any system, denoted generally by a set of coordinates, $\{\mathbf{q}\}$, immersed in an environment at a temperature, T, is not unique. If the system is not in equilibrium with its environment it will be making transitions to states closer to equilibrium, in a more or less systematic sequence. Even if it is at equilibrium at temperature $T(\neq 0)$ it still will not be in a unique microstate, but will be making transitions among what is, effectively, a family of microstates appropriate to that temperature, with the probability at any instant of being found in the microstate $\{\mathbf{q}^{(i)}\}$ given by the familiar Boltzmann distribution,

$$e^{-\beta U(\{\mathbf{q}^{(i)}\})}/Z,$$

where $U(\{\mathbf{q}^{(i)}\})$ is the energy associated with the configuration $\{\mathbf{q}^{(i)}\}, \beta = 1/k_B T$ where k_B is the Boltzmann constant, and $Z(\beta)$, the partition function, is a normalising constant. In its exploration of these accessible states the system is, of course, continuously borrowing energy from, or surrendering energy to its environment, its average energy content being a function of T. The lower the value of the temperature the smaller the size of the family, $\{\mathbf{q}^{(i)}\}$, and in the limit as T tends to zero its size, in the absence of degeneracy, would be expected to converge to a single state. This is the unique configuration of the system having the smallest value of the energy of mutual interaction and is the focus of our attention in this section. In the case of a metal, for example, starting from the liquid state we know that to attain this minimum energy state physically, i.e. to reach a pure crystalline rather than an amorphous or polycrystalline state, it is not enough just to lower the temperature, the reduction in the temperature must be carried out slowly and steadily — the material must be annealed.

If, when a system is alternating among its accessible states in equilibrium at a temperature T_1, the temperature of the environment is suddenly changed to T_2 the system will pass through a sequence of states (not characteristic of a system in equilibrium) which define some path through the phase space before it reaches the 'promised land' governed by T_2. To find the state of absolute minimum energy we might set T_2 equal to zero. In such a case, however, no energy could be temporarily borrowed from the environment, so that only configurations of lower energy could be realised en route. This is known as a *quench.* This does not mean that the final state reached is the one of globally lowest energy. If the system was in a state in the neighbourhood of a local energy minimum it would be unable to borrow the energy necessary to escape, and so would freeze into a configuration which does not correspond to the absolute energy minimum. As in life, the lowest valley cannot always be reached by going only downhill, one sometimes has to climb a peak before a path to the bottom is found. As is well known, just such a behaviour occurs in real substances — for example in glasses and in certain systems of magnetic dipoles known as 'spin-glasses' — and would occur in a computer simulation of a realistic model of such substances. To find the true minimum energy state the temperature must be reduced sufficiently slowly that the system will be able to exhaustively explore its configurations on the energy scale dictated by the prevailing temperature. It will thus be more likely to escape from configurations which correspond to local energy minima, and settle down in the family of states characteristic of equilibrium at this temperature. At the higher temperatures, changes of configurations corresponding to large decrements in the energy are encouraged with smaller scale changes being deferred until the temperature is lowered. The system must be brought down through a succession of equilibrium states.

We can easily see how the Metropolis algorithm, described in section 5.5.4, can be applied in carrying out such simulations. We select samples of an N component vector, $\mathbf{q}$, from the Boltzmann distribution by 'walking' from point to point in the configuration space (sometimes lingering), according to the following prescription. An initial configuration of coordinates, $\{\mathbf{q}^{(0)}\}$, is provided and successive states generated as follows,

$$\begin{aligned} \mathbf{q}^{(i+1)} &= \mathbf{q}^{(t)} \qquad \text{if } \frac{p(\mathbf{q}^{(t)})}{p(\mathbf{q}^{(i)})} > \xi, \\ &= \mathbf{q}^{(i)} \qquad \text{if } \frac{p(\mathbf{q}^{(t)})}{p(\mathbf{q}^{(i)})} < \xi, \end{aligned}$$

where ξ is a variate from the uniform distribution on [0,1], and the tentative target point $\mathbf{q}^{(t)} = \mathbf{q}^{(i)} + \delta$ where δ, the step in the walk, is an arbitrary displacement of the vector $\mathbf{q}^{(i)}$. Commonly in practice the only components of this displacement δ are in the coordinate(s) of one particle. For the case of the Boltzmann distribution the above prescription reduces to a comparison of $U(\mathbf{q}^{(t)})$ with $U(\mathbf{q}^{(i)})$, with automatic acceptance when the former is least and a decision to be made otherwise. Starting with an arbitrary configuration of the system at a 'high' temperature T_1, the algorithm should be executed until states characteristic of equilibrium at this temperature are being sampled. The temperature is then reduced by ΔT to T_2 and the procedure repeated, resulting in the lower energy states characteristic of T_2 — provided that the system has not got trapped in a family of configurations which correspond to a localised minimum in U. This can occur if the depth of such a local minimum is greater than the typical fluctuation in the energy at T_2. If entrapment is successfully avoided we can, by successively reducing the temperature, approach the states characteristic of an arbitrarily small $T(\neq 0, \beta \to \infty)$ and all the states sampled will be in the very close neighbourhood of the minimum energy state, although we cannot be confident of identifying it uniquely. Apart from the initial temperature, T_1, two other parameters must be decided upon in such a simulation, the step size, ΔT, and the time (or number of configurations sampled at each temperature) over which it is assumed that the system will relax to its new state of equilibrium; the combination of these two parameters is known as the *annealing schedule*. When sampling states at any value of T, a mean energy is defined, $< U >$, which systematically decreases towards U_0 as the temperature is lowered. However, it may not decrease smoothly. The rate at which it decreases, i.e. the specific heat at $T, C(T) = d < U > /dT$, contains valuable information on the range of configurations being sampled. If this specific heat suddenly rises it suggests a change in the order in the system. The specific heat, *C(T)* defined earlier, can also be related to the fluctuations in the energy of the system (section 10.3), which provides a more convenient form for determining it,

$$C(T) = k_B \beta^2 \sigma_U^2, \tag{2.1}$$

where $\sigma_U^2 =< U^2 > - < U >^2$ is the variance on U.

6.2.2 Optimisation by Simulated Annealing

In our discussion in section 6.2.1, the explicit case of an annealing fluid has been kept to the fore because of the intuitive understanding it offers of the prescribed method. In practice, however, this method of simulated annealing is of very much wider application. What we have been describing in the previous section is the determination of the coordinates of the configuration which minimises the value of a specified function of those coordinates, in this case the energy of the system, U. By defining a pseudo temperature the same method can be used to estimate the optimum configuration in any problem specified by a large number of parameters. Provided the 'cost function', i.e. the quantity to be minimised, is given, an annealing scheme can be devised which can lead to a configuration close to the absolute minimum. Conventional methods of solving such optimisation problems can run into great difficulties, because of the great number of variables and possible complexity of the cost function. Many secondary minima, on various scales, may exist in the configuration space so that finding a path to the true minimum may require an enormous amount of work. If we don't insist on the unique global minimum, the OSA method may lead to a solution of the problem adequate for practical purposes.[1] Further, if we believe we are in the immediate neighbourhood of the global minimum, a crude method, such as steepest descent, may then be used to home in on the exact location.

The limit of zero temperature in the algorithm corresponds to the method of steepest descent. The most illustrative applications of the method have been to problems in combinatorial minimisation where the parameters, which may be very numerous, assume discrete values; such an example is determining the minimum tour length in the so called travelling salesman problem. Other examples of its utility include the improvement of images and pattern reconstruction, while in computer engineering the optimum placement and wiring of components in micro electronic chips has been studied by the method. We illustrate this method of OSA by reference to some systems of interest in physics, starting with a small problem in electrostatics,[2] which has the merits of both being easily described and of serving as a prototype for any system of identical particles interacting via two-body forces. We follow this with a description of the general problem of parameter estimation using the method.

EXAMPLE 6.1: Point charges in a plane

Point electric charges, all equal and N in number, are confined to the surface of a circular disc but are otherwise free to move. How do such charges arrange themselves at equilibrium? The case of $N = 2$ is obvious, so also $N = 3$, etc.; each pair will arrange to have the largest distance possible between them, suggesting a uniform necklace of the charges around the circumference of the disc. At some stage, however, as N increases, it becomes energetically favourable, for one charge to be separated from the necklace and reside at the centre of the disc — this occurs at a value of $N = 12$. How the charges arrange themselves as N further increases is by no means obvious — although as $N \to \infty$ we expect them to tend to the distribution of a charged fluid, of total charge Q, on a disc, where the charge density is given by

$$\sigma(r) = Q/2\pi R\sqrt{R^2 - r^2}.$$

At the same time this easily understood problem serves as a prototype for the properties of multiparticle interacting systems as the temperature is reduced, i.e. prior to equilibrium ($T = 0$). We will see graphically how even such a relatively simple classical interparticle interaction, the pure Coulomb interaction, can give rise to frozen-in metastable states, and the difficulty in determining the true 'ground' state for some values of N. To use the method of simulated annealing to investigate this problem for a specified N, a function of $2N$ variables, (x_i, y_i) has to be minimised, namely the total interaction energy resulting when each charge q_i experiences a potential ϕ_i produced by the other charges. This is given by,

$$U = \frac{1}{2}\sum q_i\phi_i = \frac{1}{2}\sum_{i=1}^{N} q_i \sum_{j\neq i} \frac{q_j}{4\pi\epsilon_0 r_{ij}},$$

or, to within a constant,

$$U(x_1, y_1, \ldots x_N, y_N) = \sum_{i=1}^{N}\sum_{j>i}\{(x_i - x_j)^2 + (y_i - y_j)^2\}^{-1/2}.$$

In practice it is best to scale all lengths by the radius of the disc, i.e. set $R = 1$ — the energy will then be in units of $q^2/4\pi\epsilon_0 R$.

6.2.3 Annealing Algorithm

The steps in the solution of a general problem by simulated annealing will parallel those required in the above example, their organisation is shown in the flowchart in Fig. 6.1. Details of the steps are as follows:

(i) input the parameters of the problem and of the algorithm,

(ii) generate an initial configuration of the charges, $\{\mathbf{q}^s\}$; calculate $U_0 = U(\{\mathbf{q}^s\})$,

(iii) select an initial temperature, T_S, i.e. the parameter β,

(iv) select a sequence of successive configurations using the Metropolis algorithm, where each configuration is the same as the previous one, or

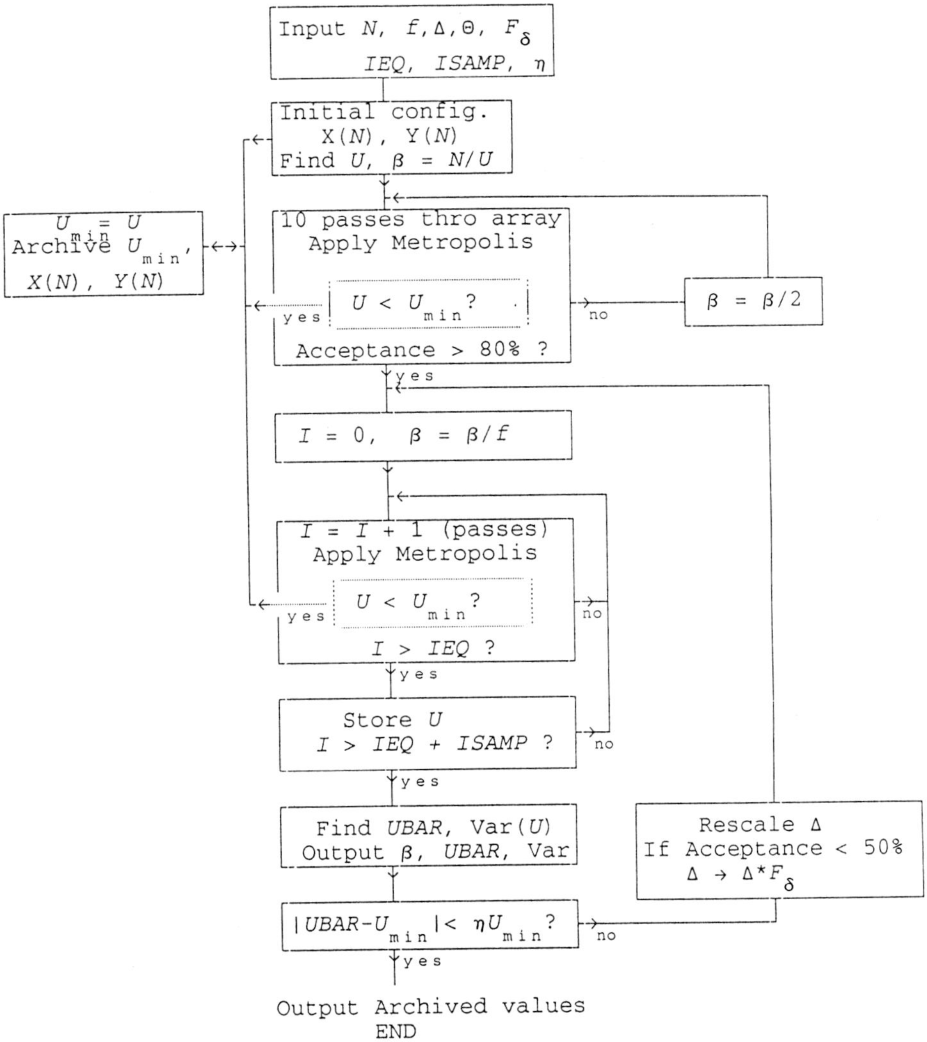

Figure 6.1: A flowchart of the basic annealing algorithm.

differs from it only in the placement of one charge. Determine what percentage of such selections are of the latter kind, i.e. give rise to a new configuration, and if it is less than some specified value ACCEP reduce β. Recycle this loop until acceptance is achieved. This gives us the starting temperature for systematic implementation of the *annealing* algorithm. The configuration of lowest energy to date should always be archived,

(v) increase β, e.g. by multiplying by a factor F_β. If the most recent percentage of 'new configurations' is less than ACCEP reduce the maximum displacement step size Δ by F_δ. Make IEQ passes (i.e. N times IEQ attempts to produce a new configuration) through the array — to attain equilibrium status at the new temperature — followed by ISAMP passes, which are used as the basis for the mean energy UBAR and its variance at this temperature. (An alternative is to calculate running average $< U >_m$ of the energy U over strings of m configurations. When these averages cease to decrease systematically the variance on the energy can be estimated), and

(vi) terminate the process if the fractional difference between UBAR and the archived minimum is less than ϵ, in which case the archived configuration is taken as the solution, otherwise return to step (v).

We will now look at the details of these steps in the context of the example.

EXAMPLE 6.1 (continued)

$N = 17$ charges are considered on a disk of unit radius. In step (ii) possibilities include arrangement on a uniform grid, or, as we adopt,

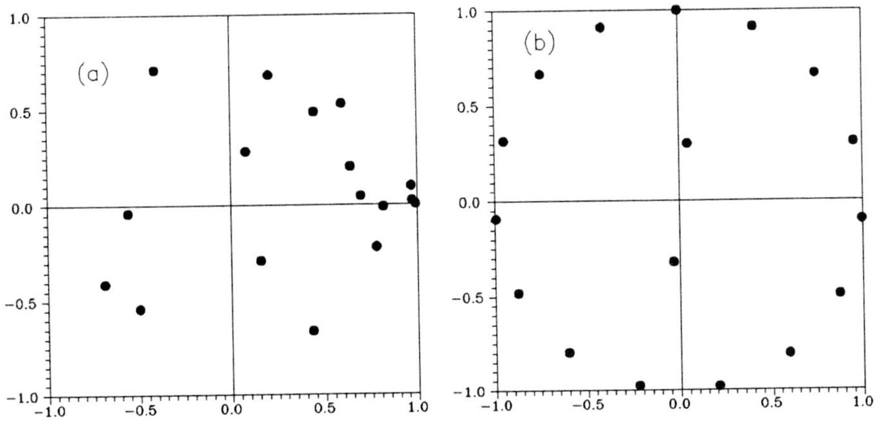

Figure 6.2: (a) The starting, random, configuration for 17 charges, and (b) the final ground state configuration.

a random distribution of the charges on the disc ($r_k = \sqrt{\xi}, \vartheta_k = 2\pi\xi$); the initial arrangement is shown in Fig. 6.2(a). The energy of this arrangement is U_0. Since U_0 is positive we can associate a characteristic temperature with this arrangement given by $U_0 \approx Nk_BT$ so in step (iii) we take as a starting value, $\beta = N/U_0$. In step (iv) $10N$ configurations were sampled and the value of β adopted as the starting value for annealing if the acceptance rate was better than 80%. Otherwise β was reduced by a factor of two before recycling. A tentative new configuration was always selected by changing the coordinates of one charge — the choice can be made at random, or a systematic pass made through the array, we choose the former — say from (x_k, y_k) to (x'_k, y'_k), where $x'_k = r'_k \cos\vartheta'_k, y'_k = r'_k \sin\vartheta'_k$, with $r'_k =\mid r_k + (2\xi - 1)\Delta \mid, \vartheta'_k = \vartheta_k + (2\xi - 1)\phi$, (always subject to the charge remaining on the disc). We took $\Delta = 0.9$ and $\phi = \pi/12$. The change in energy

$$\Delta U = U(..., x'_k, y'_k, ...) - U(..., x_k, y_k, ...)$$

$$= \sum_{j \neq k} \{(x'_k - x_j)^2 + (y'_k - y_j)^2\}^{-1/2} - \{(x_k - x_j)^2 + (y_k - y_j)^2\}^{-1/2},$$

is calculated and the Metropolis algorithm is then applied to decide on the retention of this configuration. In finding the running average, great accuracy is not called for, a value of $m \approx N/2$ might be taken,

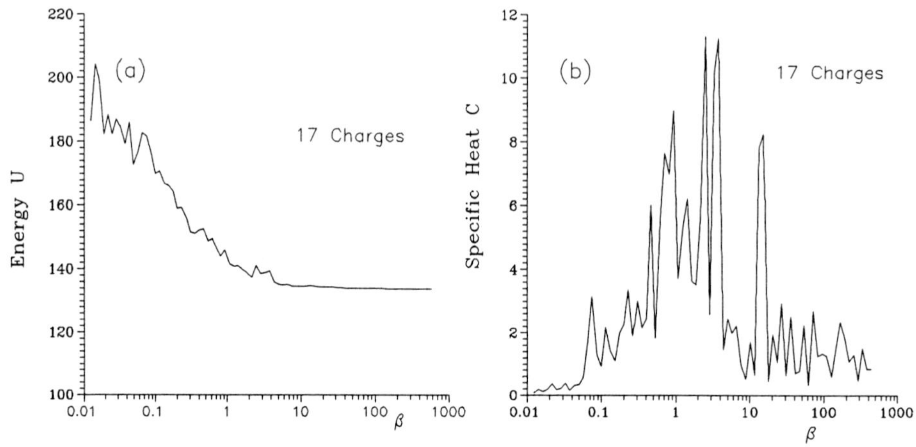

Figure 6.3: (a) The energy U, and (b) the specific heat C, of the system of 17 charges as a function of inverse temperature β.

while in the final estimates, after it is believed equilibrium has been reached, a value of $m = N$ may be suitable. In step (v) we took the temperature reduction factor $F_\beta = 1.15$, acceptable acceptance, ACCEP = 50%, and $F_\delta = 0.9$. The results below are for IEQ = 10 and NSAMP = 20 passes with $\epsilon = 10^{-5}$. Fig. 6.2(b) shows the final configuration, with two charges expelled from the ring. The approach to the minimum as a function of temperature is shown in Fig. 6.3(a), while in Fig. 6.3(b) we show the temperature dependence of the specific heat, further discussed in section 6.2.5. This latter curve shows the broad maximum characteristic of the system relaxing from a generally randomised configuration into the ordered state. The minimum energy, $U = 133.8218$, differs from that of 'meta-stable' states with one or three charges expelled by only 0.18% and 0.43%, respectively. A consequence of the small differences involved is that frequently incrementing the energy by ΔU as done above introduces round-off error in the actual energy, and it has been recalculated at each archiving. The pattern at the minimum, when, as here, the numbers on the circumference and in the interior are incommensurate is interesting. The symmetry is greatly reduced, the charges on the circumference are no longer uniformly arrayed and the pair in the interior does not lie on a diameter, but we shall leave these questions here.

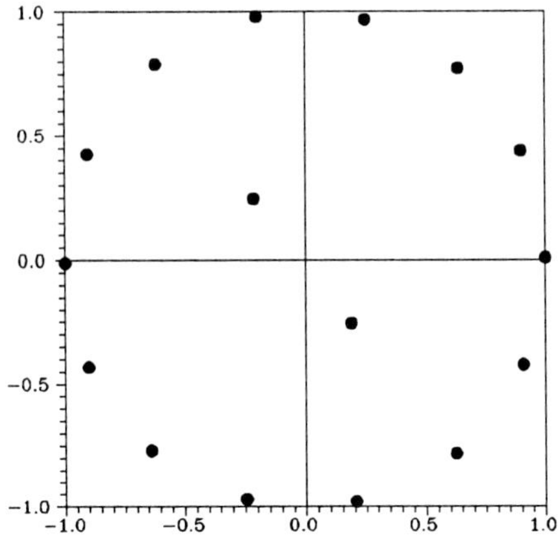

Figure 6.4: A system of 16 charges in a frozen-in, metastable, state.

6.2.4 Improved Algorithm

The example just considered gives a misleading impression of the simplicity and efficacy of the method. A case only slightly different, with $N = 16$ charges, illustrates the problems that can arise with the method. Several careful attempts at annealing ($F_\beta = 1.05, F_\delta = 0.96, \text{IEQ} = 50, \text{ISAMP} = 100$) result in the pattern in Fig. 6.4, with an energy $U = 116.532$. It is a simple calculation to show that the case of only one charge expelled has a lower energy, $U = 116.452$, but only 0.07% lower. The system has frozen into a metastable state. This is not an uncommon experience in physical systems; the system becomes isolated in a region of phase space from which escape is practically impossible. This occurs, not only because there is an energy barrier separating it from the ground state but in addition the phase space associated with fluctuations about these metastable states is much larger than in the neighbourhood of the true ground state, a higher degeneracy of the 'excited state' — the geometrical origin of this is easy to see in the present case. In some physical systems there exists another effect which makes determination of the ground state very difficult, so called *broken er-*

godicity or *frustration*, in which conflicting constraints in the interaction energy which cannot all be simultaneously satisfied occur. Examples are the so called 'spin-glasses'; to be described in more detail in chapter 10. These pose a similar challenge to optimisation by OSA. Two distinct aspects play a role in this failure to converge on the ground state, the first is the details of the annealing schedule, and the second the fact that we are focusing on one sample from what should, strictly, be an ensemble of systems. For some problems, using any practical, i.e. finite, annealing schedule an individual system may be very likely to freeze into a region of phase space — 'time averages' and 'ensemble averages' are not the same when determined by the algorithm, and neither, in general, converges to the zero temperature ground state.

There are two approaches to modifying the algorithm to reduce the likelihood of freezing:

(i) introduce deliberate bias into the phase space search so as to preferentially explore regions where the energy is low but which have a small 'natural' likelihood of occupation, or

(ii) use an ensemble of systems rather than one 'serial' system.

We will look at the second possibility first. One needs to make many independent attempts to find the minimum. For a fixed amount of computing effort intuition suggests that it may be better to explore, in parallel but less exhaustively, a series of independent systems, an ensemble, rather than focusing on one member; the reduced exploration of the individual systems may be compensated for by the opportunities to explore the phase space more widely. Basing the average, and the variance on the energy, on the samples in the ensemble will also give a more reliable indication of the thermal evolution of the system. This development will obviously be very attractive in an environment of parallel processing. In what follows we simply use a set of non-communicating members of an ensemble, all starting from the same configuration; other arrangements with communication between ensemble members, e.g. by favouring members with the lowest energies to date, can obviously be devised.

6.2.5 The Annealing Schedule

Conflicting tendencies have to be reconciled in designing an annealing schedule, and the task is very much dependent on the system concerned — essentially a matter of trial and error. A high initial temperature facilitates an extensive exploration of the space, but at the same time it means that one can easily escape from a region in which the global minimum lies. Too rapid a reduction of the temperature can cause the system to become 'supercooled', while too gentle a decrease will require unacceptable computing effort. The annealing schedule thus consists of two parts, the initial temperature T_0 of the reservoir together with its later 'time' dependence, and the degree of exploration of the space at any temperature. In our previous example we have used an exponential temperature decrease; writing a notional time t which is incremented by unity at every reduction step of the temperature, we have $T(t) = T_0 e^{-at}$, $(a = \ln(F_\beta))$. There is some theoretical work which indicates that the optimum schedule is logarithmic,[3] viz.

$$T(t) = T_0 / \ln(1+t),$$

but even if an optimum value of T_0 can be decided on, this rate of temperature decrease is likely to be too slow for practical calculations. In physical annealing we cannot easily know the thermodynamic state of the system during the process. During OSA, however, we can monitor, for example, the specific heat, and use this as feedback information to regulate the rate of temperature decrease. We want a 'quasi-static' reduction in the temperature, thus, in step (v) the decrement in temperature ΔT might be such that the associated change in the mean energy of the system is not significantly greater than the characteristic fluctuations in this energy at the prevailing temperature. We might require

$$\Delta U = \mid U(T - \Delta T) - U(T) \mid \leq f\sigma_U$$

where $f \approx 1$ is a numerical factor. Then

$$\Delta T \leq f\sigma_U / C,$$

where $C = dU/dT$ is the specific heat. Using equation 2.1 we have, in terms of $\beta, \beta \rightarrow \beta + \Delta\beta$, with

$$\Delta\beta = f/\sigma_U.$$

This is sometimes referred to as annealing with constant thermodynamic speed.

In exploring the phase space several alternatives exist to our prescription of choosing step sizes in a single coordinate from a uniform distribution in $[-\Delta, \Delta]$, although in all cases it will be necessary to reduce the characteristic value Δ as the temperature decreases.[4] An attractive possibility in this respect is to select the step sizes from a Lorentzian distribution (equation 2.4.16), with parameter $\Gamma = \Delta$. This ensures that most of the steps will be smaller than Δ but no matter how small this quantity becomes, arbitrarily large steps will sometimes present themselves, and possibly enable escape from a metastable state.

EXAMPLE 6.1 (continued)

With the modifications outlined, taking 50 members for the ensemble, $f = 1$ and $\Gamma = \Delta$ (initially $= 0.9$) and the other parameters as used earlier, (e.g. only ISAMP/50 passes through a system at a given temperature), the case of 16 charges yielded its ground state, one expelled charge at an energy of 116.4684 compared to the exact value of 116.4520. One notes that the convergence criterion (ϵ specified) is not suitable here, as many members of the ensemble freeze into the metastable state and so the mean energy does not converge to the best minimum value encountered.

We now return to the first suggestion for improvement of the algorithm, i.e. biasing of the search in configuration space. The philosophy here is that since we are only interested in the $T \to 0$ limit, the actual distribution of configurations encountered at any temperature need not be the physically prescribed Boltzmann distribution. Any reasonable distribution which scans the space will do as well. Selecting configurations from the Boltzmann distribution at a non-zero temperature does not minimise the energy of the system unconditionally, it minimises the internal energy of the system subject to the condition of equilibrium, namely that the entropy be a maximum or its negative (-S) be also a minimum. Writing a Lagrange multiplier, τ, the function being minimised is

$$U - \tau S.$$

This multiplier can be identified with the absolute temperature T, and the function which is minimised is the Helmholtz free energy, $F = U - TS$.

As $T \to 0$ this corresponds to minimisation of the potential energy, as we have assumed all along, however, minimising any other function of 'equilibrium' which tends to the same limit should be equally suitable. As we have implemented it, the Metropolis algorithm selects configurations distributed as

$$p(\mathbf{q}) \propto e^{-\beta U(\mathbf{q})},$$

which represents an importance sampling of configurations in terms of their likelihood of occurrence. Here, we are more interested in searching in energy space, especially in the region near $U_{\min}$, and our emphasis should be on exploring this region. The distribution in energy of the configurations sampled by our programme so far will be of the form

$$p(U) \propto \rho(U) e^{-\beta U}$$

where $\rho(U)$ is the density of configurations having an energy U. If this density function is vanishingly small in the neighbourhood of $U_{\min}$ the likelihood of exploring in this region is correspondingly small. This is just the case for the single charge expelled to the centre, where its room for maneuver by an amount η will be of order η^2, compared with the area of order $r\eta$ available to the two charges located at a radius r in the rival configuration. Because of our observation of the arbitrary nature of the distribution used to select configurations we can try one which will increase the probability of finding the values of U of interest. For example the distribution,

$$p(\mathbf{q}) \propto e^{-\beta U(\mathbf{q})} / \rho(U(\mathbf{q})),$$

which preferentially selects energies at the minimum value would seem an attractive candidate.

In this spirit we have modified the algorithm by replacing the ratios in the Metropolis algorithm, equation 5.6.1 by

$$\frac{r^{(i)}}{r^{(t)}}\frac{p(\mathbf{r}^{(t)})}{p(\mathbf{r}^{(i)})},$$

where the pre-factor involves the ratio of the radial coordinates of the individual charge whose move is being attempted.

Incorporating the modifications discussed earlier the program has been revised and run successfully on the 16 charge problem.

6.2.6 Further Examples

We now describe an example where actual temperature is an important par ameter in the physical problem, but the idea of dynamically simulating the behaviour of the system as a function of temperature is abandoned because of the difficulty involved. We will be satisfied with finding the ground state configuration of the system which will hold at temperatures much lower than characterise the typical energies of interaction existing in the system. The more important problem of simulating the thermal behaviour of a system we leave until chapter 10.

EXAMPLE 6.2: Polarisable fluids

We now consider an example of a thermal system, this time involving electric dipoles. Here we will also see the difficulties attendant on freezing into metastable states noted in example 6.1. The motivation for the model to be considered arises from the properties of so called *electrorheological fluids.* These are dielectric fluids which show extreme resistance to flow (very high 'viscosity') when an electric field is applied to them. Because they can respond to such a field on a fast time scale ($\simeq$ 1 msec), they are of great potential importance in the

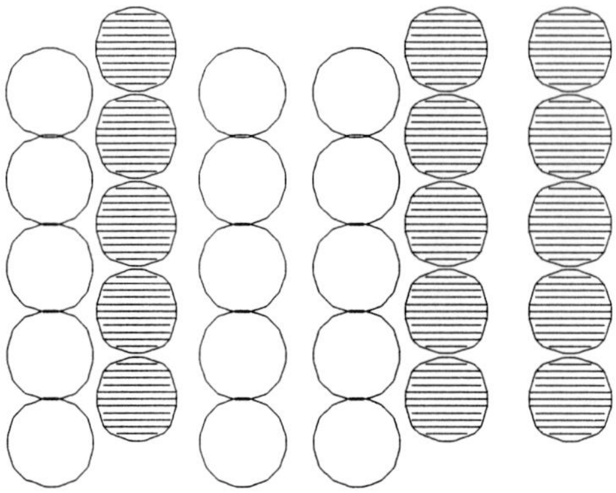

Figure 6.5: A transverse view of the presumed configuration of dipole chains. The shaded and clear molecules are just to highlight the two families of chains.

field of hydraulic control in engineering.

Such electrorheological fluids are typically colloidal suspensions of some granular material, with dimensions of the order of microns, in a fluid — an example would be silica gell in mineral oil. In the simplest model the granular material is assumed to be in the form of spheres of a polarisable substance, each of radius a.[5] When an electric field is applied these spheres acquire an induced electric dipole moment, $\mathbf{p} = \alpha\epsilon_f a^3 \mathbf{E}^{(\mathrm{pol})}$, where $\mathbf{E}^{(\mathrm{pol})}$ is the polarising field, α is the relative polarisability, $\alpha = (\epsilon_p - \epsilon_f)/(\epsilon_p + 2\epsilon_f)$, and ϵ_p and ϵ_f are the relative permittivities of the particles and fluid, respectively. Such dipoles will then experience a force due to the applied field and, in addition, inter-dipole forces, e.g. a pair of dipoles aligned end-to-end are attracted whereas a pair of parallel aligned dipoles side-by-side are repelled. If the spheres are free to move, how will their induced dipoles configure themselves? The total interaction energy, that of the dipoles with the field and of the dipoles with each other will be minimised when the latter, dipole-dipole, energy is minimised. It might then be expected that strings of particle dipoles aligned end-to-end would form but that adjacent strings would strive to be longitudinally displaced relative to each other so as to minimise the number of adjacent dipoles and thus minimise the energy of interaction. A plausible arrangement of the dipoles in the direction of the applied field is as in Fig. 6.5; two staggered interlocking arrays, shown by different symbols in the figure for clarity. Our investigation is to enquire how these chains arrange them-

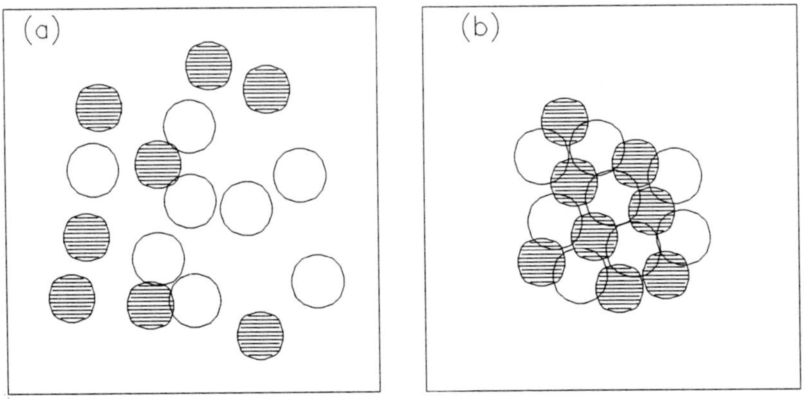

Figure 6.6: (a) A cross-sectional view of the initial configuration of chains, and (b) the final condensed state, showing a lattice structure.

selves in the transverse plane, i.e. does a cohesive configuration exist? And if so, what are the separations between the chains. As is clear, chains in each set, 'like' chains, are mutually repelling; we will see that in this configuration there is an attractive force between chains from one set with those from the other, 'unlike' chains. The minimum possible transverse separation between like chains and between unlike chains is $2a$ and $a\sqrt{3}$, respectively. If the attractive component dominates over the repulsive, the chains can be expected to coalesce at low temperatures forming an effective solid which would account for the electrorheological effect, whereas the opposite tendency will lead to dispersal of the chains. The energy of interaction, binding energy, of a chain in such a fluid is much larger than k_BT at room temperature.

The interaction energy of a system of dipole chains between plane electrodes can be derived using the fact that each dipole produces an infinite number of images in the two electrodes. The chains of dipoles are thus replaced by infinite chains. In calculating their interaction energy — the interaction energy of dipoles within a chain and with the external field is independent of configuration and need not be included — we can focus on just one layer of the two types of chains, shown in cross-section in Fig. 6.6(a), and effectively reduce the problem to one in two dimensions. It can be shown that the chain-chain interaction energy per dipole between like and unlike chains separated by a distance ρ is given by

$$U^{\text{like}}(\rho) = C\sum_{k=1}^{\infty} k^2 K_0(k\pi\rho/a)$$

$$U^{\text{unlike}}(\rho) = C\sum_{k=1}^{\infty} k^2 \cos(k\pi) K_0(k\pi\rho/a)$$

where $K_0(x)$ is the modified Bessel function, and C is a constant, (proportional to the electric field). In practice, not more than two or three terms need be carried in these summations. For a system of $2N$ chains, labelling chains of the two types by A and B, respectively, we can then write the interaction energy as

$$U = \sum_{i=1}^{N}\{\sum_{j>i} U^{AA}(\rho_{ij}) + \sum_{j>i} U^{BB}(\rho_{ij}) + \sum_{j=i}^{N} U^{AB}(\rho_{ij})\}.$$

Taking C and a equal to unity, a system of 16 chains, ($N = 8$), was modeled in this way. The chains were initially enclosed in a square of side $L = 12$. An arbitrary initial configuration satisfying the constraints of overlap inside the square was generated (shown in Fig. 6.6(a)) and its energy U calculated. As a starting value β was set equal to N/U, i.e. ($kT = U/N$). Ten complete passes through the array, i.e. 160 new configurations were generated, in each case a chain being moved in the x and y direction by an amount $(2\xi_i - 1)D$, with $D = \Delta = 1$. (If, due to the overlap restrictions, a chain fails to move after 20 attempts Δ is further reduced by a factor of 2, iteratively, until a movement is achieved — this becomes necessary in the later stages of the calculation where the possibilities for movement are more restricted.) Toroidal boundary conditions were imposed, i.e. if a tentative move of a chain took it to a point outside the square, e.g. $x' > L$, its target location was taken to be the complementary point inside the square, $x' \to x' - L$. The change in the energy, ΔU, e.g. for a change in the coordinates of chain i of type A, given by

$$\Delta U = \sum_{j\neq i}^{N}\{U^{AA}(r'_{ij}) - U^{AA}(r_{ij})\} + \sum_{j=1}^{N}\{U^{AB}(r'_{ij}) - U^{AB}(r_{ij})\},$$

is evaluated for each configuration and the Metropolis algorithm applied. If the greatest extent of the chains in the x and y directions

is less than 95% of the current length L of a side, L is reduced by a factor Fl ($Fl = 0.99$). In other words, if the chains tend to cluster the size of the containing square is reduced.
The final result is shown in Fig. 6.6(b). We see that the chains move into the central region — chain No. 1 was anchored throughout the calculation — and array themselves on a lattice structure. Overall the molecules arrange themselves on a body centred tetragonal (bct) lattice.

6.3 Monte Carlo Parameter Estimation

6.3.1 The Method of Maximum Likelihood

A very common problem in physics is the estimation of optimum parameters for a model suggested to describe a given physical system, based on a set of experimental results derived from a study of that system. In much of the statistical literature this topic is known as *regression.* In their simplest form these results are measurements at several (N) values of an 'independent' quantity, x, of a dependent quantity, y. Examples of such pairings are the temperature, T, and electrical permittivity, ϵ, wavelength, λ, and radiation intensity, I_λ, radial position of a particle in a cosmic ray shower, r, and its arrival time at a detector, t, etc.

The real difficulty in the problem arises because, in general, neither the independent quantity nor the dependent quantity is precisely defined; they contain errors, because of inadequacies in the measurement technique (especially imperfect subtraction of noise), or because they are only a finite sample from the population the model is intended to describe. Experimental information on these uncertainties may or may not be available. In practice we will assume that the errors on the independent quantity are negligible compared with those on the dependent one. The latter we will denote by the standard deviation on the quantity, so we have the data set $\{x_i, y_i \pm \sigma_i\}$, $i = 1, \ldots, N$. Finally we have the set of N values $\{y_i^{(0)}\}$ predicted as the values of the observed quantity at the corresponding values of x_i by the model. Any $y_i^{(0)}$ will be a function, not only of x_i, but of some or all of the parameters occurring in the model. It is these parameters, a total of n in number, we try to estimate from the data set.

Any one of the observed values, y_i, can be thought of as a single sample from a population of such potential observable values, a population which could be characterised by some probability density, most generally written as

$$p(y, x_i, \alpha_1, ..., \alpha_n)dy,$$

where $\alpha_1, \ldots, \alpha_n$ are the n parameters of the model. In particular cases the dependence on x_i and the parameters might just occur through the expected value, $y^{(0)}$, i.e. the density is of the form

$$p(y, y^{(0)}(x_i, \alpha_1, ..., \alpha_n)).$$

An example of this would be a constant distribution,

$$p(y)dy = \frac{1}{2\delta}dy, \qquad y^{(0)} - \delta \le y \le y^{(0)} + \delta,$$

or, more commonly, a Gaussian which in addition to the expected value involves a quantity σ which may depend on the parameters α,

$$p(y)dy = \frac{1}{\sigma\sqrt{2\pi}}e^{-(y - y^{(0)})^2/2\sigma^2}\,dy, \tag{3.1}$$

where $y^{(0)}$, and σ also, may be functions of x_i and $\{\alpha\}$. If the parameters $\{\alpha\}$ were known, the probability (in an ensemble of experiments) of getting an actual observed value y_i would also be known, viz. $p(y_i, x_i, \{\alpha\})$. The probability of getting the complete set of observed values, $\mathcal{P}(y_1, ..., y_n, x_1, \ldots, x_n \mid \{\alpha\})$, i.e. the probability conditional on knowing $\{\alpha\}$, is, assuming independent measurements, obviously proportional to

$$\prod_1^N p(y_i, x_i, \{\alpha\}).$$

The method of *maximum likelihood* asserts that the optimum values for the parameters $\{\alpha\}$ are those values which will maximise this global, or conditional, probability $\mathcal{P}(\mathbf{y}, \mathbf{x} \mid \{\alpha\})$. This function is known generally as the *likelihood* function,

$$\mathcal{P}(\mathbf{y}, \mathbf{x} \mid \{\alpha\}) = \prod_1^N p(y_i, x_i, \{\alpha\}).$$

Such a definition presumes that we have no *a priori* knowledge of the likelihood of a set of values for $\{\alpha\}$ — apart from possible general constraints, e.g. that a particular value be non-negative. If, however, we have such *a priori* knowledge on the values of $\{\alpha\}$, e.g. coming from a previous experiment or a necessary consequence of threshold levels in the experiment, we could attach some α-dependent weighting to the likelihood function. We would write

$$\mathcal{P}(\mathbf{y}, \mathbf{x} \mid \{\alpha\}) \propto W(\{\alpha\}) \prod_1^N p(y_i, x_i, \{\alpha\}). \tag{3.2}$$

When a form for the function p is assumed we can maximise the likelihood function to determine the values $\{\alpha\}$. The same result can be effected by minimising the negative of $\mathcal{P}$, or more conventionally of the logarithm of the function, i.e. we find values of $\{\alpha\}$ which minimise

$$U(\{\alpha\}) \equiv -\ln(\mathcal{P}) = -\sum_i \ln p(y_i, x_i, \{\alpha\}) - \ln W(\{\alpha\}). \qquad (3.3)$$

If the distribution $p(y, x, \{\alpha\})$ can be assumed Gaussian, i.e. equation 3.1, this function becomes, apart from a constant,

$$U(\{\alpha\}) = \frac{1}{2}\sum_{i=1}^{N}\{\ln \sigma_i^2 + \frac{(y_i - y_i^{(0)})^2}{\sigma_i^2}\} - \ln W(\{\alpha\}). \qquad (3.4)$$

Here the dependence on $\{\alpha\}$ is through the quantities $y^{(0)}(x_i, \{\alpha\})$, and possibly through σ_i if estimates of these are not available from the observations.

6.3.2 Solution by Least Squares Method

In the case that we confine ourselves only to the observations, i.e. $W = 1$, and all the σ_i are known, or do not depend on the parameters of the model, the first term in the brackets in equation 3.4 is constant and can be ignored in the minimisation process, which then reduces to within a constant to the familiar *least squares* minimisation function,

$$U = \sum_i \frac{(y_i - y_i^{(0)})^2}{\sigma_i^2}.$$

If, further, the dependence of the predicted quantities $y_i^{(0)}$ on the parameters $\{\alpha\}$ is linear, i.e.

$$y_i^{(0)} = \sum_{j=1}^{n} \alpha_j f_j(x_i), \qquad (3.5)$$

where the $f_j(x)$ are fitting functions — e.g. polynomials or circular functions — the solution which mininises $U(\{\alpha\})$ can be shown to be

$$\{\alpha\} = \mathbf{B}\mathbf{F}^T\mathbf{W}\mathbf{y}, \qquad (3.6)$$

where $\mathbf{y}$ is the array of observed values y_i, $\mathbf{W}$ is a matrix with known elements $W_{ij} = \sigma_i^{-2}\delta_{ij}$ (more generally the variance-covariance matrix of the data), $\mathbf{F}$ is the matrix with elements $F_{ij} = f_j(x_i)$ and the matrix $\mathbf{B} = (\mathbf{F}^T\mathbf{W}\mathbf{F})^{-1}$. Monte Carlo methods can be used in the solution of equation 3.6 but they are largely of academic interest and we will not dwell on them.

6.3.3 Solution by Simulated Annealing

Monte Carlo methods for minimising equation 3.3 become competitive with conventional methods for more general cases of parameter dependence than linear.[6] When the dependence of $y_i^{(0)}(x_i, \{\alpha\})$ on the parameters is not linear and the number of parameters, n, involved is not small the function $U(\{\alpha\})$ is likely to feature multiple local minima. Under these circumstances the method of simulated annealing, using $U(\{\alpha\})$ as the objective function in the parameter space, becomes a viable way for minimising equation 3.3 or 3.4. The method is in all respects similar to that described in section 6.2.2. A starting guess $\{\alpha^{(0)}\}$ for the parameters must be given, $U(\{\alpha^{(0)}\})$ found, and an annealing schedule decided on.

EXAMPLE 6.3: Crystal field parameters from EPR data

In electron paramagnetic resonance (EPR) the frequency (energy) of a transition of a magnetic moment in a known magnetic field $\mathbf{B}$ is measured, from which the energy level spectrum of the moment can be inferred. When the origin of this magnetic moment is a paramagnetic ion situated at a site in a crystal lattice there is, in addition to the purely magnetic interaction, a contribution to the energy of the moment from the interaction of the ionic charge and its associated moments (dipole, quadrupole, etc. which depend on the angular momentum configuration of the electrons on the ion) with the electric field at that location in the lattice. This latter is known as the *crystal field* energy. The deduced energy spectrum can thus also tell us something about the local environment in the crystal, e.g. the paramagnetic ion would typically be an impurity atom which serves as a probe — in theoretical analysis this interaction is formulated in terms of the symmetries of the site. The total interaction is then represented by two terms in the Hamiltonian operator for the ion

$$\mathcal{H} = g_J \mu_B \mathbf{J}.\mathbf{B} + \mathcal{H}_{CF}$$

where $\mathbf{J}$ is the total angular momentum of the ion, μ_B is the Bohr magneton, g_J is a parameter (the 'g-factor', reflecting the roles of orbital and spin angular momentum in determining the magnetic moment), and $\mathcal{H}_{CF}$ is the crystal field operator. This latter has the form

$$\mathcal{H}_{CF} = \sum_k \sum_{q \leq k} \alpha_{kq} O_k^{(q)},$$

an expansion in operators $O_k^{(q)}$ related to angular momentum states (essentially an expansion in spherical harmonics of the electrostatic potential energy) with expansion coefficients α_{kq}, which in addition to g_J, are the set of parameters of the model $\{\alpha\}$. The aim of the experiments, and this example, is to determine this family of parameters, and thus the relative importance of different aspects of the crystal field interaction. By emphasising the role of the symmetry properties of the site in determining the interaction energy the number of operators, $O_k^{(q)}$, can be minimised, and universal matrix elements between angular momentum states can be constructed — these are tabulated in the literature. As a function of the set of parameters, $\{\alpha\}$, the matrix elements of the Hamiltonian can be calculated and from them (by diagonalisation) the predicted energy levels, $\mathcal{H}_{ij} = \epsilon_i(\{\alpha\}, B)\delta_{ij}$, found. By comparing permutations of these with the observed energies the optimum set of parameters can be determined. Writing $\epsilon_{(ij)}(\{\alpha\}, B) = \epsilon_i - \epsilon_j$, a first approximation to an energy is

$$U(\{\alpha\}) = \sum_k (\epsilon_{(ij)}(\{\alpha\}, B_k) - \hbar\omega_k)^2,$$

where (i, j) is the combination of levels corresponding to the observed frequency ω_k and B_k is the value of the magnetic field at which the measurements were made. More refined functions can be constructed by allowing for the uncertainties in the magnetic fields and measured frequencies, and by making use, in addition, of the measured intensities of the transitions, I_k. The steps in the optimisation of the parameters follow the basic pattern described in example 6.1; we leave to the literature the description of a practical implementation of the method.[7]

6.4 Summary

Our concerns in this chapter have been, as is natural in such a volume, with systems of interest to physicists. The simulated annealing technique, however, has been adopted in a much wider sphere of multivariate minimisation problems, and represents a fine example of insights from thermodynamics motivating solution strategies for problems in very disparate fields. It has shown especial merit in so called *NP-complete* problems. These are problems for which it is believed that all exact algorithms for locating the minimum will require an exhaustive search in which the number of steps grows exponentially with the number of variables in the function, and no scheme can be designed which will locate the minimum in a number of steps which increases only as a polynomial in the number of variables. Many combinatorics problems are of this kind — the travelling salesman problem, scheduling problems, etc. Although the method of OSA represents only a heuristic strategy for arriving at a solution, i.e. it cannot guarantee location of the exact solution, its ease in implementation, e.g. no derivatives of the function are required, means that the quality of the minimum resulting from the method will often be acceptable when contrasted with the effort required to improve on it.

Notes

1. Kirkpatrick, S., C. D. Gelatt and *M. P.* Vecchi (1983), *Optimisation by Simulated Annealing*, Science, **220**, 671-80. and, independently by Cerny, see Cerny, V. (1988), *Methods of statistical physics and complex mathematical problems*, Eur. J. Phys., **9**, 94-100.

2. Berezin, A. (1985). *An unexpected result in classical electrostatics*, Nature, **315**, 104. The application of OSA to the problem was described by Wille, L. T. and J. Vennik(1985), *Electrostatic energy minimisation by simulated annealing*, J. Phys. A: Math. Gen., **18**, *L*1113-17.

3. Geman, S. and D. Geman (1984). *Stochastic Relaxation, Gibbs Distributions, and the Baysian Restoration of Images*, IEEE Trans. Patt. Analy. Mach. Intell., **PAMI-6**, 721-41.

4. Moving just one particle at a time is not obviously the best strategy, a correlated relaxation of the coordinates of several particles is suggested from physical considerations. Such a multi-coordinate search step, however, is not easy to design especially when the number of particles is large; a formalism for implementing correlated steps is described in Vanderbilt, D. and S. G. Louie (1984), *A Monte Carlo Simulated Annealing Approach to Optimization over Continuous Variables*, J. Comput. Phys, **56**, 259-71.

5. Halsey, T. C. and W. Toor (1990), *Structure of Electrorheological Fluids*, Phys. Rev. Lett., **22**, 2820 – 23; Tao, R. and J. M. Sun (1991), *Ground state of electrorheological fluids from Monte Carlo simulations*, Phys. Rev., **A44**, *R*6181-84 (1991).

6. The general principles of Maximum Likelihood, as well as methods for minimising the likelihood function may be found in section 7.2 of Eadie *et al.*, or chapter 9 of Frodesen *et al.* Solution by the least squares method is described in section 1.5 of MacKeown and Newman, while Monte Carlo methods for solving for the parameters can be found in sections 5.7 and 5.8 of Hammersley and Handscomb.

7. A description of crystal field theory at this level can be found in chapter 16 of Abragam, A. and B. Bleaney, *Electron Paramagnetic Resonance of Transition Ions*, (Oxford: Clarendon Press, 1970), and in Appendix

B of this work. The OSA approach to determining the crystal field parameters has been applied to data from paramagnetic ions in sites with tetragonal symmetry, Gd^{3+} ions in a CaF_2 lattice, and is described in, Puma, M., E. Laredo, A. Bello, M. E. Galavis and N. Suarez (1988), *Computer analysis of electron paramagnetic resonance data using the Monte Carlo method*, J. Phys. C: Solid State Phys., **21**, 5555-64.

Exercises

6.1 Verify that the prescription,

$$r = \sqrt{\xi}, \qquad \vartheta = 2\pi\xi,$$

where the ξ are different random numbers yields points uniformly randomly distributed on the unit disk.

6.2 Consider the one-dimensional problem in example 4.4 in chapter 4, of 5 unit charges on a line, interacting through a Coulomb like potential, where the interaction energy of a pair is $\alpha / \mid x_i - x_j \mid$, where x_i etc. are their coordinates. In addition an external potential $V(x)$ is present,

$$V(x) = 1536x^4 - 3200x^3 + 2352x^2 - 720x + 81$$

in which a charge at x_i has a contribution to its energy of $V(x_i)$. Write an OSA programme to find the equilibrium configuration of the charges, (a) for $\alpha = 0$, and (b) for $\alpha = 0.05$. As an initial configuration in each case take the coordinates of the charges randomly distributed with $0 < x_i < 1$.

[For case (b) the ground state energy is 11.034; you may also detect a stable excited state at 11.410.]

Chapter 7

MONTE CARLO INTEGRATION

> Some think it a matter of course that chance
> Should starve good men and bad advance
>
> W. B. Yeats

7.1 Introduction

One of the most straightforward applications of population sampling, discussed in chapter 5, arises in the estimation of integrals. The numerical evaluation of multidimensional integrals was one of the earliest topics in which Monte Carlo methods were employed. As it belongs more to the field of computational mathematics, our discussion will be confined to particular aspects of direct interest to physicists — variational calculations, which we will use in our example, and phase space integrals — and to aspects which illustrate techniques that can be generalised to other Monte Carlo procedures and are most easily understood in this context.[1] The techniques described here should not be adopted for the practical evaluation of low dimensional integrals; any computer library is sure to contain tested routines much more efficient for this purpose than any which could be developed on the basis of our description.

7.2 All Monte Carlo is Integration

Particular insights into Monte Carlo methods can be found by starting with integration in one dimension and generalising to many dimensions. In fact we will see that any Monte Carlo calculation can be viewed as the evaluation of a multidimensional integral. From primitive considerations, the mean value of any function *f(x)* of a stochastic variable x when this variable occurs on $[a,b]$ with a normalised frequency distribution $\pi(x)$, can be estimated by the formula

$$\overline{f} = \frac{1}{K}\sum_{1}^{K} f(x_i),$$

where the K values of x_i are selected from the distribution π. More formally this is an estimate of the integral

$$\int_a^b f(x)\pi(x)dx.$$

Taking for $\pi(x)dx$ the uniform distribution, $dx/(b-a)$, we have

$$\begin{aligned} I \equiv \int_a^b f(x)dx &= (b-a)\int_a^b f(x)\pi(x)dx \\ &\simeq \frac{(b-a)}{K}\sum_{1}^{K} f(x_i). \end{aligned} \tag{2.1}$$

This in itself is not a revelation, the only difference between it and the trapezoidal formula of conventional quadrature,

$$I \simeq \sum_{1}^{K} \Delta x_i f(x_i), \qquad \Delta x_i = (b-a)/K, \tag{2.2}$$

is that in the latter the values x_i are chosen systematically in $[a,b]$ rather than uniformly at random. What possible advantage could follow from such additional randomness? None! — at least in one, or a few dimensions — but as we shall see below, when the number of dimensions increases extension of the stochastic method remains practical when the possibility of systematic quadrature becomes hopeless. Consider for example a phase space integral for a system of $N = 50$ particles — not an enormous system — this involves integration over 150 spatial variables. Not to discriminate against any coordinate, the use of equation 2.2 would require a minimum of 3^N evaluations of the function, i.e. $\approx 10^{71}$ in our example, a totally unrealistic project.

Let us therefore return to equation 2.1. This is an estimate for any value of $K > 0$, and, in particular for $K = 1$; an unbiased estimate of I is $(b-a)f(x)$ if x is a value chosen uniformly at random on $[a,b]$. This is nothing more than the Mean Value theorem, which says that for a particular value of $x = \zeta$, in $a \le \zeta \le b$, the exact value of the integral is

$$I \equiv \int_a^b f(x)dx = (b-a)f(\zeta).$$

Reference to Fig. 7.1 will show how individual estimates made in this way may be greater or less than the true value.

The reader may already be surprised at not finding initial emphasis on derivatives and differential equations rather than integrals. Our starting with integrals, with differential equations relegated to later chapters, reflects an intimate relationship between Monte Carlo methods and integration. By direct analogy with the one-dimensional case, a function of N such variables, $f(\xi_1, \xi_2, \ldots, \xi_N)$, where the ξ_i are selected at random from the uniform dis-

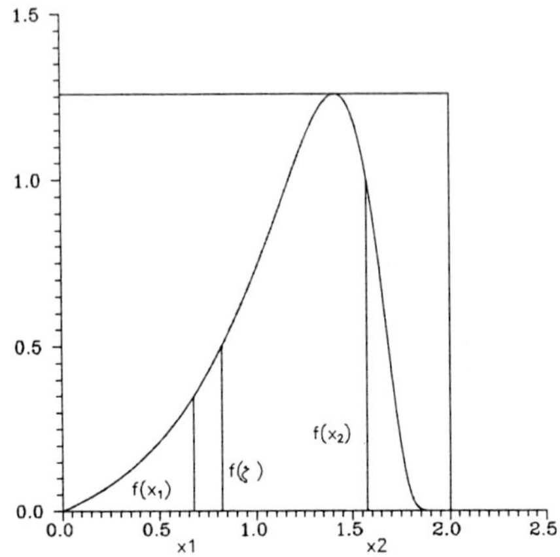

Figure 7.1: Illustrating how individual random estimates of the integral may be larger or smaller than the true value, given by $f(\zeta)$.

tribution on [0,1], — and every result of a Monte Carlo calculation is, in effect, a function of some number of uniform random variates — can be considered an estimate of the N-dimensional integral

$$I = \int_0^1 dx_1 \int_0^1 dx_2 ... \int_0^1 dx_N f(x_1, x_2, ..., x_N).$$

Thus we see that, in formal terms, any Monte Carlo calculation can be viewed as the estimation of an integral. However, we shall find that two distinct algorithms using the same set of $\{\xi_i\}$ to calculate the integral may give estimates which have very different uncertainties.

We now look successively at methods of increasing complexity for evaluating integrals, using, for notational simplicity, the case of a one-dimensional integral over a finite interval

$$I = \int_a^b g(y) dy.$$

By the transformation $x = (y-a)/(b-a)$ any such integral can be reduced to the unit interval, i.e.

$$I = \int_0^1 f(x)dx. \tag{2.3}$$

7.3 Direct Methods

7.3.1 Hit-or-miss Method

Any estimate, θ, of the integral 2.3 will depend on a sequence of N uniform random variates, i.e. $\theta = \theta(\xi_1, \ldots, \xi_N)$. The crudest of all prescriptions for employing these variates is the hit-or-miss method, essentially the acceptance-rejection method used to select random numbers from a distribution described in section 5.2.3, or indeed the Buffon's needle experiment described in chapter 1. N pairs of random numbers $(\xi_1^{(i)}, \xi_2^{(i)})$ are selected, and the number of times, ν, that the condition $f(\xi_1^{(i)}) > \xi_2^{(i)} f_{\max}$ is satisfied is noted. This is just the number of times the uniformly distributed points defined by $(\xi_1^{(i)}, \xi_2^{(i)} f_{\max})$ fall under the curve. An estimate of I is then obviously $\theta = \nu f_{\max}/N$. The quantity ν is distributed as the Binomial distribution, equation 2.4.1, with variance $\nu(1 - \nu/N)$, so the final estimate for I is $\theta \pm \sigma_\theta$, with

$$\begin{aligned} \sigma_\theta &= \frac{f_{\max}}{N}\sqrt{\nu(1-\nu/N)} \qquad (3.1) \\ &\simeq N^{-1/2}\sqrt{I(f_{\max} - I)} \\ &= \text{constant}/N^{1/2}. \end{aligned}$$

Although described for one dimension, this method is clearly applicable for any number of dimensions, and we notice that for a given accuracy the work required, i.e. the size of N, does not depend on the dimensions of the space over which the integral is carried out; for a specified accuracy we denote the sample size by $N_0 \approx 1/\sigma_\theta^2$. We will consider Monte Carlo methods much more efficient than this primitive prescription; this N-dependent feature, however, will generally be true in all cases. We can contrast this with conventional methods of numerical integration. These consist of taking a number of values along each axis of the integration region, evaluating the integrand at each point and using some algorithm, e.g. polynomial smoothing in Gauss' rule, to calculate the approximation to the integral. If n points are used to estimate a one-dimensional integral to a specified accuracy, say with absolute error $\epsilon = \sigma_\theta$ (n is less than N_0 appropriate to the hit-or-miss method, typically N_0^α where $\alpha < 0$ depends on the algorithm employed), then the number of points which must be used to estimate a d-dimensional integral to the same accuracy increases exponentially as n^d. We can thus expect stochastic methods of integration to become viable when d is larger than a certain value. A further reason for employing these methods is their utility when the region of integration, over which the integrand is defined, is very complicated. 'Brute force' specification of the function may be possible in cases where an analytical form would be wholly impractical.

7.3.2 Mean Value Method

Also known as 'crude Monte Carlo', this, most suggestive, method for estimating the integral is based on the Mean Value theorem, which says that the value of the integral is the product of the integration interval (in our case unity) and the mean value of the function on the interval. The latter can be estimated using N values, x_i, of the integration variable, and

$$\theta = < f >_N .$$

There are different possibilities for calculating $<f>_N$. The most sensible procedure would seem to be to sample the interval regularly N times at equally spaced locations — which would not require Monte Carlo methods at all! Doing so, however, would, as we observed earlier, imply using N^d points for the evaluation of a d-dimensional integral to the same accuracy. For efficient sampling of high dimensional integrands, as well as for generalisations of the method, it is better to choose the N points at random in the integration region. This gives an unbiased estimate of the mean value of $f(x)$. In one dimension we choose $x_i = \xi_i$, so

$$\theta = \frac{1}{N}\sum_1^N f(\xi_i). \tag{3.2}$$

Obviously the quantity θ in equation 3.2 is itself a random variable, any particular set $\{\xi_1^{(k)}, \ldots, \xi_N^{(k)}\}$ of the N random numbers will produce a value θ_k from a population whose mean is I. The variance on this distribution, σ_θ^2, depends on the amount of variability in the function f, and from equation 3.2 we have

$$\sigma_\theta^2 = \frac{1}{N}\sigma_f^2. \tag{3.3}$$

For N sufficiently large, and σ_f^2 finite, we know from the Central Limit Theorem that θ is Gaussian distributed. An estimate of the variance on f, s_f^2, can be made on the basis of the set of values $f(\xi_i)$,

$$s_f^2 = \frac{1}{N}\sum_1^N f^2(\xi_i) - \theta^2,$$

so we have the standard error on our estimate

$$\sigma_\theta = s_f/\sqrt{N}.$$

For a given amount of variation in the chosen samples of the function in the interval, characterised by s_f, this error in the estimate of I can be decreased by increasing N, but, because of the square root dependence, convergence to I will only occur very slowly, and beyond a certain point it will not be practical to continue increasing N. On the other hand one can show that the variance from equation 3.3 for a fixed N is always smaller than that in the hit-or-miss method, given by equation 3.1, i.e. the variance depends not only on the size of the sample of random variates used but also on their method of use. Several schemes for *variance reduction* have been proposed — control variates, stratified sampling, antithetic variates, etc., we will consider a few of them here. Further examples will be encountered in several later chapters, especially in chapter 9 in our discussion of particle transport.

7.4 Variance Reduction

7.4.1 General

The general concept of alternative methods of unbiased estimation which give rise to different variances is best illustrated by a simple mathematical example.

EXAMPLE 7.1: A very simple integral

We use N random numbers to estimate the integral

$$I = \int_0^1 x dx.$$

Using the mean value method we have the estimate

$$\vartheta_1 = N^{-1}\sum_{1}^{N}\xi_i,$$

whose expectation is $<\vartheta>= 1/2$, with a variance

$$\begin{aligned}\sigma^2_{\vartheta_1} &= N^{-2}\sum_{1}^{N}\sigma^2_{\xi_1}\\ &= N^{-1}\sigma^2_{\xi},\end{aligned}$$

which, since $\sigma^2_\xi = 1/12$, yields

$$\sigma^2_{\vartheta_1} = 1/12N.$$

Consider now the N variates used in a different way, by forming the quantity

$$\vartheta_2 = \frac{N+1}{2N}\vartheta,$$

where $\vartheta = \max(\xi_1, \xi_2, \ldots, \xi_N)$. The probability distribution for ϑ has been found in example 5.14,

$$p(\vartheta)d\vartheta = N\vartheta^{N-1}d\vartheta, \qquad 0 \le \vartheta \le 1,$$

with a mean of $\frac{N}{N+1}$ and a variance $\frac{N}{(N+1)^2(N+2)}$. Therefore we have for ϑ_2 a mean of 1/2, i.e. it is also an unbiased estimator of I, while its variance is

$$\sigma^2_{\vartheta_2} = \frac{1}{4N(N+2)}.$$

For any $N > 1$ the variance on ϑ_2 is smaller than that on ϑ_1.

We shall now look at methods for realising the possibilities illustrated by this example. The most important method, and the one most readily extended to other Monte Carlo calculations, is that of *importance sampling.*

7.4.2 Importance Sampling

The error derived in the previous section, based on a given selection of N values of the function f, depends on how these individual values are sampled to form the estimate of I. If some biased or correlated sampling of the values of $f(x)$ is made the variance on the sample of size N may be reduced. If we can then find an alternative prescription to equation 3.2 not involving significantly more work which will result in the same expectation value for such biased samples the error on the estimate can be reduced. This is the general philosophy behind variance reduction; a sample of the variables biased in a known way and having a reduced variance is used with an algorithm which makes inherent allowance for the bias to yield the same expectation value. We now look at the method of importance sampling as a way of reducing the variance of the integral, equation 2.3.

The central idea has already been encountered in section 5.2.3, one selects variates from regions in proportion to the significance these regions make to the quantity, here the integral, being sought. We rewrite the integral 2.3 in the form

$$I = \int_0^1 g(x)\pi(x)dx \tag{4.1}$$

where $\pi(x)dx$ is a probability distribution normalised on the interval of integration, and the function $g(x)$ is defined by

$$g(x) \equiv f(x)/\pi(x).$$

The integral 4.1 is obviously the expectation value of the function $g(x)$ where x is sampled from the probability distribution $\pi(x)$, i.e. $I =< g(x) >$. It can be estimated using N variates x_i sampled from $\pi(x)$ by,

$$\theta = \frac{1}{N}\sum_{1}^{N} g(x_i). \tag{4.2}$$

What determines the function π(x) to be chosen? We wish to have the variance on the estimate θ a minimum, or, since $\sigma_\theta^2 = \sigma_g^2/N$, the variance on selected values of g should be a minimum. Now,

$$\begin{aligned}\sigma_g^2 &= <(g(x) - <g>)^2> \\ &= <(g(x) - I)^2> \\ &= \int_0^1 \{\frac{f(x)}{\pi(x)} - I\}^2 \pi(x) dx.\end{aligned}$$

We note that in the case where *f(x)* is everywhere positive, if we choose a normalised version of $f(x)$, i.e. $\pi(x) = f(x)/\int f(x)$, the variance would vanish, and the integral I would be exactly estimated by θ in equation 4.2. To do this, however, presupposes we know the integral in the denominator, just the problem we are trying to solve! However the argument shows us that the closer $\pi(x)$ resembles a scaled version of $f(x)$ the smaller the variance can be expected to be. The function $\pi(x)$ from which we select the values for estimating the integral should mimic the function $f(x)$ being integrated. By so doing, values will be selected predominantly in those regions where $f(x)$ makes its greatest contribution to the integral.

Rewriting equation 4.2 in the form

$$\theta = \frac{1}{N}\sum_{1}^{N} w(x_i) f(x_i) \tag{4.3}$$

where $w(x_i) \equiv 1/\pi(x_i)$, illustrates the general principle of importance sampling. The variables are sampled from an alternative, biased, distribution, here $\pi(x)$, but the functions evaluated using these variates must be weighted by a factor $w(x_i)$ so as to remove from the final result the bias introduced. We note that equation 4.3, with the values x_i systematically chosen, is also the basis of many conventional methods for numerical quadrature of lower dimensional integrals, e.g. using Gauss' formulae which fit polynomials to the function being integrated. Technically the normalisation of $\pi(x)$ can be on a larger interval than that of integration, provided $f(x)$ is understood to vanish outside this interval. Functions $\pi(x)$, which vanish where $f(x)$ does not, are obviously excluded, while if $\pi(x)$ attains very small values relative to f some difficulty with $g(x)$ may be anticipated. Correspondingly if $f(x)$ contains singularities it will be beneficial to cancel this undesirable feature by choosing a function $\pi(x)$ which is similarly singular, and results in a finite $g(x)$ at the singular points. Choosing the uniform distribution for $\pi(x)$, of course, just returns us to the Mean Value method. With a suitably chosen $\pi(x)$ the variance on equation 4.2 will be smaller than on equation 3.2, and even though selecting from this distribution and evaluating $g(x)$ rather than $f(x)$ will involve a bit more effort it will almost surely be less than that required by the increase in N necessary in equation 3.2 to get the same accuracy.

Requiring that the importance function mimics the integrand implies that the integrand $f(x)$ is everywhere positive. If the integrand changes polarity in some sector(s) with known boundaries, the integral is just the sum of the contributions from each separate region, where the method outlined earlier can be used to calculate the negative of the integral in sectors where the integrand itself is negative. If the contributions from regions of positive and negative polarity, however, are comparable, their difference found in this way will be subject to a relatively large, stochastically induced, uncertainty. Integrands which are not positive definite, Feynman path integrals in real

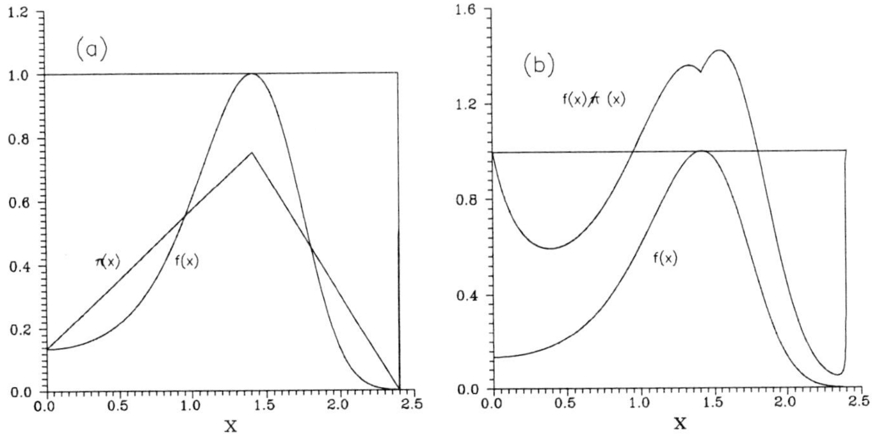

Figure 7.2: (a) The integrand $f(x)$ of example 7.2, and the crude importance function $\pi(x)$ adopted. (b) The variation of the acceptance function $f(x)/\pi(x)$.

time for example, will pose a much greater challenge to importance sampling methods when the integral is over many variables and the regions of different polarity are not transparent.

EXAMPLE 7.2: A toy problem

Before looking at useful applications of the methods discussed, it is instructive to look at a 'toy' problem, the evaluation of the one-dimensional integral

$$I = \int_0^\infty e^{-\frac{1}{2}(x^2-2)^2}\, dx,$$

which has the known value $\pi e^{-1}V(0,2)/2 = 0.9977$, where $V(0{,}2)$ is the parabolic cylinder function (Table 19.1 in Abramowitz and Stegun). The integrand is shown in Fig. 7.2(a); we take the upper limit in our evaluation equal to 2.4. Using hit-or-miss, ten evaluations of the function gave the result,

$$\theta = 1.440 \pm 0.490, \quad \text{var} = 2.400.$$

The mean value method, equation 3.2, using ten points gave

$$\theta = 1.162 \pm 0.245, \quad \text{var} = 0.600.$$

Using ten quasi-random numbers from the algorithm described in section 3.2.6 with $p = 3$ gave a result,

$$\theta = 1.030 \pm 0.229, \quad \text{var} = 0.526.$$

In using importance sampling we took two piecewise linear approximations to the function, shown in Fig. 7.2(a), i.e.

$$\pi(x) = \alpha p_1(x) + (1 - \alpha) p_2(x)$$

with $\alpha = 0.629$, and the functions $p_i(x)$ as described in example 5.5,

$$\begin{aligned} p_1(x) &= 2\Lambda_1(x - \gamma_1) \qquad & 0 \leq x \leq \sqrt{2}, \\ p_2(x) &= 2\Lambda_2(x - \gamma_2) \qquad & \sqrt{2} \leq x \leq 2.4, \end{aligned}$$

with $\Lambda_1 = 0.348$ and $\Lambda_2 = -1.029$. The corresponding function $f(x)/\pi(x)$ is shown in Fig. 7.2(b), where its variation relative to the value of the integral, the horizontal line, in the region where most variates will be selected, is contrasted with the variability of the function f itself. Using the composition method and the prescription in example 5.5, ten variates were chosen from $\pi(x)$ and used, via equation 4.3, to estimate the integral with the result,

$$\theta = 0.978 \pm 0.103, \quad \text{var} = 0.107.$$

In all four cases the estimates are within one nominal standard deviation of the true value, are unbiased as we would expect, but the variance has been reduced by more than a factor of 20, and the uncertainty therefore by almost a factor of five, between the extreme methods.

7.4.3 Correlated Sampling

The second method of variance reduction we discuss is that of *correlated sampling.* In the general case the problem to be solved may be closely related to some other problem, e.g. the result of a perturbation, and it is the difference in the solution of the two problems which is of interest. Associated with the solution of either problem there is a variance due to the stochastic method employed, which will tend to mask the small difference in the solutions. By using, as far as possible, the same set of random numbers to solve the two problems the significance of their difference may be enhanced because of approximate cancellation of random contributions. Alternatively, if a problem can be broken down into the sum of two parts, the most substantial part of which can be solved for exactly by deterministic methods, the relative variance in the final result, deriving only from that associated with the less important aspect solved for by stochastic means, may be made suitably small. These methods, where we can relate the solution of the problem in hand to that of some other problem with which it is strongly correlated, are known as correlated sampling.

We can illustrate these ideas by considering again the integral 2.3 which we might write in the form

$$I = \int_0^1 \phi(x)dx + \int_0^1 g(x)dx,$$

or more simply $I = I_1 + I_2$, where $g(x) \equiv f(x) - \phi(x)$. If the integral I_1 is exactly known an estimate of I is

$$\theta = I_1 + \theta_2,$$

where θ_2 is an estimate of I_2. If I_1 accounts for the greater part of the integral I, the variance on our estimate only arises through the variance on θ_2,

$$\sigma_\theta^2 = \sigma_{\theta_2}^2, \tag{4.4}$$

and a modest uncertainty in the estimate of the second integral may convert into a relatively much reduced uncertainty in the estimate θ. In another case it might be that the desired quantity is, like I_2, the difference between the result of two major calculations, with estimate $\theta_2 = \theta - \theta_1$ where θ, θ_1 do not necessarily involve the same number of random variates. The variance on θ_2 is then

$$\sigma_{\theta_2}^2 = \sigma_\theta^2 + \sigma_{\theta_1}^2 - 2\,\text{cov}(\theta, \theta_1). \tag{4.5}$$

If the same random numbers, as far as possible, are used in finding the estimates θ and θ_1 they can be expected to be positively correlated, i.e. $\text{cov}(\theta, \theta_1) > 0$ and the right-hand side of equation 4.5 will be smaller than would be the case if the two estimates were determined independently. Of course, if θ_1 was known exactly, equation 4.5 reduces to equation 4.4.

7.5 Variational Integrals

7.5.1 General

Monte Carlo methods have assumed increasing importance in quantum mechanics calculations as the complexity of the problems considered increases. The Variational method has been widely used in quantum theory of many-body problems to find estimates of ground state energies, and approximations to the ground state wavefunctions. The particles involved may be fermions, e.g. the many electron configuration in an atom, the many nucleons in a nucleus or an assembly of He^3 atoms, or they may be bosons, e.g. an assembly of He^4 atoms[2] — all of these have been investigated using Variational methods. The basic equation to be solved is the time independent Schrodinger equation,

$$H\psi_0(\mathbf{R}) = E_0\psi_0(\mathbf{R}),$$

where H is the Hamiltonian of the system, and, in general, $\mathbf{R} = (\mathbf{r}_1, \mathbf{r}_2, .., \mathbf{r}_N)$ for a system with N particles. The method involves guessing a function, $\phi_T(\{\alpha\},\mathbf{R})$, which contains one or more parameters, denoted by $\{\alpha\}$, and which is believed to resemble the ground state wavefunction $\psi_0(\mathbf{R})$ — in future, the parameters as arguments will be simply denoted as α. It can then be readily shown that the quantity

$$\tilde{H}(\alpha) = \int \phi_T^* H \phi_T d\mathbf{R} / \int \phi_T^* \phi_T d\mathbf{R}, \tag{5.1}$$

with $d\mathbf{R} \equiv d^3\mathbf{r}_1 d^3\mathbf{r}_2 \ldots d^3\mathbf{r}_N$, is an upper bound to the ground state energy.[3] In particular if the parameters $\{\alpha\}$ are chosen to minimise $\tilde{H}$, we have

$$E_0 \leq \tilde{H}_{\text{min}}. \tag{5.2}$$

For reasonably chosen functions ϕ_T this bound can be rather close to the true ground state energy because, in a sense described in texts on quantum mechanics, the energy is more accurate by one order than the assumed wavefunction. In common with all Monte Carlo calculations there will be a statistical error associated with a variational energy estimate in addition to the systematic uncertainty implied by equation 5.2. For higher accuracy investigations calculations like this can be used to find first estimates of the wavefunction, which can then be used as the starting point of other, more sophisticated, calculations, e.g as an importance function in the random walk methods described in chapter 11, in which only statistical errors occur. If we are satisfied with the accuracy of the variational method, the optimal wavefunction arrived at can be used to find the expectation values of other operators, e.g.

$$< A > = < \phi \mid A \mid \phi > / < \phi \mid \phi >,$$

by the method to be described.

The problem, in general, then reduces to calculating the integral 5.1 for many values of the parameters α and locating the true minimum of the integral in the parameter space — not a trivial problem when more than one parameter is involved. With as few as three particles in the system this is already a nine-dimensional integral so that Monte Carlo methods are frequently resorted to in these calculations.

Direct sampling of the functions in the integrand will be readily seen to be impractical; wavefunctions, in general, extend over wide regions of the coordinate space. Importance sampling will be essential for practical calculations. We can recast equation 5.1 into a suitable form for such calculations by multiplying and dividing by ϕ_T. Then

$$\tilde{H}(\alpha) \equiv < E_L(\alpha) >= \int \int E_L(\mathbf{R}) \pi(\mathbf{R}) d\mathbf{R}, \tag{5.3}$$

where the parameters α are suppressed. The probability distribution $\pi(\alpha,\mathbf{R})$, the importance sampling distribution, is given by

$$\pi(\alpha, \mathbf{R}) = \phi_T^* \phi_T d\mathbf{R} / \int \mid \phi_T \mid^2,$$

and the 'local' energy E_L,

$$E_L \equiv H\phi_T/\phi_T \tag{5.4}$$

is defined. If ϕ_T were exactly equal to ψ_0, $E_L(\alpha)$ would be constant, equal to E_0; in this case the variance on the estimate, equation 5.3, would vanish. So choosing ϕ_T as close as possible to ψ_0 both improves the estimate — in the sense that this upper limit is closer to the true value — and reduces its variance for a given amount of work. In most practical calculations of the integral, equation 5.3, the importance function $\pi(\mathbf{R})$ will be sampled from using the method of Markov chains.

EXAMPLE 7.3: The helium atom

As a pilot study of the method we consider a simple problem, familiar from textbooks on quantum mechanics, to which the method just discussed can be applied, namely the estimation of the ground state energy of helium-like atoms. Treating each electron as a particle of reduced mass μ moving in a central potential we have for the Hamiltonian of the system the sum of the potential energy and kinetic energy operators,

$$H = \frac{-\hbar^2}{2\mu}(\nabla_1^2 + \nabla_2^2) + V$$

with

$$V = \frac{-Ze^2}{4\pi\epsilon_0}\left(\frac{1}{r_1} + \frac{1}{r_2}\right) + \frac{e^2}{4\pi\epsilon_0 r_{12}},$$

and $\mathbf{r}_{12} = \mathbf{r}_1 - \mathbf{r}_2$. If we use the atomic unit of length, $a_0 = \hbar/\alpha\mu c$, i.e. $r/a_0 \to r$, it takes the form

$$H = \alpha^2\mu c^2\left(-\frac{1}{2}\nabla_1^2 - \frac{1}{2}\nabla_2^2 - \frac{Z}{r_1} - \frac{Z}{r_2} + \frac{1}{r_{12}}\right),$$

where $\alpha \equiv e^2/4\pi\epsilon_0 c\hbar \simeq 1/137$ is the fine structure constant. If, further, we take as our unit of energy the quantity $\alpha^2\mu c^2$, the Hamiltonian takes the form

$$H = T + V,$$

where T and V are given by

$$T = -\frac{1}{2}\nabla_1^2 - \frac{1}{2}\nabla_2^2,$$

$$V = -\frac{Z}{r_1} - \frac{Z}{r_2} + \frac{1}{r_{12}}.$$

For the case of He, $Z = 2$. Note that ∇_1^2 involves derivatives only with respect to the coordinates of electron numbered 1. With the units adopted, the spatial integration can be expected to have contributions out to distances of a few units.

The considerations in selecting a trial function ϕ_T are as follows. The spatial part of the ground state wavefunction in this system must be symmetric (with corresponding antisymmetric spin function); we can take it everywhere real and positive. In guessing a form for ϕ_T we will be guided by a first approximation of no interaction between the two electrons, when the function might be approximated as the product of two hydrogen-like functions, which for the ground state is of the form

$$\psi_0 \approx e^{-Zr/a_0} .$$

Thus we write

$$\phi_T \approx e^{-\beta r_1} e^{-\beta r_2},$$

where, with a_0 as our unit of length, $\beta \simeq Z = 2$. To account for the interelectron Coulomb interaction this form must be modified by a function to allow for the correlations in the electron positions, in particular the probability of their approaching each other should vanish. Many ways of imposing such correlations have been presented in the literature;[4] we will choose what is probably the simplest, a separation dependence r_{12}^{α} with $\alpha > 0$. Thus we can write a two parameter trial wavefunction in the form

$$\phi_T(\mathbf{R}) = C r_{12}^{\alpha} e^{-\beta(r_1 + r_2)} , \tag{5.5}$$

where C is a normalising constant which will not be required explicitly. All the elements in our multidimensional integral are now to hand, the local energy, $H\phi_T/\phi_T$, can be obtained using equations 5.4 and 5.5 — after some algebra it gives, keeping the kinetic and potential energy terms separate,

$$E_L(\alpha, \beta, \mathbf{R}) = \{-\beta^2 - \frac{\alpha(1+\alpha)}{r_{12}^2} +$$

$$\beta(r_1 + r_2)\left(\frac{1}{r_1 r_2} + \frac{\alpha}{r_{12}^2}(1 - \mathbf{r}_1.\mathbf{r}_2/r_1 r_2)\right)\} + \{\frac{1}{r_{12}} - \frac{Z(r_1 + r_2)}{r_1 r_2}\}, \tag{5.6}$$

while the importance sampling function is,

$$\pi(\alpha, \beta, \mathbf{R}) d^3\mathbf{r}_1 d^3\mathbf{r}_2 = K r_{12}^{2\alpha} e^{-2\beta(r_1 + r_2)} d^3\mathbf{r}_1 d^3\mathbf{r}_2,$$

where K, determined by normalisation, is not required in this calculation. The integral will be carried out by averaging a sample of values of E_L at configuration space points selected from the importance function π. Because the normalisation constant in this function is not readily known, the method of Markov chains (described in section 5.6.3) is used for the selection, starting with an arbitrary configuration of the particles.

For specified values of the parameters α and β an algorithm to calculate the six-dimensional integral $\tilde{H}(\alpha, \beta)$ is as follows (Cartesian coordinates will be found easiest).

(i) Choose an initial configuration, $\mathbf{R}^{(0)}$, for the two electrons, e.g. $\mathbf{R}^{(0)} = (1, 0, 0, -1, 0, 0)$.

(ii) Generate a tentative configuration, $\mathbf{R}^{(t)}$, from the previous configuration $\mathbf{R}^{(i)}$ by changing some or all of the six coordinates by amounts δ_j, e.g. $x_j \to x_j + \delta_j$. Changing just one of the coordinates is *local updating*; in the present case *global updating* is more suitable, we set $\delta_j = (2\xi_j - 1)\Delta, j = 1, \ldots 6$, i.e. the tentative positions of the electrons are rectangularly distributed in a hypercube of side Δ centered on their current configuration point. The value of Δ is considered below.

(iii) Determine the new configuration using the Markov chain method by selecting a uniform random deviate ξ and following the prescription,

$$\begin{aligned} \mathbf{R}^{(i+1)} &= \mathbf{R}^{(t)} \qquad \text{if } \frac{\pi(\mathbf{R}^{(t)})}{\pi(\mathbf{R}^{(i)})} > \xi \\ &= \mathbf{R}^{(i)} \qquad \text{otherwise.} \end{aligned}$$

A suitable value for Δ is one which allows the space region to be rapidly explored, but such that a sizable proportion of the tentative configurations are accepted, say $\geq 75\%$. An exploratory phase of the calculation needs to be carried to determine this value; in the present case a value of Δ in the range $0.3 - 0.5$ is acceptable.

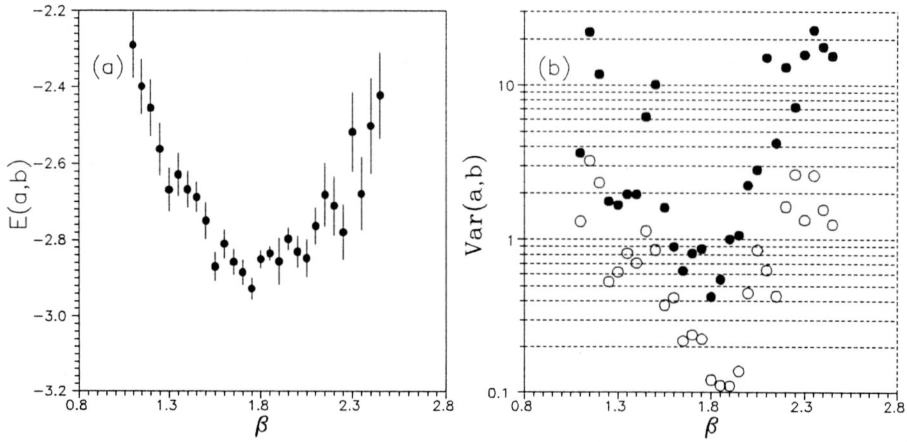

Figure 7.3: (a) The dependence of the variational energy on the parameter β (with $\alpha = 0.18$) with each point based on 300 sample electron configurations. (b) The variance on the variational energy in (a), open circles: ignoring correlations in the sample, and solid circles: taking the sampling correlations into account.

(iv) Decide whether this new configuration is to be used in averaging; to reduce the effects of correlations, only every kth sample point is used. A value of $k = 4$ is used here.

(v) Pass $E_L(\alpha, \beta, \mathbf{R}^{(i+1)})$, E_L^2, etc. to the averaging routine.

Our estimate of the Variational integral is then

$$\overline{E_L} = \frac{1}{N} \sum_i E_L(\alpha, \beta, \mathbf{R}^{(i)}),$$

with the N configurations, $\mathbf{R}^{(i)}$, selected from $\pi(\alpha, \beta, \mathbf{R})$ after the influence of the starting configuration has died away. In Fig. 7.3(a) we show the dependence on β (with α taken constant, equal to 0.18) for the case of $N = 300$. The errors here take account of the correlations described in section 2.3.3 — even when only including every fourth sample in the estimation, the effect is to increase the error by about 75%. Although this diagram clearly illustrates the method, it also emphasises how difficult it is to accurately pin down the location of the minimum; very large numbers of configuration points are required even when we employ importance sampling in the integration. Later we will see how the determination of the minimum may be speeded up by incorporating *correlated sampling* in the minimisation scheme.

At this stage we can also observe from Fig. 7.3(a) that as the minimum is approached, the standard error on the points, and therefore the variance on the samples of local energy, also decreases. This suggests investigating the variance. With E_L given by equation 5.4, we have the variance

$$\sigma_L^2(\alpha,\beta) = \int E_L^2(\alpha,\beta,\mathbf{R})\pi(\alpha,\beta,\mathbf{R})d\mathbf{R} - < E_L >^2 .$$

A first difficulty arises in estimating this quantity. Because of the correlations inherent in the Markov chain sampling method the straightforward calculation

$$\frac{1}{N}\sum_i E_L^2(\alpha,\beta,\mathbf{R}^{(i)}) - \overline{E}_L^2$$

underestimates it in general. A more reliable estimate is to take an effective variance

$$\sigma_L^2 = N s_m^2 ,$$

with s_m^2 determined by equation 2.3.15. In Fig. 7.3(b) we show these two estimates of the variance corresponding to the data in Fig. 7.3(a). Apart from the obvious uncertainty in the estimates, it is clear that the variance is more sensitive to the parameters than is the energy itself. The other advantage of the variance is that it has a well defined lower bound, namely zero. The minimum in the variance is to be expected, for it is easy to see that, were $\psi_0(\alpha, \beta)$ to be chosen for ϕ_T this quantity — always non-negative — would attain its minimum possible value, zero, for those parameter values which correspond to the ground state. In other words, when using the eigen-function of H, $\psi_0(\alpha, \beta), \sigma_L^2$ attains this minimum value for the same parameters which minimise E_L. This is not true in general for other functions, $\phi_T(\alpha, \beta)$. For an arbitrary $\phi_T(\alpha, \beta)$ the parameters which minimise $< E_L >$ are not necessarily those which minimise the variance on the integral. This is unfortunate because of the attractive properties of the variance just described. We will return to consider the variance at the end of the next section.

7.5.2 Correlated Sampling and Variational Integrals

We will now see how correlated sampling can be combined with importance sampling in our investigation of the helium atom problem, in particular in locating the minimum of the integral. We have seen above, Fig. 7.3(a), that for each parameter set $\{\alpha_i\}$ we wish to test in a search for a minimum in the local energy E_L, we must select N variates from the distribution $\pi(\{\alpha_i\}, \mathbf{R})$ to do the averaging. Each value of E_L calculated in this way is independent and the statistical fluctuations in the mean values reflect this. Each mean must be determined with a standard deviation sufficiently small that we can identify the minimum. There are two separable aspects, the determination of the parameter values which correspond to the minimum, and the accurate evaluation of the function at its minimum. It is possible to use an alternative method to identify the parameter values which correspond to a minimum, and to separately compute the value of the function at minimum accurately. This leads us to the method of correlated sampling. Using this, it is no longer necessary to calculate very accurate values of the integral at parameter values away from the minimum in order to identify its location. The technique of correlated sampling here is to use the *same* sample of configurations to calculate the integral for the different sets of parameters. This can be done by taking one set of parameters as reference values, $\{\alpha_0\}$ (e.g. corresponding to those in an, *a priori*, best known wavefunction), and the corresponding distribution

$$\pi_0(\mathbf{R}) \equiv \pi(\{\alpha_0\}, \mathbf{R}) = \mid \phi_T(\alpha_0, \mathbf{R}) \mid^2 / \int \mid \phi_T \mid^2 d\mathbf{R},$$

as the only one from which a sample will be used, and then only one sample, of size N, say. These N variates selected from π_0 will not be a fully representative sample from the probability distribution appropriate to a trial function ϕ_T containing other values of the parameter set $\{\alpha\}$, but we can correct for this by including weights, w_i, defined by rewriting the mean local energy as

$$< E_L(\alpha) >= \int E_L(\alpha, \mathbf{R}) \frac{\pi(\alpha, \mathbf{R})}{\pi(\alpha_0, \mathbf{R})} \pi_0(\mathbf{R}) d\mathbf{R} \tag{5.7}$$

with $E_L(\alpha, \mathbf{R})$ given by equation 5.4. We write

$$\begin{aligned} \frac{\pi(\alpha, \mathbf{R})}{\pi(\alpha_0, \mathbf{R})} &= \frac{\mid \phi_T(\alpha, \mathbf{R}) \mid^2}{\mid \phi_T(\alpha_0, \mathbf{R}) \mid^2} \frac{1}{\int \mid \phi_T(\alpha, \mathbf{R}) \mid^2 d\mathbf{R} / \int \mid \phi_T(\alpha_0, \mathbf{R}) \mid^2 d\mathbf{R}} \\ &= \frac{\mid \phi_T(\alpha, \mathbf{R}) \mid^2 / \mid \phi_T(\alpha_0, \mathbf{R}) \mid^2}{\int \frac{|\phi_T(\alpha, \mathbf{R})|^2}{|\phi_T(\alpha_0, \mathbf{R})|^2} \pi_0(\mathbf{R}) d\mathbf{R}} \\ &= w(\alpha, \mathbf{R}) / \int w(\alpha, \mathbf{R}) \pi_0(\mathbf{R}) d\mathbf{R}. \end{aligned}$$

Thus an estimate of equation 5.7 is

$$\overline{E}_L(\alpha) = \sum w_i E_L(\mathbf{R}_i) / \sum w_i, \tag{5.8}$$

where the same set of coordinates, $\{\mathbf{R}\}$, is used at all values of α and the weights are given by

$$w_i = \frac{|\ \phi_T(\alpha, \mathbf{R}_i)\ |^2}{|\ \phi_T(\alpha_0, \mathbf{R}_i)\ |^2}. \tag{5.9}$$

By using the same, relatively small, set of coordinates to evaluate $\overline{E}_L(\alpha)$ for a new set of parameters we ensure that the sequence of values of $\overline{E}_L$ are strongly correlated, and even though the accuracy of evaluation based on the small number of configurations in any one case is not very high, we can expect the minimum to manifest itself in these equally inaccurate values since the intra-sample fluctuations will affect them all similarly. This suggests an iterative approach; use the optimum parameters determined by the first trial function as the reference parameters for a new trial function, with further repetition if necessary.

EXAMPLE 7.3: (continued)

For illustration we return to the helium atom, taking as our reference function that corresponding to two electrons in a $Z = 2$, hydrogen-like atom, $\{\alpha_0\} = \{\alpha = 0, \beta = 2\}$ in equation 5.5. Adopting only every twentieth configuration, to reduce in-sample correlations, a reference sample of 300 configurations was built up to evaluate equation 5.8 for different combinations of α and β. In Fig. 7.4 the dependence of the energy on β when $\alpha = 0.18$ is shown by the solid squares. As can be seen, the associated statistical errors, which are obviously not relevant in estimating the minimum point, need not be accurately determined. The results can be compared with those in Fig. 7.3(a), also based on 300 configurations but not adopting correlated sampling — they also required four times as much time to compute. Finally in Fig. 7.5 we show the contours of the energy calculated from equation 5.7 in the $\alpha - \beta$ plane.

Correlated sampling is equally valuable when considering the variance, which is notoriously difficult to estimate accurately with the Markov chain method. We can write

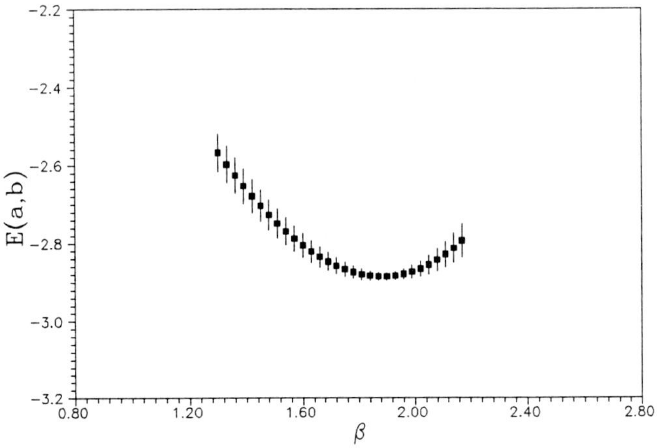

Figure 7.4: The dependence of the variational energy on the parameter β(with $\alpha =$ 0.18), using correlated sampling, each point based on 300 electron configurations.

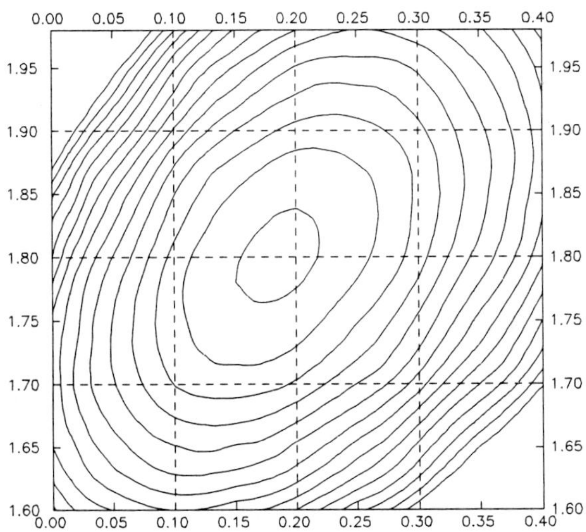

Figure 7.5: Contours of the variational energy in the $\alpha - \beta$ plane.

$$\sigma_E^2(\alpha) = \int (E_L - < E_L >)^2 \frac{\pi(\alpha, \mathbf{R})}{\pi(\alpha_0, \mathbf{R})} \pi(\alpha_0, \mathbf{R}) d\mathbf{R},$$

with the, biased, estimator,[5]

$$\begin{aligned} s_E^2(\alpha) &\simeq \sum w_i [E_{Li}(\alpha) - \overline{E}_L(\alpha)]^2 / \Sigma w_i \\ &= \sum w_i E_{Li}^2(\alpha) / \sum w_i - \overline{E}_L(\alpha)^2. \end{aligned} \tag{5.10}$$

The sensitivity of this quantity to variations in α can be expected to be greater than that of equation 5.8, but, as we noted earlier, the parameters which minimise it do not necessarily give the best upper bound on E_0 for the functional form ϕ_T chosen. We can, however, take advantage of this sensitivity in the following way.[6] A 'fitting' is done, rather than a direct estimate of the variance. Taking as a reference energy, E_r, the best known estimate of the energy, instead of equation 5.10 the quantity considered for minimisation is

$$\chi^2(\alpha) = \sum_{i=1}^{N} w_i (E_{Li}(\alpha) - E_r)^2 / \sum w_i, \tag{5.11}$$

using the same configurations and weights as before. Expression 5.11 should not be viewed as an estimate of a particular integral, rather as a fitting in the spirit of least squares minimisation. Such a minimisation lends itself to an iterative scheme as follows; starting with a trial function $\phi^{(0)} \equiv \phi_T(\alpha_0, \mathbf{R})$ and its corresponding $E_r^{(0)} =< \phi^{(0)} \mid H \mid \phi^{(0)} >$ obtained using equation 5.8, the parameter set $\{\alpha_1\}$ which minimises χ^2 is determined. This is then used to define a new trial function $\phi^{(1)} \equiv \phi_T(\alpha_1, \mathbf{R})$ and a corresponding $E_r^{(1)}$ and the prescription repeated to minimise χ^2. This can be recycled to converge on the best upper limit consistent with the trial function considered.

Dealing with the variance, or variance-like quantities such as χ^2 in equation 5.11, has real advantages. The variance on the local energy has the lower bound, exactly zero, for all states. So, by choosing suitable forms for the trial function ϕ_T (e.g. by suitable symmetry having it orthogonal to the ground state, etc.), minimisation of the variance can converge to the energy of an excited state. The quality of the wavefunction (which may be important for later estimation of expectation values of other operators), determined from minimising the variance, will often be superior to that obtained from minimisation of the energy alone. This is so, because in minimising the variance we are always minimising a sum of non-negative quantities, it can only be small when the contribution from each configuration point is small, whereas in the case of the energy, cancellation in the sum can occur from configuration points having deviations in the value of the local energy large compared to the true value and of opposite sign. Further the variance is more sensitive in general to the parameters and this facilitates the search for a minimum.

There is quite a different way of doing a Monte Carlo variational calculation, using random walks in the configuration space, which we shall see in chapter 11.

Our discussion has implied a strong positive correlation between the two problems in question. Very interesting variance reduction techniques are also possible based on considering two problems whose solutions are negatively correlated. These are known as *antithetic variable* methods, and, although very powerful in one-dimensional problems, they are not so easy to implement in many-dimensional problems.

Notes

1. A more complete discussion of the subject of this chapter can be found in chapter 5 of Hammersley and Handscomb, and in chapter 5 of Davis, P. J. and Rabinowitz P., *Methods of Numerical Integration*, (New York: Academic Press, 1975). A more physics oriented reference is, James, F. (1980), *Monte Carlo theory and practice*, Rep. Prog. Phys., **43**, 1145-89.

2. The prototype variational Monte Carlo calculation is that of W. L. McMillian (1965), *Ground State of* He^4, Phys. Rev., **138**, *A*442-51.

3. For example in D. J. Griffiths, *Introduction to Quantum Mechanics*, (Englewood Cliffs, NJ: Prentice Hall, 1994), chapter 7.

4. Whittington, S. G. and Bersohn, M. (1969), *A Monte Carlo method for quantum chemistry*, Mol. Phys., **17**, 627-31.

5. Alexander S. A., R. L. Coldwell, H. J. Monkhorst and J. D. Morgan III (1991), *Monte Carlo eigenvalues and variance estimates from several functional optimizations*, J. Chem. Phys., **95**, 6622-33.

6. Umrigar, G. J., K. G. Wilson and J. W. Wilkins (1988), *Optimized Trial wave functions for Quantum Monte Carlo Calculations*, Phys. Rev. Lett., **60**, 1719-22.

Exercises

7.1 Show that, for the same number of samples, the variance on the crude Monte Carlo estimate, equation 7.3.2 is less than that on the hit-or-miss based estimate equation 7.3.1.

7.2 On the basis of ten evaluations of the integrand in each case, estimate, along with its associated uncertainty, the simple integral

$$\int_0^1 dx\, f(x), \text{ with } f(x) = e^x,$$

(a) by crude Monte Carlo,

(b) using importance sampling with the importance function $\pi(x) = 2(1+x)/3$, and

(c) selecting the values at random on (0, 1) and using the antithetic estimator $[f(x) + f(1-x)]/2$.

7.3 Use the method of importance sampling with a sample of 20 variates to estimate the integral

$$I = \int_{-\infty}^{u} e^{-x^2/2}\, dx$$

for $u = 0.5$ using the importance function

$$\pi(x)dx = \frac{\alpha e^{-\alpha x}}{(1+e^{-\alpha x})^2} dx, \qquad -\infty < x < \infty$$

with $\alpha = \pi/\sqrt{3}$, and compare it with the known value, $I = 0.76025$ (see exercise 5.6).

Repeat the above, only using the probability distribution in exercise 5.7.

Chapter 8

RANDOM WALKS

> We may describe a Markov process picturesquely by the statement that it represents 'the gradual unfolding of a transition probability' in exactly the same sense as the development of a conservative dynamical system can be described as 'the gradual unfolding of a contact transformation'.
>
> S. Chandrasekhar

8.1 Introduction

Much of our interest in random walks will be in their application in solving partial differential equations. There are two approaches for the stochastic solution of differential equations via random walks. The most direct method depends on replacing the differential equation by a finite difference equation on a lattice, and using random walks to solve these difference equations. This approach is relatively unimportant but, for completeness sake, we will look at it in section 8.4. The second approach depends on finding a dynamic stochastic process which gives rise to the differential equation in question and finding the solution by a simulation of the process. A familiar example of this is the heat diffusion equation which is related to the dynamics of Brownian motion in this way. Studies over recent years have greatly expanded the realm of this approach and much of our discussion, in this and later chapters, will be devoted to it. Finally we shall say a few words about anomalous diffusion, and self-avoiding walks, for which no useful differential equation formalism exists. In the next section we will see an example of the relationship between a physical process and a differential equation.

8.2 Diffusion-Convection

The topic of random walks covers an enormous field; motivation via a concrete example will be helpful before embarking on it. We consider a physical process, such as the motion of positrons in a solid or the wanderings of cosmic ray particles in interstellar space, where three distinct mechanisms can be identified which effect the local density of particles — we will focus on the positron transport case. These mechanisms are: *diffusion*, which gives rise to a generally aimless drift of particles — e.g. a result of the buffeting of thermal agitation; *convection*, which superimposes a nett directed motion on this drift — e.g. as produced by an impressed electric field — and *annihilation*, which serves to remove particles from the population — e.g. as exemplified by positron annihilation on electrons in the medium. The contribution of these three processes to the change in the local density of particles (denoted by $\rho(\mathbf{r},t)$) during a time interval Δt can be written down, from first principles, as follows:

- from diffusion: $\Delta\rho =$ -div $\mathbf{j}\Delta t = \text{div}(D\,\text{grad}\;\rho)\Delta t$, where $\mathbf{j} = -D\text{grad}\rho$, is the diffusion current due to a density gradient and D is the local diffusion constant,

- from drift: $\Delta\rho =$ -div $\mathbf{j}_{\text{drift}}\Delta t =$ -div$(\mathbf{v}\rho)\Delta t$, where $\mathbf{v}$ is the local drift velocity of the particles,

- from annihilation: $\Delta\rho = -\mu\rho\Delta t$, where μ is the probability of annihilation per unit time or the reciprocal of the local lifetime against annihilation.

By combining these three contributions we find for the local time rate of change of ρ (we take the one-dimensional case for simplicity),

$$\frac{\partial\rho(x,t)}{\partial t} = \frac{\partial}{\partial x}\{D(x)\frac{\partial\rho(x,t)}{\partial x}\} - \frac{\partial}{\partial x}\{v(x)\rho(x,t)\} - \mu(x)\rho(x,t) + S(x,t). \quad (2.1)$$

In equation 2.1 we have also added in a term corresponding to the spontaneous creation of particles at x at a rate $S(x,t)$ per unit volume per unit time — such as could occur to a positron population in the presence of energetic γ-rays, or, more importantly, in a population of neutrons due to spontaneous fissions in a fissionable medium. We have assumed that D, $\mathbf{v}$ and μ depend only on position, much more general dependences, e.g. on time via a varying impressed field, or on the density ρ itself resulting in non-linear behaviour, can be visualised.

Numerous examples of phenomena involving particles behaving in this stochastic way occur; in plasma physics for example — especially in the theory of the solar wind, in high energy physics when the multiple scattering of an energetic charged particle in traversing a medium is considered, etc. The formalism is not, however, confined to the diffusion of physical particles; in the case of heat transport the thermal energy is the transported quantity. In this case the function which obeys an equation of the type shown is the temperature, $T(x,t)$; in a system not in equilibrium but at rest we have the heat conduction equation

$$\frac{\partial T}{\partial t} = \frac{1}{\rho C_p}\frac{\partial}{\partial x}\left(k\frac{\partial T}{\partial x}\right),$$

where ρ, C_p and k are the density, specific heat and thermal conductivity, respectively.

In what follows we want to show the close relationship between the type of equation displayed above, derived on the assumption of a continuous medium, and the equation obeyed by particles undergoing random walks of different kinds.[1] This comparison will not only explain how the similarity may be made the basis for a Monte Carlo approach to the solution of such differential equations, but will also provide an intuitive insight into the sometimes forbidding formalism associated with these equations.

8.3 Markovian Random Walks

8.3.1 A Simple Random Walk: The Diffusion Equation

A walker starts at $x = 0$ at time $t = 0$ and at successive times separated by an interval Δt takes discrete steps of size Δx with equal probability $T_+ = T_- = 1/2$ in either direction — the drunkard's walk. We observe that the decision on which way to step at any stage depends neither on the current location of the walker nor on where he has visited during the walk up to that time. It is the latter feature which defines the *Markov* property, the independence of the decisions on the previous history of the walk. If an ensemble of such walkers set out from the origin at $t = 0$, how would they be distributed along the line at $t = N\Delta t$? And, on average, how far would they be displaced from the origin after these N steps? From the symmetry of the problem — a walker is as likely to be at -x as at +x — the mean displacement is clearly zero; we can, however, characterise the dispersion of the walkers by the root mean square displacement, $\sqrt{< x_N^2 >}$.

The probability of the walk arriving at x at the (N+1)th step, i.e. at time $(N+1)\Delta t$ depends only on the probability of it having reached the points $x \pm \Delta x$ at the previous time step. We can write this probability, denoted by $p(x,t)$, as

$$p(x,t+\Delta t) = p(x-\Delta x,t)T_+ + p(x+\Delta x,t)T_-, \tag{3.1}$$

where we have written $N\Delta t = t$, and $T_+(T_-)$ is the probability of stepping to the right (left). For any value of t we consider allowing Δt to become arbitrarily small by taking N sufficiently large; in this case we expect the positions x to be describable as a continuum. If, with $T_+ = T_- = 1/2$, we expand the right-hand side of equation 3.1 to second order in Δx we have

$$p(x,t+\Delta t) \simeq p(x,t) + \frac{(\Delta x)^2}{2!}\frac{\partial^2 p}{\partial x^2}\Big|_{x,t},$$

or

$$\frac{\partial p}{\partial t} = \lim_{\Delta t \to 0} \frac{p(x,t+\Delta t) - p(x,t)}{\Delta t} = D\frac{\partial^2 p}{\partial x^2}.$$

We see that $p(x,t)$ obeys the diffusion equation with constant diffusion coefficient D, where

$$D = \lim_{\Delta t \to 0} \frac{(\Delta x)^2}{2\Delta t}.$$

The same equation obviously holds if the starting point of the walkers were x_0 at t_0; in this case we write the solution more generally as $p(x,t \mid x_0,t_0)$ — the *transition probability* for a walker at x_0 at t_0 to be at x at time t. If the initial ensemble of walkers were not all concentrated at the origin, but arrayed as some distribution, $\rho(x,t_0)dx$, the basis on which the diffusion equation was derived, i.e. the irrelevance of the earlier history of the walk and the non-interference between walkers, means that it is still valid for each individual walker. In this case the density at x at time t, which is just the sum of the contribution of all the individual walkers, will be given by

$$\rho(x,t) = \int dx_0 \rho(x_0,t_0) p(x,t \mid x_0,t_0).$$

As can be seen by differentiating under the integral sign, this function $\rho(x,t)$ satisfies the same equation as $p(x,t \mid x_0,t_0)$, i.e.

$$\frac{\partial \rho(x,t)}{\partial t} = D\frac{\partial^2 \rho(x,t)}{\partial x^2}. \tag{3.2}$$

The assumptions we have made here are just those adopted by Einstein in 1905 in his stochastic model of Brownian motion, and equation 3.2 is the equation he arrived at for the density of such particles.[2] The time scale on which observations of the position of a Brownian particle are made is much greater than the time interval for changes in direction of motion due to the microscopic collisions, which in turn can be thought of as being independent.

8.3.2 Monte Carlo Solution

In three dimensions the master equation for a random walk with step size Δ leads to

$$\frac{\partial \rho}{\partial t} = D\nabla^2 \rho, \tag{3.3}$$

with $D = \Delta^2/6\Delta t$. A Monte Carlo method for solving this diffusion equation, with initial conditions $\rho(r,0) = \rho_0(r)$ will now be obvious. Beginning with a large number, M, of 'particles' whose starting points are selected from the initial distribution $\rho_0(r)$, their random walk is modeled, i.e. in successive steps, corresponding to time intervals Δt, every one of these particles is displaced a distance $\pm(6D\Delta t)^{1/2}$ in one of the x, y or z directions, where the sign of the step is chosen with equal probability on the basis of a uniform random deviate ξ. A histogram of the positions of the particles after N such steps should approximate $M\rho(r,t)$ where $\rho(r,t)$ is the solution of the equation at $t = N\Delta t$. 'Approximate', first because of unavoidable statistical fluctuations dictated by a finite M, and second because the system being modeled by the random walk only mirrors the differential equation in the limit $\Delta t \to 0$. The finite time step used will mean that a truncation error is introduced. To get close to the exact solution very small step sizes may be required, and correspondingly greater computing effort.

The correspondence between the diffusion equation, equation 3.3, and the random walk described above is not unique. More general random walks (in the same approximation of ignoring terms higher than quadratic in the step size) will converge to the solution of this equation, e.g. walks where individual steps are not all of the same size, but are selected from some distribution $w(\Delta)$. Provided $w(\Delta)$ is symmetric about $\Delta = 0$ and possesses a finite second moment, it is easy to show that the analogue of equation 3.1 reduces to equation 3.2 with D given by

$$D = \frac{1}{2\Delta t}\int \Delta^2 w(\Delta)d\Delta.$$

For example in the one-dimensional case, selecting from

$$w(\Delta)d\Delta = e^{-\Delta/\Delta^*}d\Delta/\Delta^*, \qquad (D = \Delta^{*2}/2\Delta t)$$

and stepping to left or right with equal probability will solve the diffusion equation and avoid the discrete localisation of the walkers required by the simpler sampling in fixed step sizes.

Of course, analytical solutions of equation 3.3 can be found in several situations; for the one-dimensional case with an initial delta function centred on x_0, i.e. $\rho_0(x) = \delta(x - x_0)$ and where there are no boundaries which affect the walk, we can arrive at the solution for 'large' N using the Central Limit Theorem. The value of x at the time t is the result of adding to x_0 N variates from a distribution

$$w(\Delta)d\Delta = \frac{1}{2}\{\delta(\Delta - \sqrt{2D\Delta t}) + \delta(\Delta + \sqrt{2D\Delta t})\}d\Delta,$$

with mean zero and variance $2D\Delta t$. Thus at large N we have the Gaussian solution with expanding variance,

$$p(x, t; x_0, 0) = \frac{1}{\sigma(t)\sqrt{2\pi}} e^{-(x-\mu)^2/2\sigma^2}, \tag{3.4}$$

with $\mu = x_0$ and $\sigma^2 = 2Dt$. This can also be verified by substitution. Translating our origin to x_0, σ^2 is the mean square displacement from the starting point after N steps, $2Dt = N(\Delta x)^2$, or

$$\sqrt{< x_N^2 >} = \Delta x . N^{1/2}. \tag{3.5}$$

The optimum distribution, in the sense of most rapid convergence to the behaviour in equation 3.4, from which to select the path segments is then clearly the Gaussian, i.e.

$$w(\Delta)d\Delta = \frac{1}{\sigma\sqrt{2\pi}} e^{-\Delta^2/2\sigma^2} d\Delta,$$

with $\sigma^2 = 2D\Delta t$. In simulation we would take steps,

$$\Delta_i = \eta_i \sqrt{2D\Delta t}, \tag{3.6}$$

where η_i is a standardised Normal variate (see section 2.4.2).

The solution, equation 3.4, corresponding to the δ-function initial conditions, is known as the Green function for the unbounded diffusion equation;

we will, from time to time, use the alternative notation $G(x, t; x_0, 0)$ in this case. If the initial distribution is $\rho_0(x)$ the distribution at time t is just the sum of diffusion from all initial points, i.e.

$$\rho(x,t) = \int dx' \rho_0(x') \frac{1}{\sigma(t)\sqrt{2\pi}} e^{-(x-x')^2/2\sigma^2} .$$

The Monte Carlo prescription we have described is just a rather inefficient method for evaluating such an integral. Problems of a purely diffusive nature usually lend themselves to analytical methods of solution, or, at worst, difficult boundary conditions require use of finite difference or finite element methods. The more non-diffusive mechanisms contribute, the more the effort in using these methods increases; it is in such cases, as well as in problems in spaces of high dimension, that the power of the Monte Carlo method is revealed. The relationship of the method to a wide variety of random walk problems of increasing complexity is investigated later and in subsequent chapters.

8.3.3 Random Functions

The particle in the random walk follows a trajectory, however, we have not been describing this motion by a function $x(t)$, but rather, in terms of a probability distribution of the coordinate x at a specified time, like equation 3.4. For deterministic motion we would seldom use this approach – of course we could in this case write,

$$p(x, t; x_0, 0) = \delta(x - x(x_0, t)).$$

We will still find it helpful to think in terms of a trajectory, $x(t)$, for the

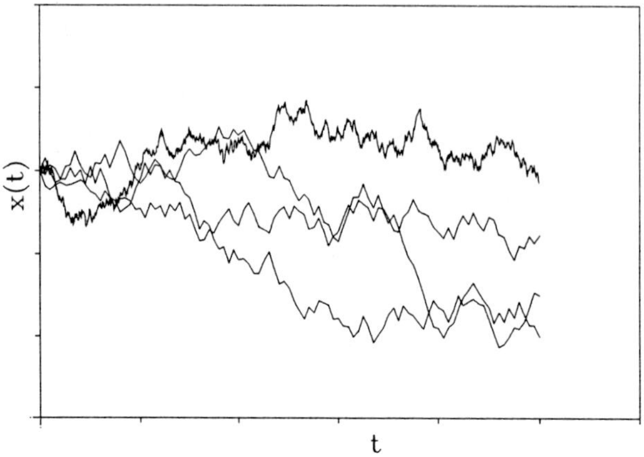

Figure 8.1: Four realisations of a random walk, each of them having one point in common. All of them have the same diffusion constant, but the uppermost in the figure has a time step ten times smaller than the others.

random walk. Such a function is a *random function.* Clearly such a function will, in general, have discontinuous first derivatives at the nodes of the walk, and in the mathematical limit of the step size tending to zero will not be differentiable anywhere. The behaviour of such paths can be appreciated from four realisations of such a function, each of them having one point in common, shown in Fig. 8.1. All of them have the same diffusion constant, but one of them — the uppermost in the figure — has a time step ten times smaller than the other three. Increments in such a function, $dx(t)$, can formally be introduced, e.g. equation 3.6 can be considered a realisation of such an increment, and integration can be defined in a consistent way.

8.3.4 Random Walks with Decay

Our first generalisation of the random walk will be to include the possibility of annihilation of the walker — the drunk in a mine field. The case where multiplication can occur (e.g. as happens to the neutrons in a fissile medium) is a simple extension of this case. We assume a constant probability per unit time, μ, for annihilation. The stepping probabilities are then $T_+ = T_- = (1 - \mu\Delta t)/2$. By repeating the analysis earlier in the chapter based on equation 3.1 we readily arrive at the equation describing the distribution of walkers,

$$\frac{\partial p(x,t)}{\partial t} = D\frac{\partial^2 p(x,t)}{\partial x^2} - \mu p(x,t), \tag{3.7}$$

with

$$D = \lim_{\Delta t \to 0} \frac{(\Delta x)^2}{2\Delta t}(1 - \mu\Delta t) = \lim_{\Delta t \to 0} \frac{(\Delta x)^2}{2\Delta t}.$$

For a walker starting from the origin at $t = 0$ in an unbounded medium, it is clear that the probability of his surviving to a time $t = N\Delta t$ doesn't depend on where he has walked up to that time and is simply $(1 - \mu\Delta t)^N = (1 - \mu t/N)^N \underset{N \to \infty}{\longrightarrow} e^{-\mu t}$. The Green function of equation 3.7 for a walker starting at x_0, i.e. the probability of being at x at the time t is then, using equation 3.4, clearly

$$p(x,t;x_0,0) = \frac{1}{\sigma\sqrt{2\pi}} e^{-\mu t - (x - x_0)^2/2\sigma^2}, \tag{3.8}$$

with $\sigma^2 = 2Dt$.

We can extend our analysis in section 8.3.1 to the case where μ is not constant, but a function of location, $\mu(x)$ — a variable density of mines. Then, provided the first two derivatives of μ are finite it is easy to show, in the same approximation viz. $\Delta t \to 0$ and $D = (\Delta x)^2/2\Delta t$ finite, that the distribution obeys the equation

$$\frac{\partial p}{\partial t} = D\frac{\partial^2 p}{\partial x^2} - \mu(x)p. \tag{3.9}$$

This is a very interesting equation, it can describe phenomena as diverse as the diffusion of a chemically reactive species or the movement of cosmic rays in interstellar space. Its solution by simulation seems straightforward: from the initial distribution $\rho(x,0)$ we select, a sample of M walkers and track each of them to their final location at $t = N\Delta t$. At every step we must check that a walker would survive, i.e. it must be annihilated — the walk abandoned — with a probability $\mu(x)\Delta t$ (or, more rigorously, continued with probability $e^{-\mu(x)\Delta t}$), otherwise a step, selected optimally from a Gaussian using equation 3.6, is taken to the left or right with equal probability. As before, a histogram based on the positions of the walkers at t is the solution. A difficulty with this simple method in practice is that, for $t \geq O(\mu^{-1})$, almost no walkers will survive, no information will be available from which a distribution at t may be constructed.

The survival probability of any one walker (denote its particular walk by λ) is the product of its surviving every step, i.e.

$$\begin{aligned} w^{(\lambda)} &= \prod_{\{i\in\lambda\}} e^{-\mu(x_i)\Delta t} \\ &= e^{-\Sigma\mu(x_i)\Delta t} && (3.10) \\ &= \exp\{-\int_{\lambda 0}^{t} \mu(x(t))dt\}. && (3.11) \end{aligned}$$

This surrender of information is a familiar feature of random walk simulations, it gives rise to wasted effort and greatly degraded statistics with the result that the variance on the finally computed quantity is great. Various methods can be devised to reduce the variance, as we have seen in chapter 7, and will describe in more detail in chapter 9, and here we conveniently introduce the method of *survival biasing*. This is implemented as follows;

when carrying out the walk no annihilation is permitted at any step. This distortion of the walk is compensated for, however, by attaching a 'weight', equal to the survival probability on that step, i.e. $e^{-\mu(x_i)\Delta t}$, to the walker. The total weight of the walker at time t is the product of all these partial weights accumulated in the course of the walk (it having started with a weight of unity), i.e. equation 3.10. The final 'histogram' is no longer based on integers — the number of walkers which end up in a bin — but rather the total sum of the weights of all the, many more, walkers which end up in that bin, $\sum_\lambda w^{(\lambda)}(x)$. If the initial conditions correspond to a δ-function source at the point x_0, we see that the solution at x at any later time t, i.e. the average probability of a walker being at x, is just the average over a population of walks starting from x_0 of the weight $w^{(\lambda)}$ given by equation 3.10 or 3.11, i.e.

$$p(x,t;x_0,0) = < \exp\{-\int_0^t \mu(x(t))dt\} >_\lambda \,. \tag{3.12}$$

In other words, it is the average of the exponential term weighted by the probability of particular paths, λ. Analogous averages over paths play an important role in Feynman's formulation of quantum mechanics, which we will turn to in chapter 11.

At a more elementary level, new insights into the formalism of quantum mechanics, and the basis for important Monte Carlo methods for solving quantum problems, can be obtained by noting also that equation 3.9 is mathematically identical with the time dependent Schrodinger equation for the state of a particle moving in a potential having the shape of the function $\mu(x)$. This equation is of the form

$$\frac{1}{i}\frac{\partial \Psi}{\partial t} = \frac{\hbar}{2m}\nabla^2\Psi - \frac{V(r)}{\hbar}\Psi.$$

Thus if we allow ourselves to think in terms of an imaginary time, *it*, we see that Schrodinger's equation is identical with equation 3.9 in which we can identify a diffusion constant D with $\hbar/2m$ and a lifetime, $\tau \equiv \mu^{-1} = \hbar/V(r)$. This latter quantity in a Monte Carlo approach to solving Schrodinger's equation will represent capture or branching in the execution of a random walk, depending on whether the potential $V(r)$ is positive or negative. Consideration of imaginary time is not just an intellectual extravagance, we will see how to employ it in the practical solution of Schrodinger's equation using random walks, the important method of Quantum Diffusion Monte Carlo in chapter 11.

8.3.5 The Fokker-Planck Equation

We now consider a more general form of the random walk where the stepping probabilities are not equal in either direction, and, more generally, may depend on position. We write

- $T_+(x)$: the probability that a particle at x will step to $x + \Delta x$ during Δt, with an analogous definition of $T_-(x)$.
- $1-\omega(x)$: the probability that the particle will remain at x rather than take a step at Δt — corresponding to trapping.

The equation governing the probability of finding a particle at x after $(N+1)$ steps of the walk may be readily deduced, viz.

$$p(x,t+\Delta t) = p(x,t)[1-\omega(x)]+p(x-\Delta x,t)T_+(x-\Delta x)+p(x+\Delta x,t)T_-(x+\Delta x)$$

where we have again written $N\Delta t = t$. Proceeding as before to expand the right-hand side to order $(\Delta x)^2$, assuming the positions form a continuum, we find, in the limit

$$\frac{\partial p}{\partial t} = \frac{\partial^2}{\partial x^2}\{D(x)p(x,t)\} - \frac{\partial}{\partial x}\{b(x)p(x,t)\} - \mu(x)p(x,t), \tag{3.13}$$

where

$$\begin{aligned} D(x) &= \{T_+(x) + T_-(x)\} \lim_{\Delta t \to 0} \frac{(\Delta x)^2}{2\Delta t}, \\ b(x) &= \{T_+(x) - T_-(x)\} \lim_{\Delta t \to 0} \frac{\Delta x}{\Delta t}, \\ \mu(x)\Delta t &= \omega(x) - \{T_+(x) + T_-(x)\}. \end{aligned}$$

We have also provided for the possibility that the probability of stepping plus the probability of lingering do not add up to unity, i.e. there is a mechanism corresponding to absorption or decay such that particles disappear at x at a rate $\mu(x)$ per unit time — or with a characteristic lifetime $\tau(x) = 1/\mu(x)$. In the above construction we necessarily have μ positive, the case of negative μ, corresponding to particle multiplication, as well as a source of spontaneous creation of walkers during the walk can also be easily accommodated in the execution of the walk.

As before, equation 3.13 also holds for walks in which the individual steps in the walk, Δ, are selected from some suitable distribution $w(x,\Delta)d\Delta$, with the correspondences

$$\begin{aligned} D(x) &= \frac{1}{2\Delta t}\int_{-\infty}^{\infty} \Delta^2 w(x,\Delta)d\Delta, \\ b(x) &= \frac{1}{\Delta t}\int_{-\infty}^{\infty} \Delta w(x,\Delta)d\Delta = < \Delta > /\Delta t. \end{aligned}$$

The simple case considered corresponds to

$$w(x,\Delta) = T_+(x)\delta(\Delta - \Delta x) + T_-(x)\delta(\Delta + \Delta x).$$

It is helpful, by analogy with the physically derived equation 2.1 to rewrite equation 3.13 in the form,

$$\frac{\partial p}{\partial t} = \frac{\partial}{\partial x}\{D(x)\frac{\partial p}{\partial x}\} - \frac{\partial}{\partial x}\{v(x)p\} - \mu(x)p, \tag{3.14}$$

where

$$v(x) = b(x) - \frac{\partial D}{\partial x}.$$

These relations enable us to relate the stepping probabilities in the walk, T_+, T_- and ω to the physical mechanisms underlying the problem.

Equation 3.14 or 3.13 is a one-dimensional form of the Fokker-Planck equation describing diffusion-convection with the possibility of escape or annihilation or multiplication allowed for. Such equations are encountered in a large number of fields, typified by the example of positron diffusion mentioned at the beginning of the chapter, but of no less importance are examples like neutron transport in a nuclear reactor, or the acceleration of cosmic rays in the shells of supernovae, the migration of vortices in a viscous fluid or the transfer of heat in a body.

For the simplest case where $\mu = 0$ and D, b are constants the solution for a walker starting from x_0 at $t = 0$ is

$$G_0(x,t;x_0,0) = \frac{1}{\sigma\sqrt{2\pi}}e^{-\{x-(x_0+vt)\}^2/2\sigma^2} \tag{3.15}$$

with mean $< x >= x_0 + vt$ and variance $\sigma^2 = 2Dt$, where $v =< \Delta > /\Delta t$ and $D =< \Delta^2 > /2\Delta t$.

8.3.6 Fokker-Planck Based Variable Selection

If we consider equation 3.14 with D a constant and $\mu(x) = 0$, i.e.

$$\frac{\partial p}{\partial t} = \frac{\partial}{\partial x}\{D\frac{\partial p}{\partial x} - v(x)p\}, \tag{3.16}$$

it clearly has a time independent solution,

$$p(x,t) = p(x) = e^{\;D^{-1}\int v(x)dx}\;.$$

This observation suggests an alternative random walk (Markov chain) method of selecting variates from a distribution of the form

$$p(x)dx = e^{-f(x)}dx. \tag{3.17}$$

We write

$$v(x) = -D\frac{df}{dx},$$

and carry out a random walk to simulate equation 3.14. With $\mu = 0$ we can take $\omega(x) = 1$. In addition to the diffusive term, which is modeled by a step $\eta_i\sqrt{2D\Delta t}$ (equation 3.6), a nett drift $\{T_+(x) - T_-(x)\}\Delta x = v(x)\Delta t$ should be added at each step. Points visited in this walk, started from an arbitrary point, will, after a time long enough such that the influence of the starting point has decayed, be distributed according to the distribution 3.17. The steps in the Markov chain are given by

$$\Delta_i = \eta_i\sqrt{2D\Delta t} - D\frac{df}{dx}\Delta t. \tag{3.18}$$

Since the time scale doesn't come into it, other than for discarding a number of steps at the beginning of the walk, we can set $\Delta t = 1$. D is still a free parameter with the dimensions of the square of the step size.

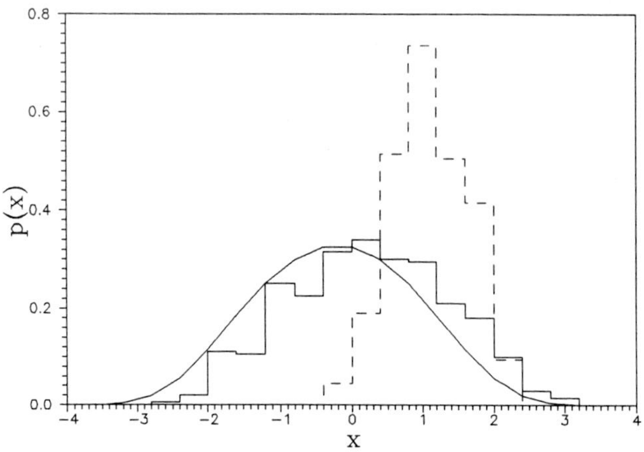

Figure 8.2: Variates from the distribution $p(x)$ in example 8.1 selected using a random walk starting at $x = -3$, and histogrammed for comparison with $p(x)$, solid curve. The dashed histogram corresponds to taking $D = 0.01$, while the solid histogram is the result with $D = 0.20$. The first 200 variates generated were discarded in each case.

EXAMPLE 8.1: Selection from $e^{-0.5\cosh x}$

As a simple example we illustrate selection from the un-normalised distribution

$$p(x)dx = Ce^{-0.5\cosh x}\,dx,$$

using the algorithm 3.18. We have $df/dx = 0.5\sinh x$, and so the random walk is executed with step size

$$\Delta x = \eta_i\sqrt{2D} - 0.5D\sinh x.$$

The locations of the walker in a sequence of 500 steps, following on from an initial discarded 200 steps which were started at $x = -3.0$, are histogrammed and compared with $p(x)$ in Fig. 8.2 for two cases. The dashed histogram corresponds to taking $D = 0.01$, while the solid histogram is the result with $D = 0.20$. The comparison shows how a compromise value of D must be chosen, as small as possible so that the underlying theory holds (D here is really the product $D\Delta t$), but large enough so that the exploration of the space is exhaustive within the chosen number of steps. For our small number of steps, 500, the value of 0.01 for D is too small for this latter condition to be satisfied.

In the literature, the algorithm incorporating equation 3.18 or versions of it is known variously as the *Langevin* method since it is an exact parallel of Langevin's equation,

$$\frac{dq_k}{dt} = -\frac{\partial H}{\partial q_k} + \eta_k(t)\sqrt{2}, \tag{3.19}$$

where $\eta_k(t)$ is an uncorrellated standardised Gaussian random force. *Smart Monte Carlo* and *Force bias Monte Carlo* are variants on this method.

8.3.7 Continuous Time Random Walks

A particular feature of our treatment of random walks is the lack of symmetry between time and space. In section 8.3.2 we considered variety in the size of a spatial step, characterised by $w(\Delta)$, and saw that the basic features of the walk were unchanged. Analogous variability in the time step can also be introduced, each step in the walk is taken after a random time τ. In the general case there is a probability distribution $w(\Delta, \tau)d\Delta d\tau$ representing the probability that a step in the size range $\Delta, \Delta + d\Delta$ is taken in a time interval $d\tau$ at a time τ since the previous step. For the case where the steps Δ and τ are independent, $w \approx w(\Delta)\psi(\tau), \tau \geq 0$. Time intervals τ are selected from a distribution $\psi(\tau)$, these walks are known as *continuous time random walks*, (CTRW). The formal theory of CTRW has been worked out in the literature,[3] but the basic properties of such walks are most easily seen by simulation. Provided the distributions in Δ and τ are well behaved (have finite low order moments) the distribution of walkers starting from x_0 are distributed according to the moving Gaussian, equation 3.15, with

$$\sigma^2 = 2Dt, \qquad v = < \Delta > / < \tau > , \qquad D = < \Delta^2 > /2 < \tau > . \tag{3.20}$$

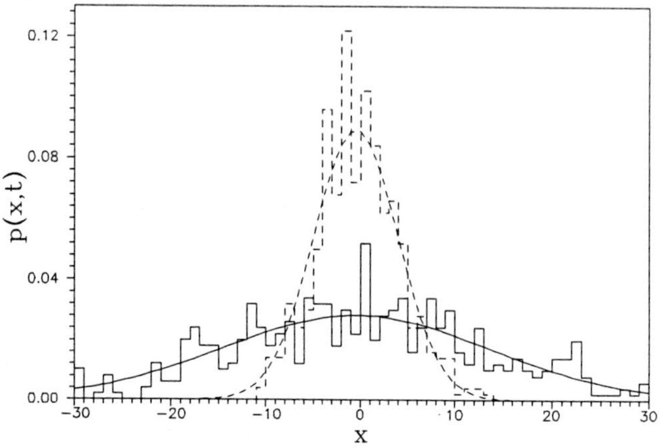

Figure 8.3: The distribution of 500 walkers which started from the origin, at later times $10 < \tau >$, dashed curves, and $100 < \tau >$, solid curves, compared with equation 3.15 using the relations in equation 3.20. The steps were selected from exponential distributions, $w(\Delta) = 0.5e^{-|\Delta|}, \psi(\tau) = e^{-\tau}$.

EXAMPLE 8.2: A simple CTRW

We simulate the distribution of 500 walkers starting from the origin with both time step and space step selected from an exponential, $w(\Delta) = 0.5e^{-|\Delta|}, \psi(\tau) = e^{-\tau}$, with $< \Delta >= 0, < \Delta^2 >= 2, < \tau >= 1$, respectively. Fig. 8.3 shows the distribution of walkers at times $10 < \tau >$, dashed plots, and $100 < \tau >$, solid plots, and compares them with equation 3.15 using the relations in equation 3.20, illustrating the difference from a regular random walk.

Walks of this type are relevant, for example, to the diffusion of vacancy defects or interstitial atoms in crystalline or amorphous media. Their simulation is of much greater relevance for the cases where either Δ or τ arises from a distribution whose first or second moment is not finite; we describe these interesting cases in section 8.6

8.4 Random Walks: Finite Difference Methods

8.4.1 Introduction

Our description of how random walks can be utilised to find solutions to differential equations has been based on the primitive origins of the equation. We can also work in the opposite direction, i.e. starting from the fully fledged equation we can derive a random walk prescription which will generate its solution. One way of doing this is to recast the equation in terms of discrete values of the function, the method of *finite differences.* We will see later in the chapter the close relationship between this method and the random walk approach which we have developed up till now. The relationship is not only of interest in itself, but opens the way to exploiting techniques developed in the context of finite differences in the application of Monte Carlo methods, e.g. the Multigrid method and methods of over-relaxation.

We will consider again the equation

$$\frac{\partial \rho}{\partial t} = \frac{\partial^2}{\partial x^2} D\rho - \frac{\partial}{\partial x} b\rho, \tag{4.1}$$

with $\rho(x,0)$ specified — an initial value problem. As we saw in section 8.3.5, this arises from the master equation

$$p(x,t+\Delta t) = T(x-\Delta x)p(x-\Delta x,t) + T(x+\Delta x,t)p(x+\Delta x,t) + [1-\omega]p(x,t)$$

in the limit $\Delta t, \Delta x \to 0$, where the stepping and lingering probabilities are related as $\omega(x,t) = (T_+ + T_-)$. If we treat the $(x,\ t)$ space as a grid of points with spacings Δx and Δt, with $x = i\Delta x$ and $t = j\Delta t$ and use the notation $p(x,t) = p_{i,j}$ etc., the contributions to the function p at $(x, t+\Delta t)$ can be viewed schematically as in Fig. 8.4(a). The master equation has the form

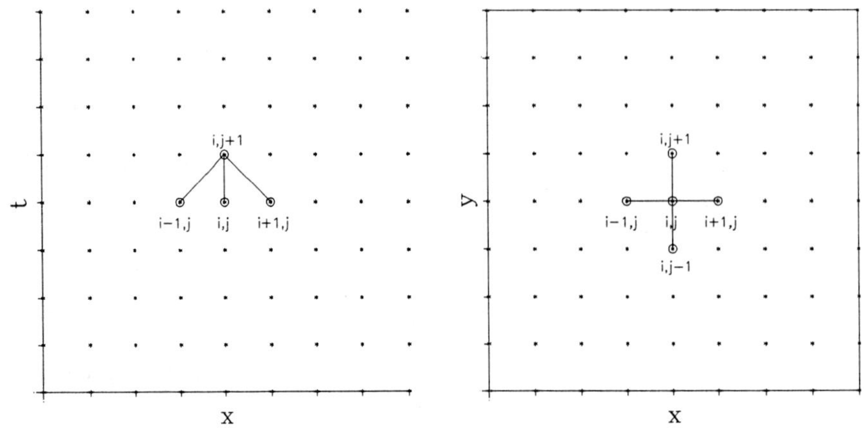

Figure 8.4: (a) Discrete grid with $x = i\Delta x$ and $t = j\Delta t$ on which the differential equation is approximated. (b) The function at the mesh point (i, j) can be expressed in terms of its values at its four neighbouring mesh points and the value of the function q at the point.

$$p_{i,j+1} = T_{+i-1}p_{i-1,j} + T_{-i+1}p_{i+1,j} + [1 - \omega_i]p_{i,j} \tag{4.2}$$

with $\omega_i = T_{+i-1} + T_{-i+1}$. The value $p_{i,j+1}$ is determined by the three points all at the earlier time t. We have also seen that the quantities $p_{i,j}$ are approximations (truncated forms) to discrete values of a function $\rho(x, t)$ which satisfies the differential equation where

$$\begin{aligned} D(x) &= (T_+(x) + T_-(x)) \lim_{\Delta t \to 0} \frac{(\Delta x)^2}{2\Delta t}, \\ b(x) &= (T_+(x) - T_-(x)) \lim_{\Delta t \to 0} \frac{\Delta x}{\Delta t}, \end{aligned}$$

and the initial conditions, $p_{i,0} = \rho(i\Delta x, 0)$ is specified. For illustration all our discussion will be constrained to two-dimensional problems, where the Random Walk method will not be competitive with other non-stochastic methods in practice. We now look briefly at the finite difference formulation of equation 4.1.

8.4.2 Finite Difference Schemes

The derivatives of a function, denoted generally by $F(x)$, at a point can be written, formally, in terms of values of the function at the point and at some neighbouring point(s). By expanding the function F about the point $x+\Delta x$ it is trivial to show that

$$\frac{dF}{dx} = \frac{F(x+\Delta x) - F(x)}{\Delta x} + O(\Delta x). \tag{4.3}$$

By combining this expansion with one about $x - \Delta x$ it is easy to show that a better expression is

$$\frac{dF}{dx} = \frac{F(x+\Delta x) - F(x-\Delta x)}{2\Delta x} + O(\Delta x^{2}), \tag{4.4}$$

better in the sense of having a higher order truncation term. By applying equation 4.4 to the function $F'(x) \equiv dF/dx$, we have

$$\frac{d^2F}{dx^2} = \frac{dF'}{dx} = \frac{F'(x+\Delta x/2) - F'(x-\Delta x/2)}{\Delta x} + O(\Delta x^{2}),$$

or

$$\frac{d^2F}{dx^2} = \frac{F(x-\Delta x) - 2F(x) + F(x+\Delta x)}{(\Delta x)^2} + O(\Delta x^{2}). \tag{4.5}$$

The starting point of the method of finite differences is the replacement of derivative terms by truncated forms of these or similar expressions; this is followed by solving for values of the functions involved in them. Because the 'error' terms have been ignored in this replacement the values of the function so determined from the difference equations will not be the same as the values of the function F which obeys the original differential equation. We in general distinguish between the two functions by using lower case for the former function, e.g.

$$\frac{d^2F}{dx^2} \to \frac{f(x-\Delta x) - 2f(x) + f(x+\Delta x)}{(\Delta x)^2}. \tag{4.6a}$$

Partial derivatives can be replaced in the same way. For example, the partial derivative with respect to x of a function $F(x,t)$ at a point $x = i\Delta x, t = j\Delta t$, using the notation earlier introduced, may be replaced as

$$\frac{\partial^2 F(x,t)}{\partial x^2} \to \frac{f_{i-1,j} - 2f_{i,j} + f_{i+1,j}}{(\Delta x)^2}. \tag{4.6b}$$

8.4.3 Initial Value Problems

We are now in a position to write the finite difference form for equation 4.1. In doing so we will use equation 4.3 in replacing the time derivative, (thus incurring a truncation $O(\Delta t)$), but the more accurate expression equation 4.4 in replacing the first order space derivative. Writing p for the truncated form of ρ we have (we assume that neither D nor b depend on t)

$$\frac{p_{i,j+1} - p_{i,j}}{\Delta t} = \frac{D_{i-1}p_{i-1,j} - 2D_i p_{i,j} + D_{i+1}p_{i+1,j}}{(\Delta x)^2} - \frac{b_{i+1}p_{i+1,j} - b_{i-1}p_{i-1,j}}{2\Delta x},$$

or

$$p_{i,j+1} = T_{(+)i-1}p_{i-1,j} + T_{(-)i+1}p_{i+1,j} + [1 - \omega_i]p_{i,j}, \tag{4.7}$$

where $T_{\pm}(x)$ and ω are given by

$$T_{\pm}(x) = \frac{\Delta t}{(\Delta x)^2} D(x) \pm \frac{\Delta t}{2\Delta x} b(x),$$
$$\omega = T_+ + T_- = 2D(x)\frac{\Delta t}{(\Delta x)^2}.$$

We have arrived back at equation 4.2, and, if we take $\Delta x^2 = 2D\Delta t$, with the same stepping probabilities as we derived on the basis of the random walk in section 8.3.5. The Finite difference prescription for 'marching' the solution from one time slice to the next in a deterministic way is exactly paralleled by the stochastic procedure we obtained on the basis of a random walk. The replacement of the differential equation made above is not unique, and is well known not to be a very effective one. The requirement that at any point, $T_{\pm} \leq 1$, yields a constraint on Δx,

$$\Delta x \leq 2D/b,$$

while requiring $T_{+} + T_{-} \leq 1$ yields the important condition

$$\Delta t \leq \frac{(\Delta x)^2}{2D(x)}. \tag{4.8}$$

This means that the time steps employed must be an order smaller than the space steps, which in turn should be small enough to result in a negligible truncation error, i.e. that the solution p be close to the desired function ρ. The condition in equation 4.8 is well known in the theory of finite differences as a necessary condition for the solution of the finite difference equations to converge.

We have illustrated the relation between random walks and the finite difference formalism; the weakness of the random walk method for problems in low dimensions is seen from its equivalence to the least sophisticated of the finite difference schemes. In such problems finite difference techniques of systematically marching through the array, using more accurate finite difference replacements will be more practical than solution by the stochastic method.

We see that taking the equality sign in equation 4.8 corresponds to the absence of a lingering term in the random walk approach, and since to do so would permit the largest possible value of Δt consistent with convergence to be used it suggests itself as the most obvious choice. The case of $\omega \neq 1$, i.e. where some memory of the function at the point at the previous time slice is retained, is known as *under-relaxation* and is of some theoretical interest.

8.4.4 Boundary Value Problems

In this section we contrast the finite difference and random walk methods for solving the equation,

$$\frac{\partial^2 \Phi}{\partial x^2} + \frac{\partial^2 \Phi}{\partial y^2} = q(x, y),$$

in a two-dimensional region bounded by a curve $\Gamma(x, y)$ with $\Phi(x, y)$ specified on the boundary Γ, and given there by a function $F(s)$ where s denotes a coordinate on the boundary — a boundary value problem. To find a random walk prescription we transfer the equation to a discrete mesh with step sizes Δx and Δy. Using equation 4.6(b) we can write an equation for ϕ, the truncated function corresponding to Φ, at each mesh point in the form

$$\phi_{i+1,j} + \phi_{i-1,j} + \lambda^2 \phi_{i,j+1} + \lambda^2 \phi_{i,j-1} - 2(\lambda^2 + 1)\phi_{i,j} = (\Delta x)^2 q_{i,j},$$

where $\phi_{i,j}$, etc. correspond to the value of the function at the point $x = i\Delta x, y = j\Delta y$, and $\lambda = \Delta x / \Delta y$. By rearranging this expression we see that the function at the mesh point (i, j) can be expressed in terms of its values at its four neighbouring mesh points and the value of the function q at the point (see Fig. 8.4(b) — where it is assumed that the boundary Γ coincides with the mesh). Locally relabelling the points east, north, west and south of the point (i, j) $\phi_1, \phi_2, \phi_3, \phi_4$, respectively, and the point (i, j) itself by the subscript *0*, we can write

$$\phi_{i,j} \equiv \phi_0 = \sum_1^4 T_k \phi_k + \mu q_0, \tag{4.9}$$

where

$$T_1 = T_3 = 1/2(\lambda^2 + 1), \quad T_2 = T_4 = \lambda^2/2(\lambda^2 + 1), \quad \mu = -(\Delta x)^2/2(\lambda^2 + 1).$$

The possibility of using a random walk to solve equation 4.9 arises from the fact that $\sum T_k = 1$, so that, apart from the contribution $\mu q_{i,j}$ the expression represents a weighted average of the value of ϕ at the four points surrounding (i, j). Thus an estimate of $\phi_{i,j}$ is simply

$$\phi_m + \mu q_0,$$

where m is one of the four values of k selected at random according to the probabilities T_k. This is an unbiased estimate in the sense that the mean of a large sample of such estimates would converge to the value of equation 4.9. In general, of course, such an estimate is purely formal if we don't know the value of ϕ_m. The value of ϕ_m, however, is also determined by an expression like equation 4.9 with q_0 replaced by q_m, so that an estimate of ϕ_m can be made by selecting one of the values, call it ϕ_n, surrounding the point m at random according to the probabilities T_k. Inserting our estimate $\phi_m = \phi_n + \mu q_m$, our estimate of $\phi_{i,j}$ then becomes

$$\phi_n + \mu(q_0 + q_m).$$

If the value of ϕ_n is not known, i.e. if the point does not lie on the boundary Γ, the procedure can be repeated. In this way a single estimate of $\phi_{i,j}$ is made by carrying out a random walk starting from the point (i, j), at every point stepping with probabilities T_k, and accumulating the score until the boundary is reached. If this occurs after M steps, our estimate is

$$F_M + \sum_{k=0}^{M-1} \mu q_k,$$

where F_M is the value of the function F at the boundary point and the summation runs over all M-1 points visited in the course of the walk (some of which may occur multiply). Many such estimates, each based on an absorbing walk, would be required just to solve the finite difference equation at the one point (i, j). Although refinements can be made to greatly improve the efficiency of the random walk,[4] the method remains, in general, more a classic illustration of the principles of Monte Carlo estimation and of the relationship between stochastic processes and potential theory than a viable procedure for solving such equations.

8.5 First Passage Times

Many parameters of a random walk apart from its distribution and moments can be studied, its mean maximum extent in a time t is an example. One of the more important parameters is the *first passage time.* Referring to Fig. 8.5 a particular random walk starting from x_o at $t = 0$ defines a random variable at any later time t, $x(t)$, which is distributed as $p(x, t; x_o)dx$. At the same time it defines a multivalued random variable $t_i(x), i = 1, 2, \ldots$, being those successive values of t for which the walker is located at x. By focusing on the case $i = 1$, i.e. the earliest time at which the walker reaches x (or the time necessary to arrive at x), we have a random variable $t_1(x)$ known as the *first passage time* to x. For all walks starting at $(x_o, t = 0)$ this quantity

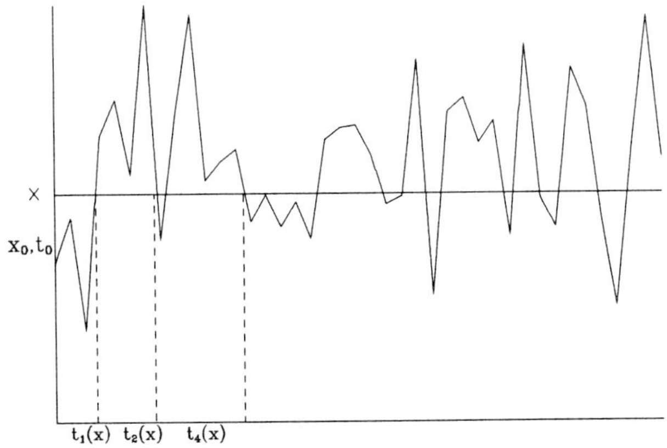

Figure 8.5: A random walk starting from x_o at t_0, defines a random variable $t_1(x)$, the first passage time at x.

will, in general, take on different values, i.e. it will have a distribution, which we denote by $f(t_1, x; x_o)$, such that

$$f(t_1, x; x_o)dt_1$$

is the probability that the time required for a walker to arrive at x for the first time will lie in the interval $t_1, t_1 + dt_1$. So, in general, one can talk of the mean time required to reach x for the first time, its median, variance, etc.

Schrodinger was one of the first people to investigate the form of this distribution, while Chandrasekhar applied it in a discussion of the escape of a star from a cluster. If the velocity of a star in a cluster can be considered as a random variable, due to its collisions with other members of the cluster, it may at some stage attain a value corresponding to the escape velocity, in which case it will evaporate from the system. Such problems, as well as others in which the attainment of a threshold by some randomly varying quantity is important, lead to a discussion of this first passage time distribution. If the velocity of the walker is constant the passage time is a measure of the distance travelled by the walker; if *en route* some process is occurring, such as the fragmentation of a cosmic ray nucleus in collisions with interstellar gas, the distribution of first passage times gives the lifetime distribution of the species. Frequently in mechanical systems stochastic disturbances occur, and failure of the system, corresponding to a response amplitude V exceeding a critical value V_c, can be viewed as a first passage problem. Determining the probability thus requires knowledge of the cumulative distribution function

$$\int^t f(t_1, V_c; 0)dt_1.$$

For any specified random walk, with given boundary conditions, the investigation of the distribution by simulation seems obvious, and is indeed usually necessary, since not only are analytical results restricted to a few special cases, but in the general case the threshold to be passed may itself be a function of time. By way of example we will look at a toy problem of trapping — the diffusion of impurity atoms in a solid would be a physical analogue.

EXAMPLE 8.3: Mean first passage time on a circular random walk

As an example of the use of Monte Carlo methods in random walks this short exercise illustrates the idea of first passage times, and, because its exact solution is known, serves as a probe of correlations in the sequence of uniform numbers used. This problem uses the numbers produced by a random number generator to simulate a random walk over sites, one of which is a trap. Starting from a given site the mean time to reach the trap is estimated and compared with the known solution to this problem. Significant disagreement (in the absence of programming errors) must be due to lack of randomness in the numbers used in constructing the walk.

We consider a simple random walk on N discrete sites, one of which is

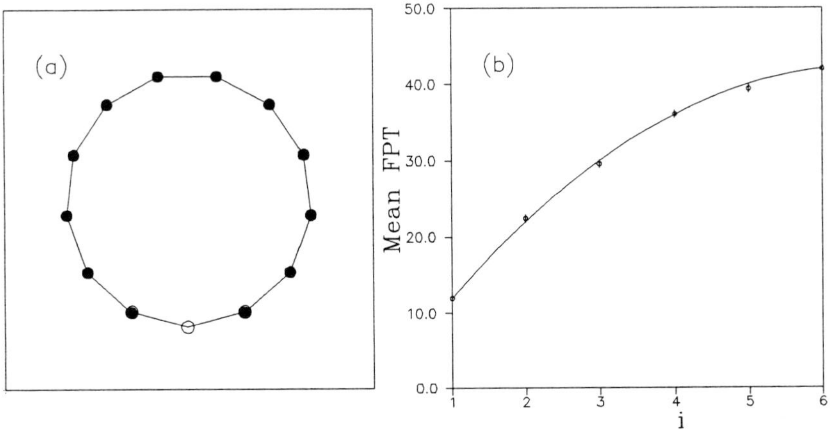

Figure 8.6: (a) Thirteen sites arranged on a circle, with the open symbol indicating the trap. (b) The mean time to capture, with its standard error, is shown for sites 1—6, with the theoretical values connected by a curve.

absorbing, on a circle — or equivalently the line with periodic boundary conditions — and determine the mean first passage time $< t_{1i} >$ to absorption for walkers starting at each of the sites i. We take the case of $N = 13$ and number the sites, starting with the one nearest the trap, $i = 1, 2, \ldots, 12$; the arrangement is shown in Fig. 8.6(a). A walker steps with equal probability to the left or right every unit of time until it reaches the trap. Based on samples of 4000 walkers starting from each site the mean time to capture, with its standard error, is shown for sites $1-6$ in Fig. 8.6(b). The exact solution for this problem is known, $< t_{1i} >= i(N - i)$. For ease of comparison with the simulated data we have plotted this as a continuous function of i in the figure. The agreement with the simulation results is a further test of the quality of the random number generator used in this calculation.

Analytical methods can be used to find mean first passage time distributions for some walks;[5] here we just quote the result for the most common case. For the case of diffusion with a constant drift velocity, $v \geq 0$, in an infinite medium (i.e. corresponding to equation 3.14 with D constant and $\mu = 0$) it can be shown that the first passage time distribution has the form

$$f(t_1, x; x_0)dt_1 = \frac{| x - x_0 |}{\sqrt{4\pi D t_1^3}} e^{-(x - x_0 - vt_1)^2/4Dt_1} dt_1,$$

with mean and variance ($v > 0$),

$$\mu = (x - x_o)/v, \qquad \sigma^2 = 2D(x - x_o)/v^2.$$

It is interesting to note that for pure diffusion, $v = 0$, this reduces to the Levy distribution (equation 2.4.19), for which both the mean and variance are infinite — a famous result of Polya's — although the walker is sure, eventually, to reach the point x. When the drift velocity is negative however, $v < 0$, the probability of ever reaching x from $x_o (x > x_0)$ is less than unity, explicitly it is

$$e^{v(x-x_0)/D}.$$

8.6 Anomalous Diffusion

Referring to section 8.3.7, it is clearly possible, and in practice just as simple, to simulate a random walk, selecting steps from a distribution $w(\Delta)$ or from a time distribution $\psi(\tau)$ which have divergent first or second moments,[6] such as the power law distribution, equation 2.4.18, with $1 < \gamma < 3$ or the Lorentz distribution, equation 2.4.16. Walks based on these distributions do not describe models of classical diffusion where the dispersion of the walkers is $\propto t^{1/2}$, and their properties are remarkably different — but they do have relation to some physical phenomena. We will start with the case of a regular distribution in spatial steps, $w(\Delta)$ with moments $< \Delta >$ and $< \Delta^2 >$, and a pathological time step distribution $\psi(\tau)$. Walks of this type can be related to anomalous diffusion (dispersive transport) of charge carriers in an amorphous material or to the relaxation of electric dipoles in a dielectric via transport of mobile defects. The passage of electric current in many materials can be viewed as a random walk, simple diffusion of the carriers biased by an applied electric field. In some materials, however, the mechanism is more complex; from the point of view of the carrier, electron or hole, the medium is a warren of potential wells into which it can fall, and remain trapped until an escape opportunity arises. This can occur through thermal excitation or by tunneling to a nearby well. The time required for the occurrence of one of these steps, the time step in the random walk, is a random variable distributed according to some distribution. In a normal random walk for which $< \tau >$ is defined the mean location of the walker (starting from the origin at $t = 0$) would be $< x >= vt$ with $v =< x > / < \tau >$, equation 3.15. In anomalous diffusion the relation found is $< x >\simeq t^\nu$ with $\nu < 1$. This can be accounted for by a 'long tailed' distribution in the time steps, e.g. the power law distribution, equation 2.4.18,

$$\psi(\tau)d\tau = (\gamma - 1)\tau_0^{\gamma-1}\tau^{-\gamma}d\tau, \qquad \gamma > 1, \quad \tau \geq \tau_0. \tag{6.1}$$

The mean only exists for values of $\gamma > 2$ so for $1 < \gamma < 2$ we have $< \tau >= \infty$.

EXAMPLE 8.4: A Comparison with normal diffusion

In this example we compare the distribution of walkers, starting from the origin, for the case where the time steps are from a regular distribution, $e^{-\tau}, < \tau >= 1$, and from the distribution, equation 6.1, with $\gamma = 3/2$ and $\tau_0 = 0.06$. We consider a drifting walk with the space steps selected from $e^{-\Delta}, < \Delta >= 1, < \Delta^2 >= 2$. The distribution of 5000 walkers at times $5 < \tau >$ and $70 < \tau >$ are shown in Fig. 8.7(a), the regular case by the dashed histograms and the anomalous diffusion by the solid curves. The reluctant diffusion in the latter case can be contrasted with the propagating Gaussian characterising reg-

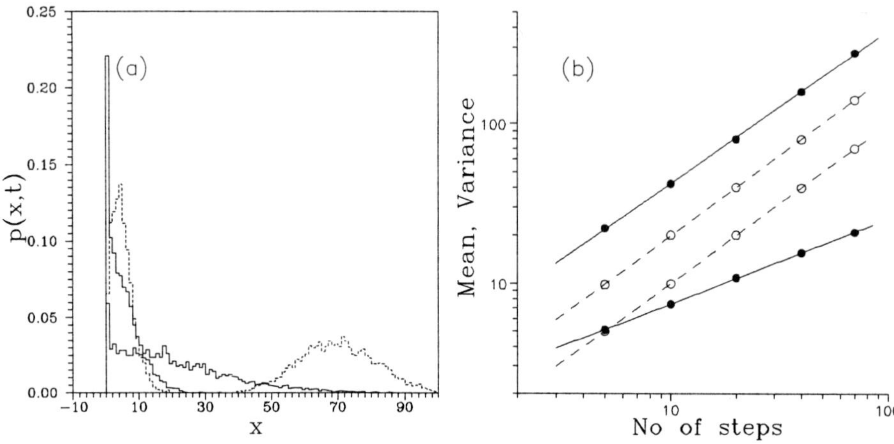

Figure 8.7: (a) The distribution of 5000 walkers at times $5 < \tau >$ and $70 < \tau >$ for the case of a simple random walk — dashed histograms, and for the case of anomalous diffusion — solid histograms. (b) The mean displacement, and variance of the walkers for the simple random walk— the dashed curves, and for anomalous diffusion — solid curves.

ular diffusion. The mean displacement, and variance of the walkers for the two cases are compared in Fig. 8.7(b), the dashed curves refer to the regular walk. For the anomalous diffusion we see $< x > \simeq x^{\alpha}$ and $< x^2 > \simeq x^{\beta}$ with $\alpha = 0.53 \simeq \gamma - 1$, and $\beta = 1.84 \simeq 2(\gamma - 1)$ compared with the regular values of 1.0 and 1.0, respectively. The theoretical values quoted are valid at large times.

We turn now to the case of a pathological distribution in the spatial step size — walks based on such distributions are known as *Levy walks.* The qualitative difference of these walks can be seen in Fig. 8.8 where we show examples of a 60 step random walk in two dimensions for the case where the steps sizes are selected from a standardised Gaussian, and a Levy walk with steps from a 'similar' Lorentz distribution with $\Gamma = \sqrt{2/\pi}$, (Fig. 2.5). Note the difference in the scale for the two parts of the figure.

Our specification of a walk so far in the chapter allowed for the stepping probabilities to be an arbitrary function of x; it can be made more

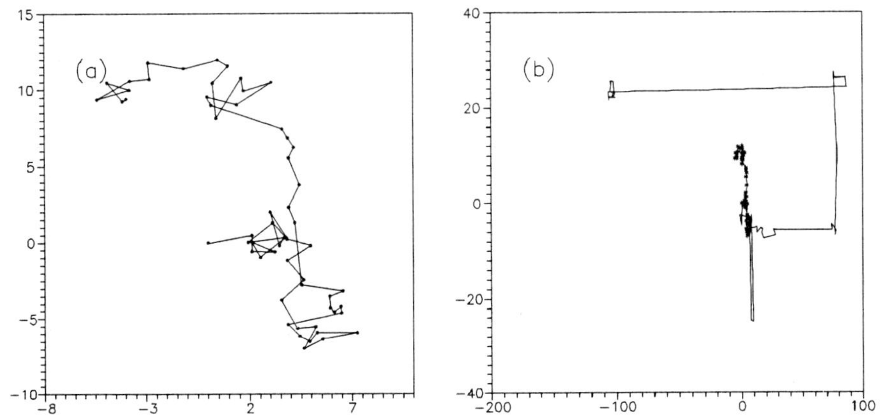

Figure 8.8: Examples of a 60 step random walk in two dimensions for the case where the steps sizes are selected: (a) from a standardised Gaussian, and, (b) from a 'similar' Lorentz distribution with the same intensity at the mode.

general by introducing the notion of *memory*. Stepping probabilities then depend not only on the present position of the walker, but also on where he last came from. Some correlation exists between successive steps and a persistence, or inertia, is imposed on the motion. In this case we can define two transition functions, $p_+(x,t)$, the probability of a walker arriving at x at time t and travelling to the right (i.e. arriving from the left) and $p_-(x,t)$ the corresponding probability of arriving from the right. The total probability of reaching the point x at time t is then the sum of these two values, and gives rise to differential equations, such as Kelvin's equation and the Telegrapher's equation. Their extension to imaginary time leads to Dirac's equation, which, in principle, could also be solved using Monte Carlo simulation, although practical methods have not so far been developed.

8.7 Non-Markovian Walks

8.7.1 Introduction

It is easy to visualise more complex behaviour of the stepping probabilities in a walk than considered earlier. Very interesting cases arise when the memory extends back more than one step in the walk. For example, the probability of stepping to a point might depend not only on the position of the walker, but on whether and how often that point has been visited earlier. Such walks, where some or all of the previous history of the walk enters into decisions for future steps, are non-Markovian, and their theoretical description is much more difficult than the cases so far considered. Here the value of the Monte Carlo approach when compared with analytical methods is more obvious than in the case of the random walks studied so far, and the availability of computer methods has enabled the properties of many new types of walks to be studied in recent years. As in the case of Markovian walks, we will be interested in the general properties of the walks and in their possible applications to physical problems. In addition, however, we will be preoccupied to a considerable extent by the development of algorithms to generate samples of these walks — a task which is trivial for their Markovian counterparts where direct mimicry of their definition enables practical algorithms to be constructed. The logistics of data handling, updating a walk and testing for memory constraints, are much more demanding than in the simple random walk case. We shall be concerned with two generic classes of these interacting walks, *self-repelling walks* (SRW) and *self-avoiding walks* (SAW). Among the fields where concepts based on these walks find applications are the statistical mechanics of polymer molecules, the general theory of phase transitions and critical phenomena and the properties of certain four-dimensional quantum field theories.

Before looking at these we remind ourselves of the properties of the pure random walk of N steps, each of length h, in d dimensions, starting from the origin, with the kth step arriving at $\mathbf{r}_k$. The number of configurations for which the end walker lies in a volume element $d^d\mathbf{R}$ at $\mathbf{r}_N = \mathbf{R}$ from the start of the walk, for large N, is Gaussian distributed,

$$p(\mathbf{R})d^d\mathbf{R} = Ce^{-\mathbf{R}^2/b^2} d^d\mathbf{R},$$

where $b = 2h^2N/d$ and $C = C(d,b)$ is a normalisation constant. The mean square end-to-end separation is $< R^2 >= h^2N$. In other words,

$$\sqrt{< R^2 >} = hN^\nu, \tag{7.1}$$

with $\nu = 1/2$, independent of the dimension d. In the case of a polymer chain, this end-to-end distance is not something which can be easily measured, however the *radius of gyration*, S_N, of the chain, defined as the mean square displacement of a molecule from its centre of mass coordinate $\mathbf{R}_{CM}$, can be determined from scattering experiments. For a random walk it is defined by

$$S_N^2 = \frac{1}{N}\sum_{i=1}^{N} \mid \mathbf{r}_i - \mathbf{R}_{CM} \mid^2, \qquad \text{with } \mathbf{R}_{CM} = \frac{1}{N}\sum_{i=1}^{N} \mathbf{r}_i. \tag{7.2}$$

This can be written in the alternative form

$$S_N^2 = \frac{1}{N}\sum_{i=1}^{N} \mid \mathbf{r}_i \mid^2 - \mid \mathbf{R}_{CM} \mid^2,$$

or as the average squared internal separation of all sites,

$$S_N^2 = \frac{1}{2N^2}\sum_{i=0}^{N}\sum_{j=0}^{N} \mid \mathbf{r}_i - \mathbf{r}_j \mid^2 .$$

For a simple random walk in three dimensions, the mean value of this quantity is simply related to the end-to-end distance,

$$< R_N^2 > / < S_N^2 >= 6. \tag{7.3}$$

For walks confined to a lattice with coordination number q (in two dimensions $q = 4(5)$ on a square (triangular) lattice, and in three dimensions $q = 6(12)$ on a simple cubic (face centred cubic) lattice), the total possible number of distinct walks of N steps is clearly $c_N = q^N$. The determination of quantities analogous to $p(\mathbf{r}), < R_N^2 >, c_N$, etc. for non-Markovian walks relies heavily on computer simulations.

8.7.2 Self-repelling Walks

The simplest non-Markovian behaviour arises when we introduce some dependence of the stepping probability not only on the location of the walker at time t, but on the extent to which his neighbouring points have been already visited up to that time (or on the local density of walkers if many are considered). The Markov condition, namely that the probabilities determining the choice of a step at any stage in the walk be independent of the past history of the walk, is no longer valid. The dynamics of these walks is much more complicated than the random walks discussed in section 8.3. A quantity which characterises walks of a given type, a critical exponent, is the index ν in the root mean square end-to-end extent of the walks for large values of N in equation 7.1; we are interested in how it depends on the dimension of the space (we recall that for a pure random walk $\nu = 1/2$ in all dimensions).

As usual it is simplest to look at the case of a discrete walk in one dimension. We denote by n_i the number of previous visits to the site i in the course of N steps, then a model walk might feature transition probabilities for the next step from site i of the form

$$T_{\pm} = Z^{-1} e^{-g n_{i\pm1}} ,$$

where the normalisation factor Z is just $(e^{-g n_{i-1}} + e^{-g n_{i+1}})$. The constant g can be positive or negative. We will concentrate on positive values of g, which is then a measure of the reluctance of the walker to revisit sites. If one considers many walkers simultaneously present the pure random walk behaviour is unaffected since they do not mutually interact; here we have effectively introduced a repulsive interaction between them. The case $g = 0$ is a singular case. It clearly corresponds to the Markovian walk for which we know $\nu = 1/2$ although the assymptotic behaviour of the walks do not tend to this value as $g \to 0$. Similarily the case $g \to \infty$ might be thought to correspond to the SAW discussed in the next section for which $\nu = 1$ but there is a subtle difference between the two cases. We should point out that in the literature these walks are sometimes referred to as *true self-avoiding walks* (TSAW). Although very interesting in themselves, such walks do not seem to have much application in physics. The simulation methods are basically the same as for SAWs, which we now consider.

8.7.3 Analogue Generation of Self-avoiding Walks

Simpler in many respects than the self-repelling walk is the self-avoiding walk, where the walker is forbidden from visiting any site more than once. In one dimension such a walk is trivial, the only stochastic element involved is in the choice of the first step from the origin; the length of the walk after N steps is simply $N\Delta x$ and the index ν has the value of unity.

On a two-dimensional square lattice we find that walks of N steps ($N > 11$) are of two kinds, those which can continue to $N + 1$ steps and those which have reached an impasse — they cannot be continued because all possible steps would lead to a site which has already been visited. In

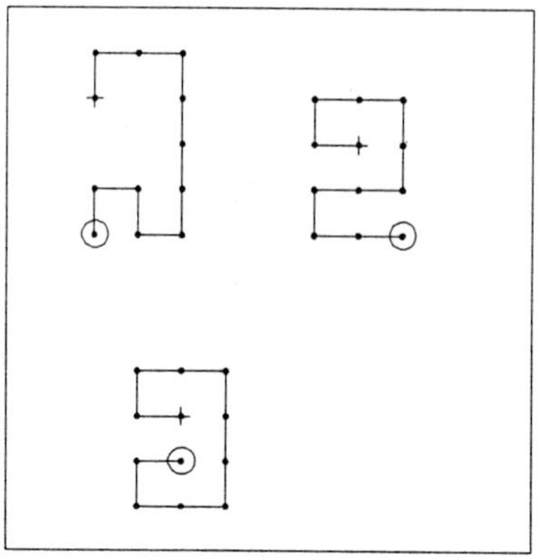

Figure 8.9: Three examples of self-avoiding walks of 11 steps each starting from the origin, denoted by •.

Fig. 8.9, for the case of $N = 11$, starting at the origin, we show one example of the former and two examples of walks which cannot continue. Similar configurations can occur for walks in higher dimensions although the relative probability of the two types of configuration will be dimension dependent. There is yet one more walk of a similar type, namely one where points can be revisited but only from a direction which does not backtrack, i.e. intersections of the walk can occur but a step must not overlap with the immediately previous step. These are known as *self-avoiding trails*.

Prior to describing the application of SAW models for the configurations of polymer molecules we will look at the computational aspects of their simulation. In a pure random walk the number of choices available to the walker prior to a step is q, the coordination number of the lattice, i.e. the number of first nearest neighbours; for a square lattice this is just $q = 2d$ in a space of dimension d. At any point in the SAW this number is at most q-1 since the walker cannot backtrack to a site from which it has arrived. The computer generation of an unbiassed sample of SAWs of N steps is easily described; for simplicity our discussion refers to the two-dimensional square lattice. From the origin use a uniform random number ξ to step to one of the four nearest neighbour sites with equal probability, from there with equal probability step to one of the three adjacent unvisited sites. Continue in this way, selecting one of the three neighbouring sites with equal probability and stepping to it if it has not already been visited, until N steps have been completed. Strictly speaking the choice of backtracking should also be included, but because its effects on generating all walks of N steps is the same, ignoring it introduces no bias in the sample generated. If a site, selected before N steps have been taken, has already been visited — it can occur for N as small as three — the walk must be abandoned and a new walk started from scratch. The probability of reaching such a situation is, of course, dependent on N (and on the dimensionality of the space — it is zero in one dimension, and in dimensions greater than two it would intuitively be expected to be less probable). This phenomenon is known as attrition. The probability that a walk which has completed N steps will be able to continue decreases exponentially with N, i.e. of M_0 walks started the number on average attaining N steps (at large N) is

$$M_0 e^{-N/\kappa} \tag{7.4}$$

— the constant $\lambda \equiv \kappa^{-1}$ is known as the attrition constant. On a two-dimensional square lattice $\kappa \simeq 7.75$, while on a similar lattice in three dimensions $\kappa \simeq 15.4$. The chance of completing a 100 step walk in either case is very small and the possibility of generating a large enough sample to study their properties is remote. It is also clear the the number of possible walks of N steps is much less than q^N. In fact it is believed that the total number of possible walks of N steps is given by

$$c_N \approx \mu(q)^N N^{\gamma-1} \qquad N >> 1 \tag{7.5}$$

where γ is a critical exponent which depends only on the dimension, and $\mu(q)$ is an effective coordination number of the lattice.

We will investigate successively some of the many more efficient proposals for generating a sample of SAWs, of length N, noting, for reference, that a method to do this which is transparently both totally unbiased and totally inefficient, is to generate random walks of length N and extract the set of them which have no intersections. The most obvious procedure is to include only those sites which have not been previously visited in the selection process for the next step; this is a method of enhancing the survival probability of the walker. Indeed just this technique, called *survival biasing*, is an important mechanism of variance reduction in particle transport calculations, as we will see in the next chapter. If at the end of the jth step q_j of the neighbouring sites are unvisited, one of them is chosen with probability $1/q_j$ and the step made in that direction. If $q_j > 0$ termination of the walk for which the probability is $1 - q_j/(q-1)$ is not entertained. This method is known as *inversely restricted sampling.* Two difficulties arise in this method. First it may turn out that before N steps have been executed we find $q_j = 0$. Compare the two examples in Fig. 8.9, in which case the walk can continue no further, and second, even if the walk survives for N steps it cannot be taken as a random sample from the population of N step SAWs because of the biased way in which step selection is made.

Eventual entrapment puts a limit on the length of walks which can be generated in this way, the attrition constant on a square lattice in two dimensions is $\lambda \simeq 1/600$. Another technique from particle transport simulations, *splitting*, can be adopted here. Rather than waste all the effort which has gone into generating a relatively long walk which may come to an impasse before reaching N steps, a number of copies of its early development, say a segment of length N_0 steps (e.g. with N an integer multiple of $N_0, N_0 = N/k$) are used as starting segments for generating other walks, some of which may survive to a total of N steps. Clearly any number of sample of walks generated in this way are not independent — since they all consist of the same N_0 initial steps. This splitting can be carried out again on walks which survive to $2N_0$ steps, duplicate copies being used in turn as the starting segment of $2N_0$ steps for a number of walks. This 'favouring of survivors' can be continued with succesive successful additions of N_0 steps, and will result in more N step walks than would a comparable amount of work starting each new walk from scratch. A suitable value for N_0 would be of the order of the attrition length κ, about 15 in three dimensions. More details on selecting the parameters in splitting can be found in section 9.3.4. The degree of correlation among the walks generated in this way is not easy to allow for in later statistical analysis.

As in many other places in this book, the problem of biased sampling can be overcome by attaching a compensating weight to the walk, at each step the walk should acquire a weight $q_j/(q-1)$, so that the weight of the final walk, which must be used in any estimation of characteristics of these walks such as their mean end-to-end length, is $w = \prod_{j=2}^{N} \left(\frac{q_j}{q-1} \right)$. The sample of walks produced in this way is from a population other than the one we wish, with the discrepancy compensated for by weighting each walk. The larger N, the more this discrepancy grows, and at large N the variation in the weights across a generated sample becomes very great, many of them very small and a small, and poorly defined, number of them very large. These latter then dominate in the estimation of any quantity based on the sample, which can lead to very unreliable estimation. Both limitations mean that the method of inversely restricted sampling is of limited usefulness when we wish to generate walks with large values of N.

There are many variations on the methods described so far, but for algorithms which generate samples of long SAWs efficiently we must turn to Markov chain methods — dynamical methods (more correctly pseudo-dynamic, since in the case of a polymer no pretence is made that the true real time dynamics of the walk is simulated).

8.7.4 Markov Chain Generation of SAWs

The basis for Markov chain methods is that if we have one sample of a SAW of length N — a straight line is such a sample — we can use it as an arbitrary starting configuration from which to generate other walks of this length by carrying out successive incremental changes in the configuration according to some transition probability. As in all sampling using Markov chains (see section 5.5), provided suitable transition probabilities which ensure that all configurations in the population are accessible (ergodicity) are used, valid samples can be generated. However, the initial members of the sample, in general, are not representative of the population and should be ignored; thereafter the sample manifests corelations which are characterised by some finite correlation length. Although, as we said, an arbitrary starting configuration can be used, the size of the sequence which must be discarded, if any, can be greatly reduced by starting with a single sample walk gener-

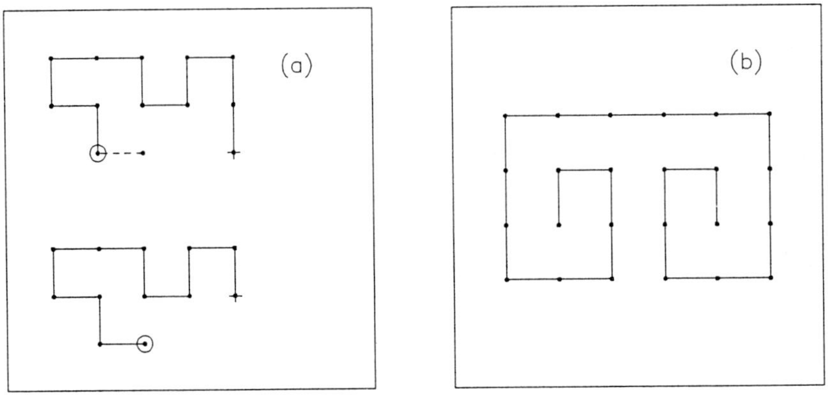

Figure 8.10: (a) Illustration of a reptation move, the lower walk is transformed into the upper one by removing the leading step and attaching it at the tail at a random, and acceptable, orientation giving rise to a new configuration. (b) An example of a 23 step SAW which could never be generated by the reptation algorithm.

ated using one of the less efficient methods described earlier. The essence of the algorithm is: from any walk choose a new configuration of the steps in it, and if the resulting walk has no intersections take it as the next member of the sample, otherwise count the original walk as the 'new' member. There are many prescriptions for revising a walk, we will consider just two examples, *reptation* and *pivoting*.

In the reptation or slithering snake algorithm, one end of the walk, the head or the tail, is chosen at random and the bond there removed and transferred to the other end where it is added at an arbitrarily chosen orientation, and a decision on its acceptability made. Fig. 8.10(a) illustrates a move in two dimensions giving rise to a new configuration. Here we say a few words about the data management problem in the simulation of SAWs. Two distinct activities have to be provided for, the updating of a configuration and its testing for self- avoidance. The simplest way to store a configuration is as a circular list, $\{\mathbf{r}_i\}, i = 0, 1, \ldots, N$, with, in three dimensions, $\mathbf{r}_i = (l_i, m_i, n_i)$

where l_i, m_i, n_i are the lattice coordinates of node i in the walk. A flag, or label, k, is maintained which indicates the address, k, of the head node $\mathbf{r}_k$; initially $k = 0$. When a link is transferred from one end to another, rather than move the whole chain along in memory, which would require N operations, we simply replace the coordinates of the removed node by those of the new node, and at the same time revise the flag corresponding to the memory address of the head of the walk, $k \to (N+k+1\pm1) \bmod (N+1)$. The easiest, but crudest, way to check for non-avoidance is to maintain an $M \times M \times M$ array, the only non-zero elements of which correspond to the nodes occupied by the walk — similar to the method described for percolation in chapter 4. When a tentative node arises only this memory location need be checked, and it and the vacated node updated if accepted. Clearly, M must be larger than the dimensions of the region over which the configuration is likely to wander in the course of the simulation, e.g. $M \approx 2-5N$. For storage reasons this method is impractical for large N, especially in three dimensions, and more sophisticated algorithms, hash tables, must be used (see for example Knuth's book).

The speed with which the configuration space is explored using the slithering snake algorithm can be judged by noting that an end bond at any instant, as time goes by, diffuses into the walk, so that after about the order of N^2 attempted moves the walk will have lost all memory of its starting pattern, i.e. we will have arrived at an independent configuration. This very long correlation length in the sequence of generated walks is one of the disadvantages of this algorithm. An even more serious limitation of the algorithm is that it is not ergodic, all the walks of N steps cannot be reached from a given walk using it. This can be clearly seem from the example of a walk from the population of $N = 23$ step SAWs in two dimensions shown in Fig. 8.10(b) which could never be generated by this algorithm, and which, if used as a starting configuration, could never generate other samples. This non-ergodicity can be corrected for by using more complex moves in the chain, we will leave it here, having seen how such chain methods are used, and at the same time learned a lesson of caution in using them.

An approach, which on initial appearance does not suggest efficiency but turns out to be highly efficient for large N, and whose ergodicity can be ensured, is the *pivot* algorithm. Here a Markov chain move consists of gross but identical reorientations of all the steps in the walk beyond an arbitrary point along it. As usual, the resulting walk is tested for intersections and if such occur the original version is retained as the 'new' walk. To ensure ergodicity of the method the reorientations should consist of (at least some of) the transformations of the symmetry group of the lattice. The method is best explained by an illustration in two dimensions, for which the most general description is as follows. With probability $1/N$ select one of the nodes, k, on the walk, other than the tail, as a pivot point. Select with probability $1/7$ one of the elements of the symmetry group other than the identity and, taking the point k as origin, apply the transformation to all the steps beyond k. The elements of this group apart from the identity are: reflections about the two orthogonal axes, reflections about the two diagonal directions, clockwise and anti-clockwise rotations of $\pi/2$, and a rotation of π. Not all of these need be used, nor need they be taken with equal probability, but for ergodicity there are certain requirements — we leave the precise details to the literature.[7] Clearly the testing of self-intersection plays a major role in this algorithm. While it is true that many of the attempted moves will be rejected due to self-intersection, when a different walk is accepted it will often be radically different from its predecessor. This means that a sequence containing a relatively small number of accepted walks is enough to overcome the correlations which are so persistent in the earlier methods, and to yield effectively independent samples from the population. It can be shown, and is seen from simulations, that if an optimum method of checking for avoidance is used, the required number of attempted moves to generate an effectively independent walk is of order $(N \ln N)$, to be compared with $O(N^2)$ for reptation described earlier, and close to the limiting theoretical efficiency of $O(N)$.

There are methods of inferring the asymptotic properties of SAWs from the results of simulations based on walks of relatively modest length, which make use of scaling properties of these chains.[8] These methods, based on renormalisation group analysis, are beyond the scope of this introductory presentation.

8.7.5 Applications to Polymer Chains

Results of simulations in two dimensions confirm theoretical expectations for the critical exponents ν and γ in equations 7.1 and 7.5, viz.

$$\nu = 3/4 \qquad \gamma = 43/32,$$

and at the same time emphasise that one has to go to quite large values of N to reach the assymptotic region where these relationships hold. This region is approached as

$$< R^2 > \approx N^{2\nu} \left(1 + \frac{b}{N^{\Delta}} \right),$$

where b and Δ are parameters, themselves of importance in the theory of critical phenomena. The ratio in equation 7.3 takes the value 7.13.

The work required to probe the assymptotic region in three dimensions, where no reliable theoretical estimates have been obtained, becomes even greater. An important result of simulations in this case is that $\nu = 0.588 \pm .001$, significantly different from the value of $3/5$ arising from simple theory, and long suspected to be true.

The most direct application of SAWs is to the simplest type of linear polymer chain molecules in a good solvent, where each individual chain is free to disport itself without interference from the others. In the ideal case, where no interaction exists between molecules in the chain except that giving rise to the chain structure, nor any interaction with the solvent molecules, the self-avoidance can be alternatively described as the *excluded volume effect.* Although monomers in such a chain do not lie on a lattice, and the bonds in practice are not rigid, the basic features of the molecule can be described by the simulations described in the previous section. These simulations can also, with only a little additional work, be generalised to a continuum with the monomers modeled by hard spheres separated by bonds of fixed length, as they also can to more complex chains, star polymers, etc. The results can be compared with the parameters of real polymer molecules determined, for example, using light scattering.

The characterisation of the interaction between two monomers along a polymer chain by an excluded volume is only a first approximation to the actual interaction between such monomers. In addition to the short range repulsion allowed for in this way, there is typically a van der Waals type attraction for moderate monomer separation, resulting in an energy of interaction of each monomer with a number of others in the chain. This means that different configurations of an N step chain in general have different energies associated with them. At a finite temperature T the probability of occurrence of the different configurations is no longer equal, but depends on the interaction energy of the molecule, E, with frequency of occurrence given by the Boltzmann distribution (with $\beta \equiv 1/k_B T$, for more detail see section 10.2)

$$p(\{\mathbf{r}^{(i)}\}) \propto e^{-\beta E(\{\mathbf{r}^{(i)}\})} . \tag{7.4}$$

When working on a lattice it is usual to approximate this interaction as being of fixed strength ($-\epsilon$, with $\epsilon > 0$) and only occuring between non-bonded nearest neighbours. The thermal properties of polymers can be studied by simulating walks which are not only self-avoiding but are also selected from this distribution. In using the method of inversely restricted sampling any N step SAW configuration generated has a 'natural preference' proportional to $e^{-\beta E}$, and so should be assigned a weight equal to this quantity, in addition to any other weight required to remove bias introduced in its generation. The number of nearest neighbour interactions introduced by adding the jth step in the walk is $(q - 1 - q_j)$, where q_j is the number of neighbouring sites not visited at that stage along the walk. The energy of the final N step configuration is then,

$$E^{(i)} = -\epsilon \sum_{j=2}^{N} (q - 1 - q_j).$$

Carrying out simulations at a temperature T, a sucessfully completed walk will carry a weight $W^{(i)} = w^{(i)} e^{-\beta E^{(i)}}$ where $w^{(i)}$ is any weight introduced by survival biasing, etc. A property of the polymer, e.g. its mean square end-to-end distance, would then be estimated on the basis of m generated walks as

$$< R^2(N,T,\epsilon) >= \sum_{i=1}^{m} W^{(i)} R^{(i)2} / \sum_{i=1}^{m} W^{(i)}.$$

In the case of dynamic simulation the Markov chain method can be readily modified to ensure that the generated configurations are not only self- avoiding but form a sample (highly correlated) from the distribution 7.4, without the necessity for any correcting weights. The method exactly parallels that described in section 6.2 for selecting configurations at a temperature T. The initial configuration has an interaction energy $E^{(0)}$. Successive tentative configurations produced by some move, after passing the self-avoidance test are subject to a further acceptance test based on the effect of the move on the energy of the walk. A move is accepted with probability $\min[1, e^{-\beta\Delta E}]$, where ΔE is the energy difference between the tentative configuration and the existing configuration.

For a given value of ϵ, the effect of the attraction between monomers is to counteract the expansive tendency due to the excluded volume, with the former becoming less and less important as the temperature increases. As the temperature is decreased however, a stage is reached where the van der Waals attraction just compensates for the repulsion induced by the excluded volume effect, so that the average length of the molecule, and its other properties, are very much the same as those for a pure, Markovian, random walk, e.g. $\nu = 1/2$. The critical temperature at which this occurs is known as the Θ−temperature; it is an important parameter characterising the region over which a 'collapse' transition occurs, separating the relatively mobile chain from a more compact object. It is interesting that the method of inverse restricted sampling discussed above but *without* including the weights for survival bias basically samples the Boltzmann distribution, equation 7.4, at an effective temperature which is very close to the $\Theta-$temperature.

In the following chapter we will see how more general random walk techniques can be used in the study of very complex physical systems.

Notes

1. Chandrasekhar, S. (1943), *Stochastic Problems in Physics and Astronomy*, Rev. Mod. Phys., **15**, 1-89, is the classic reference for random walks in physics.

2. Einstein's work on Brownian motion may be found in Einstein, A., *Investigation on the Theory of Brownian Motion*, (London: Methuen, 1956).

3. CTRW are described in Montroll, E. W. and M. F. Schlesinger, *On the Wonderful World of Random Walks*, in J. L. Lebowitz and E. W. Montroll (Eds.), *Nonequilibrium Phenomena II From Stochastics to Hydrodynamics* (Amsterdam: North Holland, 1984). They also discuss walks where the steps Δ and τ are coupled and their distribution cannot be factored as here.

4. Many suggestions to optimise the step size in the random walk have been proposed, see Shreider (1964), and Shoichiro Nakamura, *Computational methods in Engineering and Science: with applications to fluid dynamics and nuclear systems* (New York: Wiley, 1977).

5. First passage time distributions are described in M. C. Wang and G. E. Uhlenbeck (1945), *On the Theory of Brownian Motion II*, Rev. Mod. Phys., **17**, 323-42. and in more detail in Siegert, A. J. F. (1951). *On the First Passage Time Probability Problem*, Phys. Rev., **81**, 617-23.

6. A lot has been written in recent years on the theory and simulation of anomalous diffusion; notable for its relevance to physical phenomena is the very readable article *Time-Scale Invariance in Transport and Relaxation* by H. Scher, M. F. Shlesinger and J. T. Bender in Physics Today, Jan. 1991, 26-34. For an interesting discussion of the relevance of Levy flight to the intermixing of powders, e.g. in the manufacture of pharmaceutical compounds, see chapter 7 of Bryan H. Kaye, *A Random Walk Through Fractal Dimensions* (New York: VCH Publishers, 1989).

7. See, for example, section 9.4.3 of N. Madras and G. Slade, *The Self-Avoiding Walk* (Boston: Birkhauser, 1993).

8. See, for example, K. Kremer, A. Baumgartner and K. Binder (1981), *Monte Carlo Renormalization of Hard Sphere Polymer Chains in Two to Five Dimensions*, Z. Physik B - Condensed Matter, **40**, 331-41.

Exercises

8.1 A-one dimensional walk quite different from any described in this chapter is to imagine equal steps to left or right as introducing a unit resistor, so that a segment which has been traversed n times has a resistance $1/n$. Carry out a simulation to find the distribution of the resistance between the begining and end points of a walk of N steps, and the average of this distribution. Investigate this average as a function of N for $N < 500$.

Repeat the above for the resistance between the extreme ends of N step walks. (For applications of this, see J. R. Banavar, A. B. Harris and J. Koplick, Phys. Rev. Lett., **51**, 1115-18, 1983.)

8.2 Starting from a specified $x(0)$ the position of a walker at time t is given by

$$x(t) = x(t-1) + \eta\delta$$

where η is a random variate from a standardised Gaussian.

(a) Take $\delta = 0.2$ and $x(0) = 0.01$ and plot such a walk for 500 steps.

A trajectory which is completely deterministic — no random numbers involved is

$$x(t) = x(t-1) - \mu \sin(2\pi x(t-1)).$$

(b) Take $\mu = 0.732, x(0) = 0.01$ and present a walk, as in (a).

(c) Repeat (b) but with $\mu = 0.750$.

The similarity between case (c), a regime of deterministic chaos, and (a) suggests that the phenomenon of chaos may a possible source of random numbers, a subject which has still not been fully explored — but touched on in exercise 3.6.

8.3 Verify by direct substitution that equation 8.3.4 is a solution of the equation 8.3.2.

8.4 Show that equation 8.3.7 follows from taking $T_- = T_+ = \frac{1}{2}(1 - \mu\Delta t)$.

8.5 In the simulation of polymer chains at a finite temperature the following distribution for the azimuthal angle φ of a bond may arise,

$$p(\varphi)d\varphi = Ke^{\lambda \cos\varphi} \sin \varphi d\varphi, \qquad 0 \leq |\ \varphi\ | \leq \pi,$$

where $\lambda = \beta V$ and V is an interaction energy. Show that the normalising constant is $K = \lambda/2\sinh\lambda$, and that samples of the variable can be selected as

$$\varphi = \pm\cos^{-1}\{1 + \lambda^{-1}\ln[1 - \xi(1 - e^{-2\lambda})]\}.$$

8.6 Show that for a simple random walk in three dimensions the mean square length $< R^2 >$ and mean square radius of gyration, $< S^2 >$ defined by equation 8.7.2 are related by $< R^2 > / < S^2 > = 6$.

8.7 Develop a programme to estimate the reciprocal of the attrition constant, κ, defined by equation 8.7.4 for (a) a two-dimensional square lattice, and (b) for a simple cubic lattice.

Chapter 9

PARTICLE TRANSPORT

> The original method of programming might be likened to a giant plugboard, that is to say, to a can of worms.
>
> N. Metropolis

9.1 Introduction

Particle transport is the field which gave rise to much of the modern Monte Carlo paradigm, and it was in this field that many of the earliest developments in computing evolved — as Metropolis reminds us — and in which many of the 'tricks of the trade' evolved.[1] It is still the field where it finds greatest application, having the advantages of being relatively easy to implement and of giving good insight into the evolution of the system under study, especially as regards fluctuations. The problems can be of great variety; on the grand scale it may concern the design of a shield for a nuclear reactor and, at the other extreme, the determination of expected properties of a certain configuration of semiconductor components in an electronic device. In both of these cases a pilot study through direct experimental exploration would be very costly in time and money, the former for obvious reasons of size and the latter because of the many *a priori* plausible material combinations and the exacting conditions of purity and cleanliness under which such studies must be carried out. Results of simulation can be very valuable in these design studies. Resort also may be had to Monte Carlo methods because of great complexity in the underlying physics in the problem, or because of geometrical complexity, e.g. inhomogeneity, in the system being studied. Examples of topics where the methods are used include studies of transport of solar radiation in a cloud field, which can be very inhomogeneous, or of the albedo radiation from a highly irregular surface like that of the ocean, radiative transport in astrophysics or in laboratory fusion, radiation damage to an environment (e.g. spacecraft shielding), radiological dose calculation, charge transport in semiconductor devices, etc. As examples we describe four concrete problems of contemporary application.[2]

When a high energy photon, electron or hadron, typically with energy in the 100s of GeV, is incident on a large mass of sufficient thickness, most of its energy (i.e. excluding that part converted into neutrinos) appears as excitation or ionisation in the medium, deposited there by the myriad of secondary particles produced in the cascade resulting from the collision. By interleaving the medium with detectors such as plastic scintillators and sampling the energy deposited at several levels in the medium, with suitable correction for any energy escaping, a reasonable estimate of the energy of the primary particle can be made. Instruments based on these principles are known as ionisation calorimeters; calorimetry of this type has become an important tool in experiments in high energy physics. For the design and calibration of such devices, very extensive Monte Carlo simulations of their response to different primary particles will be required.

When an energetic primary cosmic ray (nucleus, photon or electron) is incident on the earth's atmosphere a cascade of particles and electromagnetic radiation of great complexity results. These cascades, 'extensive air showers' (EAS) contain information in several different channels — radio waves, optical (Cerenkov and fluorescence radiation), gamma rays, leptons and hadrons — and the only practical way to predict the characteristics of these components, especially their variability, is through Monte Carlo simulation.

The design of a barrier to act as a shield from a beam of energetic γ-rays or other energetic particles can be carried out by simulating the response of materials of different thicknesses when irradiated with such particles — we shall focus on this problem in our description of the general problem.

The most famous problem in this category is the behaviour of a population of neutrons in a nuclear reactor assembly. The competition between fission, escape and absorption will determine the state of the system, whether it be runaway, decay or criticality. We shall describe this problem in greater detail later in the chapter.

The quantity sought, the 'score', in each case will be quite different, e.g. the energy deposited by ionisation at different depths in the calorimeter, maybe the intensity distribution of Cerenkov light at ground level in the EAS, the average number of particles which escape per incident particle in the shielding problem, and the energy released per unit time in the reactor. It might even refer to the transporting particles themselves, such as their drift velocity or diffusion coefficient. Notwithstanding these differences, the solutions have much in common. In particle transport we model the dynamics of individual quanta and particles in the system. There are various levels

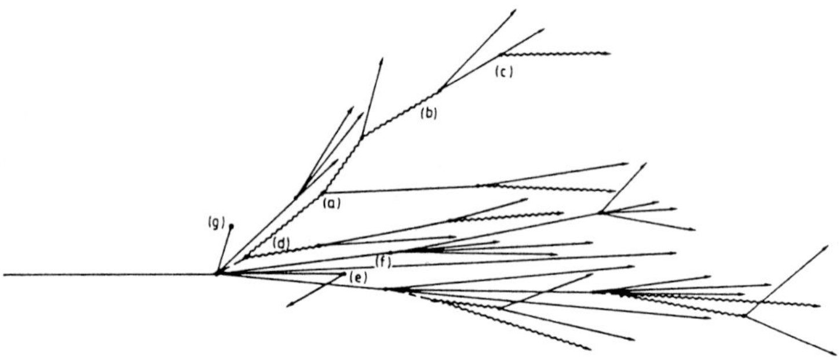

Figure 9.1: A schematic diagram of a cascade produced by a nucleon-nucleus collision at high energies. The nodes represent possible processes: (a) Compton scattering, (b) pair production, (c) electronic Bremsstrahlung, (d) π^0 decay, (e) charged pion decay — $\pi \rightarrow \mu\nu$, (f) charged pion interaction, and (g) negative pion absorption (after MacKeown and Newman, 1987).

of complexity, depending, for example, on whether the coupling between the particles in question can be ignored — failing which simultaneous simulation of a whole ensemble of particles is required — and whether a classical trajectory-like description is adequate, or quantum motion must be imposed from the very beginning. Neutron transport in a reactor would represent the simplest category, while an accurate simulation of electron transport in a miniature semiconductor device might require the more sophisticated treatment. Our discussion will deal mainly with the simpler case.

EXAMPLE 9.1: A particle cascade

The main features of these calculations can be seen in Fig. 9.1 where we show, schematically, a possible sequence of events when an energetic cosmic ray nucleon strikes the atmosphere. Different particles are created, each of which can undergo certain physical processes, e.g. photons may undergo Compton scattering on an electron in the material, process (a) in Fig. 9.1, or, if energetic enough, may give rise to an e^+e^- pair, process (b). Electrons may produce γ-rays by Bremsstrahlung, process (c). High energy hadrons in a collision will produce a multiplicity of secondary particles, e.g. π^o mesons, which decay rapidly into

$\gamma\gamma$, process (d), charged pions which decay to muons, process (e), or collide with nuclei to produce further secondaries, process (f). In the case of fission, the secondaries will be neutrons whose fate may be absorption, (which may also happen to low energy π^-, process (g)). Depending on the nature and the energies of the particles involved, many of these processes may be of negligible importance; several others, such as positron annihilation or photon absorption by the photoelectric effect, may occur but in any case the general 'tree-like' structure of the transport will remain.

9.2 Structure of the Calculation

9.2.1 General

All particle transport problems are essentially solutions of a Boltzmann- or Fokker-Planck-like equation, similar to that encountered in our discussion of random walks — they are just one great branching random walk.[3] The configuration space in which the walker moves will, in general, have more than just the three spatial dimensions, and so, apart from the location $\mathbf{r}$, additional labels will have to be attached to it, e.g. its type (electron, photon, pion, etc.), energy E, direction of motion (ω, *a* unit vector), etc. The positions of the nodes will be determined by stochastic transition probabilities, and their complexity by the detailed physics of the underlying processes involved. Carrying out an analogue simulation of the system has the merit over, say, finite difference methods, that it yields not only the average value of the solution based on representative samples of the stochastic population but can also give estimates of the variability in this population. In many cases such information will be of direct experimental interest.

In the simulation the positions of the nodes on the tree are determined by some functions (cross-sections) of the parameters of the incoming particle and the composition of the medium in each case, and will have to be sampled from the appropriate distributions. Many other such samplings will have to be made at each node to determine the multiplicity and nature of the outgoing particles, their direction of motion, energy, etc. When more than one particle comes out from a node, i.e. an inelastic interaction, all but one must be stored, since only one can be followed at a time. To keep the storage requirements to a minimum it is advisable to process first the particle which is likely to produce the least complicated branch; in general it will be the secondary with lowest energy. Secondaries from an interaction should thus be stored such that they will be accessed in order of increasing energy.

In practice one distinguishes between hadronic cascades and electromagnetic cascades, although they are often physically interrelated, and for either case, between low energy particles (for hadrons, e.g. neutrons, this means kinetic energies ≤ 1 GeV while for electrons/photons it means energies ≤ 1 MeV) and high energy particles. Accurate simulation of the low energy particles is very demanding of effort; an electron of 1 GeV needs only about 20 significant collisions on average in water before its energy is decreased to a few MeV, a further thousand-fold drop in energy to a few keV will involve in excess of 10^4 scatters.[4] For problems which are sensitive to the details at such low energies, e.g. medical radiation dosimetry, it will almost certainly be more economical to use, or graft on, thoroughly tested library programmes for the calculations, e.g. the National Bureau of Standards' ETRAN programme or the EGS4 code developed by the Stanford Linear Accelerator Center.[5] Such programmes will depend on an extensive, updated, data base of cross-sections and angular distributions. Code for problems involving high energy particles can be developed without too great complications, with quasi-averaged low energy behaviour grafted on at later stages as necessary.

The stages of the calculation can be characterised as follows.

(i) Definition of the geometry of the problem and specification of the parameters of the source particle, if unique, or its sampling from a source distribution.

(ii) Specification of cross-sections, multiplicity distributions, energy spectra, etc.

(iii) Construction of sampling algorithms for the distributions, specification of the logical structure of the 'tree'. Displacement coordinates are best specified in a Cartesian frame, using direction cosines.

(iv) The incorporation of variance reduction techniques; we defer discussion of these until section 9.3.

(v) Decision on how the result is to be scored; a direct method will automatically suggest itself but, in conjunction with (iv), more subtle methods based on potential signal rather than analogue signal may be desirable.

We illustrate the basic approach by considering a photon of energy E_γ incident on a block of material, important, for example, in the design of a shield for a radiation source or in the estimation of radiation dosage in a radio biological simulation. Successive interactions of electrons and photons arising in the cascade are individually sampled and reported, as though observers were assigned to travel with each such particle. The basic techniques, variate sampling and variance reduction, have been encountered in previous chapters, although we will develop the latter aspect in greater detail in the context of particle transport.

For the primary photon the relevant processes are the photoelectric effect, Compton scattering and pair production. The probability of a particle interacting in a path dl may be written as $\mu_t dl$, where the total interaction coefficient μ_t may depend both on the parameters of the particle (energy, etc.) and the nature of the medium and, in general, has contributions from the different processes. In the present case,

$$\mu_t(E) = \mu_P(E) + \mu_C(E) + \mu_{pp}(E),$$

where the terms refer to the three processes mentioned. In general, $\mu_i(E) = n_i\sigma_i(E)$, where σ_i is the cross-section for the process and n_i is the density of relevant targets; in the case of Compton scattering, for example, $n_C = Z\rho N_A/\mathcal{M}$ where ρ and $\mathcal{M}$, respectively, are the density and molecular weight of the medium, Z the number of target electrons per atom and N_A is Avogadro's number. If depths are measured in terms of amount of matter (gcm^{-2}), the relevant quantity is μ/ρ in $cm^2\text{g}^{-1}$.

9.2.2 Differential Sampling

The simplest prescription which suggests itself for selecting an interaction point is to divide the medium into successive segments Δl and test for interaction on each segment, using a uniform random number ξ as follows:

(a) if $\mu_P\Delta l > \xi \quad \longrightarrow$ photo-electric,

else $(\mu_P + \mu_C)\Delta l > \xi \quad \longrightarrow$ Compton scattering,

else $(\mu_P + \mu_C + \mu_{pp})\Delta l > \xi \quad \longrightarrow$ pair production,

else $l \to l + \Delta l$, update particle parameters for any changes on

Δl, e.g. energy loss, call a new ξ and go to (a).

Apart from the truncation error involved in using this expression, this would be a very tedious and slow procedure; it can, fortunately, be superseded in most cases. Conditions where this is not so would be the presence of a 'substructure' which is not being explicitly sampled, e.g. the presence of multiple scattering or a magnetic field in the case of a charged particle.

9.2.3 Integral Sampling

Even in a medium of inhomogeneous composition we can sample tentative interaction points directly. The probability that a particle will have interacted on traversing a path l is one minus the product of probabilities of its surviving over all the path segments up to l, i.e.

$$\begin{aligned} P(l) &= 1 - \prod_i e^{-\Delta l_i/\lambda_i} \\ &= 1 - e^{-\sum \Delta l_i/\lambda_i}, \end{aligned}$$

where $\lambda_i (= 1/\mu_t(l_i))$, the interaction length, may depend on the segment, either because the medium (density of targets is not constant) or the parameters of the particle are changing (the latter for example through slowing down by energy loss to ionisation, or, in a semiconductor simulation, due to acceleration in an electric field). To select a distance to interaction, l, using the inversion method (see section 5.2.1) we write

$$P(l) = 1 - exp\{-\int_0^l dl'/\lambda(l')\} = \xi,$$

which yields for l the solution of the integral equation,

$$\int_0^l dl'/\lambda(l') = -\ln \xi. \tag{2.1}$$

Of course, for constant λ, $(= 1/\mu_t)$, it yields the familiar

$$l = -\lambda \ln \xi. \tag{2.2}$$

These results hold for the tentative occurrence of any particular interaction, with λ replaced by $\lambda_i (= 1/\mu_i)$, the earliest of which to occur when all are sampled giving the actual interaction point. If the motion between interactions is not rectilinear, e.g. motion of a charged particle in a magnetic field, the selection will have to be made from the corresponding distribution in time intervals between interactions and the relevant trajectory constructed from the dynamics.

EXAMPLE 9.2: Discontinuous medium

For stacked parallel plane slabs of material each characterised by a constant interaction length λ_i, solution of equation 2.1 corresponds to finding k, the smallest integer such that

$$\sum_{i=1}^{k} l_i/\lambda_i > -\ln \xi,$$

where l_i is the thickness of the ith slab along the line of travel. One then finds

$$l = -\lambda_k \ln \xi + \sum_{1}^{k-1} (1 - \lambda_k/\lambda_i) l_i.$$

EXAMPLE 9.3: Linear medium

For a linear variation of the interaction length, often a reasonable approximation,

$$\lambda(l) = \lambda_0 + bl,$$

equation 2.2 can be solved to yield,

$$l = \lambda_0(\xi^{-b} - 1)/b.$$

In general, however, it will not be so easy to solve equation 2.1, and to avoid this difficulty an elegant trick, the introduction of *self-scattering*, is sometimes used.[6] We introduce a constant, μ^*, with a value not less than the maximum value of $\mu_t(x)$ in the medium, and define an additional, fictitious, process, self-scattering, with a probability coefficient $\mu_0(x)$,

$$\mu_0(x) = \mu^* - \mu_t(x).$$

Occurrence of this process is a null event, the particle proceeds from the simulated interaction point with properties identical to those it had before the 'interaction'. The total probability coefficient for all interactions is then constant,

$$\mu^* = \mu_0 + \sum \mu_i,$$

and selection of the interaction point simply given by equation 2.2 with $\lambda = 1/\mu^*$; self-scattering being selected with probability μ_0/μ^*. The possibility of self-scattering will be the last interrogated since it requires evaluation of all the other terms in μ_t. Of course, effort is wasted in selecting these self-scattering events, and it should be kept to a minimum by selecting μ^* as close as possible to the maximum value of $\sum \mu_i$, but the effort of invoking it may be more than compensated for by not having to solve the integral equation. Self-scattering can also be of value in other circumstances, e.g. if it is required to know the extent to which a certain region where no interactions can occur (a void) is traversed by particles. In such a case the void can be replaced by material and all interactions occurring in it monitored and treated as self-scatters.

Assuming we are dealing with a source photon, using equation 2.1 or 2.2 an interaction node is selected; if it lies beyond the boundary of the medium suitable scoring is done and the particle is abandoned. Otherwise, the particular type of interaction is selected based on the relative probabilities, by dividing the unit interval (0, 1) into subintervals of length μ_i/μ_t (μ_i/μ^* if self-scattering is included) associated with each process, the one in which a uniform random number falls determining the outcome of the collision. It will be helpful to arrange this testing in order of processes of decreasing importance — if such a distinction is clear — especially if self-scattering is included, in which case the various μ_i will be newly computed. By way of example let us assume that a Compton scattering occurs — this is an important process in many practical problems, e.g. in the design of a satellite borne detector of cosmic X-rays, in which it plays a decisive role. The total cross-section for the process is

$$\sigma_C(\epsilon) = \frac{3\sigma_T}{4}\{\frac{2(1+\epsilon)^2}{\epsilon(1+2\epsilon)} - \frac{1+3\epsilon}{(1+2\epsilon)^2} - \frac{\epsilon+2}{2\epsilon^2}\ln(1+2\epsilon)\},$$

where ϵ is the energy of the incident photon in units of the electron rest mass, $\epsilon = E_\gamma/mc^2$, and $\sigma_T = \frac{8\pi}{3}(\frac{e^2}{mc^2})^2$ is the Thompson cross-section. The directions and energies of the outgoing electron and photon (see Fig. 9.2) have to be selected. The energy of the scattered photon as a fraction of the incident energy is denoted by $z = E'_\gamma/E_\gamma$ and the cosine of its angle of scattering by $q \equiv \cos\vartheta$. These two quantities are uniquely related through the Compton relation,

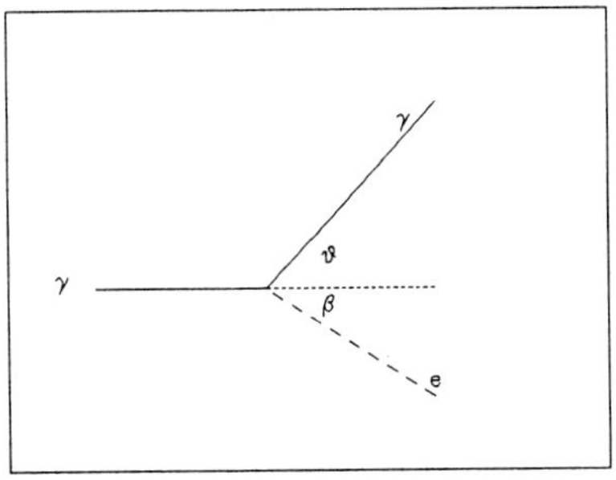

Figure 9.2: The geometry of Compton scattering by an electron.

$$z = \frac{1}{1+\epsilon(1-q)},$$

or

$$q(\epsilon, z) = 1 + \frac{1}{\epsilon} - \frac{1}{\epsilon z'}. \tag{2.3}$$

To select the scattered photon, we can do so either in terms of its scattering angle or its energy, and use equation 2.3 to find the other quantity. In terms of energy the probability distribution for the secondary photon is given by the Klein-Nishina formula as follows,

$$p(z)dz = K(\epsilon)\{z + z^{-1} + q^2(\epsilon, z) - 1\}dz, \qquad \frac{1}{2\epsilon+1} \leq z \leq 1, \tag{2.4}$$

where $K(\epsilon)$ is a normalising constant. This distribution can be sampled from using decomposition.

EXAMPLE 9.4: Klein-Nishina distribution

It was observed by Kahn that the distribution, equation 2.4, can be decomposed into a sum of two probability distributions, each of which can be readily sampled from using von Neumann's method, as follows

$$p(z) = K(\epsilon)\{\beta_1 Q_1(z)\pi_1(z) + \beta_2 Q_2(z)\pi_2(z)\},$$

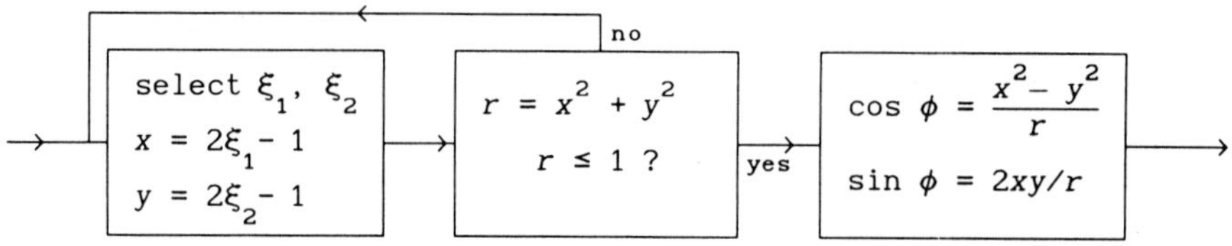

Figure 9.3: An algorithm to select the sine and cosine of a polar angle.

with

$$\beta_1 = (2\epsilon + 1)/(2\epsilon + 9), \beta_2 = 8/(2\epsilon + 9),$$

$$Q_1(z) = 4(z - z^2), Q_2(z) = (z + q^2)/2,$$

$$\begin{aligned} \pi_1(z)dz &= \frac{dz}{2\epsilon z^2} \qquad 1/(2\epsilon + 1) \le z \le 1, \\ \pi_2(z)dz &= \frac{2\epsilon + 1}{2\epsilon} dz \qquad 1/(2\epsilon + 1) \le z \le 1, \end{aligned}$$

The 'channel' to be selected from is chosen with probability β_i, and thereafter the value of z from π_i. The distributions π_1 or π_2 can be sampled from by inversion giving $z = 1/(2\epsilon\xi + 1)$ and $z = (2\epsilon\xi + 1)/(2\epsilon + 1)$, respectively. The azimuthal angle of the scattered photon φ, can be chosen directly as $2\pi\xi$. In practice, however, it is the sine and cosine of this angle which will be required, when forming direction cosines; these quantities can be generated directly.

EXAMPLE 9.5: Selecting geometric functions

The sine and cosine of a uniformly distributed polar angle, ϕ, can be selected by using a rather efficient rejection algorithm due to von Neumann, which is shown in the flowchart in Fig. 9.3. Points, uniformly distributed over the unit square, are generated and those which fall inside the unit circle only accepted. Because the annulus of area is proportional to $\Delta\phi$, these points select ϕ uniformly distributed on $[0, 2\pi]$; the final expressions follow from trigonometry. The method has an efficiency of $\pi/4$ and the merit of not calling any high level functions.

The energy and direction of the outgoing electron are determined by the kinematics, and are given by

$$E_e = (1 - z)E_\gamma,$$

and

$$\tan\beta = (1+\epsilon)^{-1}\text{cotan}(\vartheta/2).$$

If the γ-ray being processed is not the primary particle — assumed to define the z-axis, the two directions selected will then have to be converted to direction cosines in the reference coordinate frame. One of these particles will then be stored and the other followed. This must be continued until all particles have come to rest, or been absorbed, or escaped from the shield.

The straightforward way of scoring is to augment the score, S, by unity every time a particle, photon or electron, exits the shield. If, after tracking N incident photons through the shield the accumulated score is S, an estimate for T, the transmission coefficient is, trivially,

$$T = S/N.$$

Although formally a straightforward calculation, an investigation of the uncertainty in this answer will reveal the impracticality of such an approach. Let us assume that a standard deviation uncertainty of 10% would be acceptable in the estimated value of T. The number S is a variate from a Binomial distribution (section 2.4.1) where the number of trials is N and the probability of success is T. Thus the variance on it is

$$\sigma_S^2 = NT(1-T).$$

which, since $T \ll 1$

$$\simeq NT.$$

Or, using our estimate for T, we have $\sigma_S^2 = S$, in other words S is Poisson distributed. Thus the variance on T is

$$\sigma_T^2 = N^{-2}\sigma_S^2$$
$$= T/N.$$

This N^{-1} dependence of the variance on the sought for solution is universal in the type of analogue simulation we have been describing and, as we shall see, will often impose severe constraints on the practical solution of these problems. We then have the standard error on our estimate of the transmission coefficient,

$$\sigma_T = \sqrt{S}/N.$$

By our requirement that this not exceed $T/10$, we have

$$\sqrt{S}/N < S/10N,$$

or, $S > 100$. To be useful a shield would typically attenuate to better than one part in 10^6; in such a case we would need to track $10^6 S$, i.e. of the order of 10^8 primary histories — and to improve the accuracy in T from 10% to 1% would require 100 times as many. Each of these in turn would involve tracking an enormous number of particles, the great majority of which, in contrast to a similar investigation of an ionisation calorimeter, make no contribution to the score S. Simulation of particle histories in the way described, where the particle configurations are image analogues of a physical situation, is called an analogue Monte Carlo calculation. We shall look at some of the ways in which the amount of information from a simulation can be increased by introducing variance reduction techniques, leading to so called non-analogue Monte Carlo, but before doing so return briefly to an alternative algorithm for event sampling.

9.2.4 Joint Sampling

Our description of event location and energy sampling may give a misleading impression of simplicity, which often does not hold in practice. Frequently the interaction probability for a process, $\mu(E)$, is not easily available in closed form, but presented in terms of a differential form — as we would have seen if pair production had been chosen for illustration above. For example, in terms of the energy, ϵ, of a particle emerging from an interaction we would have,

$$\mu(E) = \int \phi(E, \epsilon) d\epsilon,$$

with the differential spectrum ϕ specified. In other words we have an explicit expression for the joint probability of the particle interacting in dx at x and giving rise to an outgoing particle with energy in $\epsilon, d\epsilon$,

$$p(x, \epsilon) dx d\epsilon = e^{-\mu(E)x} \phi(E, \epsilon) dx d\epsilon.$$

An algorithm exists for sampling both x and ϵ from the joint distribution which we now describe. The un-normalised distribution $\phi(E, \epsilon)$ is selected from by combining the methods of superposition (section 5.2.4) and von Neumann's method. The function ϕ is written as

$$\phi(E,\epsilon)d\epsilon = \sum_k a_k(E)Q_k(\epsilon)\pi_k(\epsilon)d\epsilon, \qquad k \geq 1,$$

where any E dependence in the functions $Q, Q \leq 1$, and π have been suppressed for clarity. Except in the ideal case, where all $Q(\epsilon) \equiv 1$, we have on integration, and defining $\mu^*(E) \equiv \sum a_k(E)$,

$$\mu^*(E) > \mu(E).$$

The algorithm depends on using the explicitly known 'pseudo interaction' coefficient $\mu^*(E)$ to locate a tentative interaction point, and then making a decision on whether to accept it biased in such a way as to compensate for not using the correct interaction coefficient $\mu(E)$. The philosophy behind it is closely related to the method of self-scattering described in the previous section. The steps in the algorithm are as follows:[7]

(i) with the above decomposition of $\phi(E,\epsilon)$ form $\mu^*(E) = \sum a_k(E)$,

(ii) use this interaction coefficient to select a tentative interaction point, i.e. $x_t = -{\mu^*}^{-1} \ln \xi$,

(iii) with probability a_k/μ^* select channel k, and a variate ϵ_t from π_k,

(iv) if $Q(\epsilon_t) > \xi$, where ξ is another uniform variate, the selection is accepted, i.e. $x = x_t$ and $\epsilon = \epsilon_t$, otherwise,

(v) advance the particle by a distance x_t and return to step (ii).

The justification for the algorithm is as follows. It is clear that when a pass is accepted the energy is correctly sampled. As far as selection of the location is concerned, we note from section 5.2.4 that the average number of passes required to locate a true interaction is $\Sigma a_k/\mu = \mu^*/\mu$, thus the probability per pass is μ/μ^*. The probability of requiring exactly k passes is the probability of (k-1) unsuccessful attempts times this probability, i.e. $\mu/\mu^*(1-\mu/\mu^*)^{k-1}$. The probability of travelling a distance in x, dx between interactions is

$$\begin{aligned} p(x)dx &= \sum_{k=1}^{\infty}(\text{probability of requiring } k \text{ passes}) \times \\ &\qquad (\text{probability that the sum } \sum_{t=1}^{k} x_t \text{ lies in } x, dx) \\ &= \sum \mu/\mu^*(1-\mu/\mu^*)^{k-1} p_k(x)dx. \end{aligned}$$

However, x, being the sum of k variates independently drawn from the exponential distribution $\mu^* e^{-\mu * x}$ follows the gamma distribution, equation 2.4.10, i.e.

$$p_k(x)dx = \frac{\mu^{*k}}{(k-1)!} x^{k-1} e^{-\mu * x} dx.$$

Therefore, writing $j \equiv k-1$, we have

$$\begin{aligned} p(x)dx &= \mu e^{-\mu * x} \sum_{j=0}^{\infty} \frac{(\mu^* - \mu)^j}{j!} x^j dx \\ &= \mu e^{-\mu * x} e^{(\mu * - \mu)x} dx \\ &= \mu e^{-\mu x} dx, \end{aligned}$$

as required.

EXAMPLE 9.6: Pair production

In a medium with nuclear charge Z, pair production by a γ-ray of energy E_γ will result in one particle (e^- or e^+) with energy ϵ^+ and the other carrying the remainder. The interaction coefficient, using variables $\epsilon \equiv E_\gamma/mc^2$ and $z \equiv \epsilon^+/E_\gamma$, in units of inverse radiation length is given by

$$\mu_{pp}(\epsilon) = \int \phi(\epsilon, z)dz.$$

$\phi(\epsilon,z)$ can be decomposed as[7]

$$\phi(\epsilon, z) = \beta_1 Q_1(z)\pi_1(z) + \beta_2 Q_2(z)\pi_2(z),$$

where

$$\beta_1 = \frac{a(Z)}{3}\left(2 - \frac{b(Z)}{12}\right) \qquad \beta_2 = \frac{a(Z)}{9}\left(1 + \frac{b(Z)}{12}\right),$$

independent of γ-ray energy with

$$a(Z) = 1 - \frac{f_c}{\ln 183Z^{-1/3}} \qquad b(Z) = [\ln(183Z^{-1/3}) - f_c(Z)]^{-1}$$

$$f_c(Z) = (\alpha Z)^2 \sum_1^{\infty} \frac{1}{k(k^2 + (\alpha Z)^2)},$$

where α is the fine structure constant. The probability functions are

$$\pi_1(z)dz = dz, \qquad \pi_2(z)dz = 12(z - 0.5)^2 dz, \;\; 0 \le z \le 1,$$

which can be selected from as $z = \xi$ and $z = [1 + (2\xi - 1)^{1/3}]/2$, respectively, while the rejection functions $Q(z)$ are given by

$$Q_1(z) = 1 - \left(\frac{4}{b(Z)} - \frac{1}{6}\right)\begin{cases} [2.914\chi - 0.447\chi^2] & \chi \le 1 \\ [4.184\ln(\chi + 0.952) - 0.280] & \chi > 1 \end{cases}$$

$$Q_2(z) = 1 - \left(\frac{4}{b(Z)} + \frac{1}{3}\right)^{-1}\begin{cases} [3.898 - 0.9805\chi^2] & \chi \le 1 \\ [4.184\ln(\chi + 0.952) - 0.280] & \chi > 1 \end{cases}$$

These functions depend on a shielding parameter,

$$\chi(\epsilon, Z, z) \equiv 136Z^{-1/3}\frac{1}{\epsilon z(1 - z)},$$

and are valid when the Landau-Pomeranchuk effect is not important.

9.3 Variance Reduction

9.3.1 General

To whatever extent inter-event fluctuations may be of interest in a simulation, it is invariably the average behaviour, as in the previous case, which is of foremost concern. Its uncertainty should be as small as possible. The variance on the score can, of course, always be reduced by running more histories, however the figure of 10^8 in the example of the shield for only a 10% accuracy in the final result is a warning not to pin too great a hope on such a line of approach — the sample size on which the average must be estimated will be severely constrained by the speed of the computer and the amount of time available.

It is easy to observe a lack of discrimination in a Monte Carlo programme of the type described so far; often at least as much effort is devoted to stages of a history which are unimportant in determining the final score, as is devoted to the important parts of the cascade. In some sense the calculation is not as 'efficient' as possible. *Variance reduction* represents techniques aimed at improving this efficiency. A natural definition of efficiency suggests itself when we note that, in general, the variance on the score in an analogue Monte Carlo calculation is inversely proportional to the number of histories, N, used in making the estimation. As the time required for the calculation is directly proportional to N, we have the product

$$\text{variance} \times \text{time} \approx \text{constant}.$$

The smaller this product, the more efficient the calculation. A variance reduction technique is a prescription which will reduce this product; it may simply reduce the time, as in the case of *roulette* described in section 9.3.8 or reduce the variance as in the *splitting* technique, or reduce both. It is not always helpful to follow the history of every particle in the transport. The general principle behind the methods used usually involves a modification of one or more of the distributions sampled from in the simulation in such a way as to arrive at an unbiased estimate for the answer but with a reduced variance on it. To maintain the average score unbiased a compensating weighting of individual particles and histories will have to be introduced. This concept of a *weight* which every particle carries, and which, in general, will vary as the particle propagates, is central to the methods. When particles are produced in an interaction they must be assigned a weight which is the product of the weight of the incident particle with any weight peculiar to that interaction. For example, in the case of a thick shield, discussed earlier, most histories will result in no particle exiting at the end face, only the rare cases of one (and much rarer cases of more than one), exiting contributing to the score. Thus we seek some way of emphasising cases where particles, possibly in fractional numbers, could exit and contribute to the statistics on the score, but without, at the same time, introducing any bias into the estimate of the mean. As we will see further on, in a calculation in which the direct analogue of the physical process is abandoned in favour of one employing distorted distributions accompanied by compensating weights, it is possible to considerably increase the efficiency of a calculation. Some of these methods have surfaced at odd places in previous chapters; here we shall present a more complete, and more coherent, account in the context of particle transport.

9.3.2 Biased Source Sampling

Since the number of histories to be followed will usually be decided in advance, the source (primary) particles, if they derive from a distribution (e.g. an energy spectrum) may be systematically sampled. This will remove some of the sources of variability in the final result. This procedure of *stratified sampling* needs no introduction of weights, but, as with all variance reduction methods, distorts the fluctuations in the system. In addition, or alternatively, the incident particles may be selected from a source biased in favour of particles likely to contribute to the score. For example the particles might be selected from an energy spectrum $n^*(\epsilon)$ richer in high energy particles than the real spectrum $n(\epsilon)$, and a weight $w = n(\epsilon)/n^*(\epsilon)$ attached to the incident particle at the beginning of the simulation.

9.3.3 Potential Scoring

Potential scoring, or non-analogue scoring, in which no modifications are made to sampling distributions can be, in some problems, a valuable technique for variance reduction. There are two distinct schemes, *collision estimation* and *next event estimation.*

Collision Estimation

In conventional scoring of energy deposition by a particle of energy E on absorption in a region V_i, the score is increased as

$$S_i \longrightarrow S_i + E,$$

whenever such an absorption event is sampled. This will occur with a probability proportional to the absorption coefficient $\mu_a(E)$. In an actual history the number of such absorptions in the volume V_i may not be very many. On the other hand, many particles may have interacted in this volume without being absorbed. This total number of collisions can be used to estimate the energy deposition by assuming in every collision an energy deposited equal to E, scaled by the ratio of the probability of absorption to the total interaction probability. Thus, in every collision we augment the score according to

$$S_i \longrightarrow S_i + \frac{\mu_a(E)}{\mu_t^*(E)} E.$$

This is known as the collision estimator and because it is based on a much larger sample its variance will be smaller than in the analogue case.

Next Event Estimation

So called to distinguish it from analogue absorption, which is a *last event* estimator, this technique will be essential if the number of collisions in the volume region in a realisable simulation is itself very small, either because the volume is so small (e.g. a probe region) and/or very far away from the bulk of the cascade. Information on the response of a detector in such a region can still be obtained by basing an estimate on the collisions of particles in all other regions. If it is sufficient just for a particle to traverse a region to score in an analogue calculation, e.g. detection by ionisation loss in a probe, we can proceed as follows. At any node in the transport tree the potential contribution of every outgoing particle to the score in the region of interest, a distance R away at a direction ϑ, can be written down. It is

$$\Delta S = wp(\cos\vartheta)\Delta\Omega \exp\{-\int_0^R \mu_a dl'\},$$

where w is the weight of the outgoing particle, $p(\cos\vartheta)$ is the angular distribution of the outgoing particles, suitably normalised, $\Delta\Omega$ is the solid angle subtended by the region at the node, and the integral in the exponential is carried out along the line connecting the node to the region of interest.

By far the most important variance reduction methods depend on the notion of *importance sampling.* We now describe some of these methods.

9.3.4 Splitting

Probably the easiest of the variance reduction techniques to understand is that of splitting, originally due to von Neumann. At one or more depths in the shield penetration calculation each particle passing that level would be replaced in the simulation by ν particles, all having identical properties — energy, direction, etc. — to the incident particle but each assigned a weight equal to the weight of the incident particle multiplied by ν^{-1},

$$w_{\text{out}} = w_{\text{in}}\nu^{-1}.$$

This will obviously be helpful in promoting the penetration of the cascade, especially by a particle which survives with rather high energy to a great depth in the shield but which may still only have a relatively small probability of giving rise to a transmitted particle. The splitting technique can be incorporated in a systematic way into a calculation but obviously cannot be arbitrarily frequent. If, after a distance l, a ν-fold splitting is introduced, the fraction of events starting from there on average is $\nu e^{-\mu_a l}$, which, after another distance l becomes $\nu e^{-2\mu_a l}$. Splitting again gives $\nu^2 e^{-2\mu_a l}$ particles. If this is continued at j equally spaced intervals, we will have on average $(\nu e^{-\mu_a l})^j$ events to process. This average should not become significantly greater than unity in order not to introduce bias in favour of the particular beginning of this history, i.e. we require $\nu e^{-\mu_a l} \leq 1$. If the splitting is carried out after every absorption mean free path, i.e. $l = 1/\mu_a$, a suitable value for ν would be 2. Splitting criteria might also depend on location rather than energy; particles in the neighbourhood of an 'important' region — like a detector — might be processed in this way. Care must be taken when such weighted particles are scattered backwards across a splitting level to invert the procedure. The particle is annihilated with probability $1 - \nu^{-1}$, otherwise its returning weight is augmented by a factor ν.

9.3.5 Survival Biasing

A more sophisticated technique for achieving effects similar to splitting is that of *survival biasing*, also known as *implicit capture*. The possibility of selecting absorption as a choice for a particle in an interaction is suppressed. In every interaction only those processes which do not lead to absorption of the incoming particle are sampled, but all particles emerging from the interaction are assigned a weight w,

$$w = (1 - \mu_a(E)/\mu_t(E)),$$

i.e. $w_{\text{out}} = w w_{\text{in}}$ where w_{in} is the weight of the in-going particle, and μ_a and μ_t are the absorption and total interaction coefficients of the particle, respectively. For example, in the case of low energy photons μ_a might refer to the photoelectric effect. In this way all produced particles make a, possibly very small, contribution to the final score.

Both the survival biasing and splitting methods are generally applicable when events with small probability of occurrence are of interest, e.g. for particles in the tail of an energy distribution exceeding a threshold for some reaction. They also find application in other random walk type calculations, as in the simulation of non-intersecting polymer chains.

EXAMPLE 9.7: A toy shielding problem

To illustrate some of these ideas we consider the following one dimensional toy shielding problem. The absorption coefficient and total interaction coefficient for a particle in a medium are $\mu_a = 0.3$ and $\mu_T = 1.0$, respectively, and when absorption does not occur the particle continues from the interaction point in the same direction. For a shield of thickness, D, in units of μ_T^{-1}, we know the result for the transmission probability — it is simply $T = e^{-\mu_a D}$.

Taking $D = 30$ a direct simulation of the passage of $N = 20000$ particles gave two particles output, which using the methods described earlier gives a transmission coefficient $T = 1.00 \pm 0.71 10^{-4}$ — more helpfully the one standard deviation error can be replaced by the 68.3% fiducial limits as defined in section 2.4.1, to give the estimate $1.00^{+1.32}_{-0.65} 10^{-4}$ for T. Suppressing absorption and attaching the weight $(1 - \mu_a/\mu_T)$ gives all N particles out, their average weight gave $T = 1.27 \pm 0.034 10^{-4}$. Both values are, of course, consistent with the known value of $e^{-9} = 1.23 10^{-4}$. Using the number of interactions processed in each case, n_{int}, as a measure of the work done we find the 'efficiency' $(n_{\mathrm{int}} \sigma_T^2)^{-1} = 1.0 10^4$ and $1.2 10^5$, respectively for the two methods, reflecting an improvement of more than an order of magnitude from the introduction of survival biasing.

9.3.6 The Exponential Transformation

A generalisation of survival biasing, this technique ensures that particles have the opportunity to reveal themselves in regions (of phase space) where their presence is of interest, e.g. deep in a shield. By distorting not only the outcome at a node but also the interaction length, or interaction probability μ_t which locates it, we can choose where in the medium, on average, a particle interacts. In the analogue problem the probability of a particle interacting in dl after traversing a path length l is $\mu_t e^{-\mu_t l} dl$. Such an interaction might correspond to a certain amount of energy being deposited in dl. If we arbitrarily change the interaction probability to μ^*, we must ensure that such physical conditions are unaffected. This we do by attaching a weight w to the particle so that

$$w\mu^* e^{-\mu^* l} dl = \mu_t e^{-\mu_t l} dl,$$

i.e. the weight must be a function of the path length traversed, l, and is given by

$$w = w(l) = \frac{\mu_t}{\mu^*} e^{-(\mu_t - \mu^*)l} .$$

EXAMPLE 9.8: Weight verification

It is worthwhile to verify that the weight assignment gives the correct probability for exiting the medium. The weight builds up cumulatively. The probability of exiting the medium after a path L is not $w(L)e^{-\mu^* L} = {}^{\mu}t/\mu^* e^{-\mu_t L}$. We can write it as one minus the accumulated number of 'interactions', i.e.

$$1 - \int_0^L w(l) e^{-\mu^* l} \mu^* dl$$

$$= e^{-\mu_t L},$$

as it should be.

By making μ^* less than μ we can 'delay' the interaction and thus ensure greater penetration of the particles, as required earlier in the case of the shield, while by taking μ^* greater than μ_t we can force the interaction to occur early in the medium — something we might wish to do if studying surface effects in an irradiated medium. In a pure one-dimensional problem we could write $\mu^* = \mu_t - c$, where c could be a constant or it could be a function of the energy of the particle. A positive value for c — which, to avoid negative interaction probabilities must not exceed the minimum value of μ_t possible in the history — will correspond to a 'stretching out' of the cascade, while a negative value for c will correspond to early occurrence of the interactions. In either case the weight acquired by a particle in travelling a distance l is

$$w = \frac{e^{-cl}}{(1 - c/\mu_t)}.$$

In a problem like the evaluation of a shield it will be further helpful to bias the motion of particles towards directions normal to the faces of the medium, i.e. along the normal direction. A convenient way of doing this is to write $c = c_o \cos\vartheta$, where ϑ is the angle the particle track makes with the normal. In other words we simulate particles propagating with interaction length λ^*

$$\lambda^* \equiv 1/\mu^* = \frac{\lambda}{1 - c_0\lambda\cos\vartheta},$$

with acquired weight

$$\frac{e^{-c_0 l\cos\vartheta}}{(1 - c_0\lambda\cos\vartheta)}.$$

The particular case of $c = \mu_a$ corresponds to survival biasing but with the nodes located on the basis of $\mu^* = \mu_t - \mu_a$, rather than μ_t.

Special care must be taken with this technique. The term in the denominator of the weight can give rise to particles arriving in a region with weights of great variability, e.g. a particle arriving at a given point on the x-

axis without deviation from that axis direction and one which gets there after several scatters. This is not something guaranteed to reduce the variance. Combined with splitting and roulette in experienced hands, however, the exponential transform is an important element in particle transport codes.

9.3.7 Interaction Forcing

This is another technique to allow for the failure of interesting interactions to occur in an important region, and the consequent scanty and uneven flow of information from the region — a particular organ in an inhomogeneous human phantom for example. As with the exponential transform, this method consists in altering the distribution of the interaction points, in this case from the usual exponential to a truncated exponential, and weighting the outgoing particles from the interaction suitably; carefully employed, it can result in the variance on the final score being reduced. If a particle at a point is a distance d short of a boundary, beyond which interactions are not of interest, we can sample its interaction from the distribution

$$\begin{aligned} p(x)dx &= Ke^{-x/\lambda}dx \qquad & 0 \le x \le d, \\ &= 0 & x > \text{ d}. \end{aligned}$$

with $K^{-1} = \int_0^d e^{-x/\lambda}dx$. Thus we select the path as

$$x = -\lambda \ln(1 - \xi/K\lambda).$$

With this selection prescription the total probability of interacting in the distance d is unity, so to restore average behaviour we must attach a weight to out coming particles equal to the unforced probability of interacting in that region, i.e.

$$w = \int_0^d e^{-x/\lambda} dx/\lambda = 1/K\lambda.$$

This can be a valuable technique in high energy physics experiments in determining the response of a particle detector such as a gas proportional counter or thin plastic scintillator to γ-rays, most of which in an analogue simulation would ordinarily pass right through so that the response would be estimated on very poor statistics.

9.3.8 Roulette

Often known under the more colourful name of *Russian Roulette*, and sometimes as thin sampling, this technique is directed towards cutting back on the computation time in the simulation. It strives to reduce the work spent on branches in a history which *a priori* are unlikely to contribute much to the final score, e.g. numerous low energy particles in the early stages of a cascade in the shielding problem. It may also be required as an antidote to splitting, discussed earlier, from which a proliferation of particles, each eventually with extremely small weight, may occur. The goal is to have an algorithm in which the number of particles being processed at any stage of the calculation is proportional to the expected contribution of such particles to the final score.

A 'renormalisation' of the particle population is carried out at stages during the simulation. For example, at some level, particles are discarded from future tracking with a probability α, but each of the fraction $(1 - \alpha)$ which are retained continue with an assigned weight of $1/(1 - \alpha)$. The parameter α could be a function of energy, e.g. $\alpha = 0$ for $E > E_{\text{ref}}$, where E_{ref} is some reference energy, and $\alpha = 1 - E/E_{\text{ref}}$ for lower energies so that all high energy particles are followed but only a fraction of the, very numerous, low energy ones are. In the presence of splitting, the weight of a particle can be taken as the parameter on which to base roulette, to avoid wasting time following particles whose weight has become so small that they could only make an insignificant contribution to the score. A weight management scheme can be devised, e.g. a weight cut-off $w_{\rm min}$ can be defined. A suitable prescription is as follows, when a particle's weight w falls below this specified $w_{\rm min}$, compare $w/0.5$ with a variate ξ from the uniform distribution. If $\xi > w/0.5$ abandon the particle, otherwise continue it in the simulation with a weight equal to 0.5.

9.3.9 Correlated Sampling

It may be that the purpose of a simulation is to study the effect of a small perturbation to a system. This could be determined by making two simulations, with and without the effect taken into account, independently, and taking the difference. The variance on this result, is the sum of the variance on the two independent simulations, and may be relatively very large compared with the effect itself. Alternatively, we may be in a position where the solution of the unperturbed problem is known exactly, or with great accuracy; it would then seem to be advisable to try to make use of this information in finding the solution. An interesting example of how correlated sampling can greatly reduce the effort in minimising the value of a multidimensional integral has been noted in section 7.5. Correlated sampling means using the same set of random variables, as far as possible, to solve the two problems, yielding results x and y, say, respectively. Then although the variance on each separate calculation will not be affected, because of the approximate cancellation of random contributions — 'positive correlation' — or, formally from the relation for the variance of a quantity $z = x - y$,

$$\mathrm{var}(z) = \mathrm{var}(x) + \mathrm{var}(y) \text{ -2 } \mathrm{cov}(x,y),$$

the variance on the difference can be much smaller than the 'macro-variances'. Examples of problems where such positive correlation can be identified include the case of the photon incident on a shield. The solution for the average number of particles crossing any level in a semi-infinite medium can be found analytically, though not easily. The solution of the shielding problem differs from this only in that it does not have contributions from particles scattered back from beyond the depth of the shield, as may occur in the semi-infinite case. This analytical solution can serve as a valuable control variate in the shielding calculation.

There are two other, very interesting and much written about techniques of variance reduction, *adjoint biasing* and *antithetic (anticorellated) sampling.* In simple one-dimensional examples these can be seen to be very powerful. Unfortunately, implementing them when the phase space has more than one dimension is usually very difficult — although we note that anticorellated sampling has been used in simulating charge scattering in semiconductors as a way of cancelling the effects of thermal fluctuations when estimating the drift velocity — and so, though in principle very powerful, they find little application. We shall not discuss them further, but refer the reader to Hammersley and Handscomb.

The degree of variance reduction introduced in a simulation must reflect a compromise between demands on computing time, complexity of programming and acceptable limitations on output information. Obviously all the methods described, by definition, distort the variance in the calculation, and will be very difficult to impossible to use if the degree of fluctuation in the transport is of interest, as it often will be since it is an aspect for which Monte Carlo methods are almost the only way of arriving at reliable information. Improvement of efficiency in such calculations is a much more challenging task.

9.4 Semiconductor Device Simulation

The nature of the problems here are quite different from the example discussed in section 9.3, e.g. a study of the current pattern in a device in constant or time varying fields, the characteristics of the signal from a semiconductor detector for ionising radiation, or the effects of radioactivity or cosmic rays in causing rare disruptions in the operation of semiconductor devices. Frequently, in these cases, it is enough to model a steady state. This can often be done based on a single particle's history followed for a long time, rather than using the more time consuming ensemble simulation necessary in other problems. The transported particles are electrons or holes, with associated effective masses, while the principal interaction mechanisms involved are scattering by phonons (acoustic and optical), and by ionised impurities. The presence of an electric field means that the energy of the carriers is continuously changing with time, and it is more direct to use time rather than traversed distance as the variable in locating the interaction nodes. The distribution is

$$p(t)dt \quad = \quad \mu_T \frac{dl}{dt} \exp\{-\int_0^t \mu_T dl/dt'dt'\}dt$$

$$= \Gamma(\epsilon(t)) \exp\{-\int_0^t \Gamma(\epsilon(t'))dt'\}dt,$$

where the interaction rate $\Gamma \equiv \mu_T(\epsilon)v(\epsilon)$, v is the particle's velocity and ϵ its energy. The selection of interactions by solution of the corresponding integral equation 2.1 is prohibitively difficult in general, selection using the method of self-scattering is usually used. A pseudo interaction rate Γ^* is used, which must be larger than the largest actual value at any point on the trajectory, but, for reasons of efficiency, should differ from this largest value by as little as possible. The value of Γ^* does not have to be constant, and its value can be optimised by an iterative procedure. The particle's path is then calculated over the selected time segment, and further selection made to determine its scattering. Depending on what quantity is of interest the simulations can be of great complexity, e.g. for consistency it is usually necessary to solve Poisson's equation in parallel with the simulation at regular intervals to update the values of the electric field; full expositions can be found in research monographs.[8]

Except when dealing with very small systems (dimensions $\leq 1\mu m$), where quantum transport theory may be required, the methods of attack are generally the same as those outlined so far. In cases where quantum effects are important, even though the central concept of a trajectory — location and momentum — is excluded, Monte Carlo methods may still be useful. In this case the description in terms of discrete points in phase space is replaced by a more general probability density matrix, with a considerable increase in the complexity of the formalism.

9.5 Reactor Criticality

In nuclear reactor design two topics are especially amenable to analysis by Monte Carlo methods, shielding from the neutron and gamma ray fluxes, and calculation of the rate of energy release from fission in the assembly. The methodology and many of the techniques for the former problem have formed the bulk of this chapter so far; we now look briefly at the latter, the central problem in the field. Neutrons are the particles of interest here, and we are interested in two things, the conditions under which their distribution will assume a steady state, e.g. the configuration of control rods which correspond to a critical assembly, and the actual neutron population in the reactor under these conditions — this is known as the *criticality* problem. A reactor is composed of three types of materials, absorbing, moderating, i.e. causing neutrons to slow down, and fissionable material. The rate of fissions at any time, and thus the energy release, will depend on the instantaneous density of neutrons throughout the assembly and their energy spectrum, $\rho(\mathbf{r}, E, t)$, where $\rho(\mathbf{r}, E, t)d^3\mathbf{r}dE$ is the number of neutrons at t with energies in E, dE in the element of volume at $\mathbf{r}$. Competition between absorption, escape from the assembly and multiplication of neutrons in fissions causes this neutron population to evolve with time. If, independent of the distribution at an initial time — which might just be a single neutron from a spontaneous fission or an injected start-up pulse — the population approaches a time independent steady state $\rho_c(\mathbf{r}, E)$ we talk of the critical state, and it is the energy release in this state as well as its maintenance which is of interest. The principal processes determining the transport of neutrons are (i) elastic scattering, (ii) inelastic scattering (nuclear excitation, possibly followed by deexcitation by fission), (iii) nuclear capture, and (iv) fission. Assuming the physics of all these processes is known an ensemble of neutrons can be simulated in a specified array configuration geometry and the spatial density of neutrons monitored as a function of time. It is usually not practical to keep track of the growth or decay of the whole neutron population in the assembly. To understand how the scoring is done it is helpful to look at the underlying theory of the critical state.

The theoretical development is most easily done using the directional flux density $\psi(\mathbf{r}, E, \omega, t)$ rather than the number density, where, formally,

$$\psi(\mathbf{r}, E, \omega, t) \equiv v d\rho(\mathbf{r}, E, t)/d\Omega,$$

where $v = (2E/m_n)^{1/2}$ is the neutron velocity, directed along a direction specified by the unit directional vector ω with components $(\sin\vartheta\cos\phi$, $\sin\vartheta\sin\phi, \cos\vartheta)$ and $d\Omega = \sin\vartheta d\vartheta d\phi$ is an element of solid angle. The function ψ satisfies the Boltzmann transport equation[9] which can be written, when spontaneous fissions are negligible, in operator form

$$\frac{\partial\psi}{\partial t} = \mathbf{L}\psi.$$

Here $\mathbf{L}$ is a time independent, linear integro-differential operator which accounts for neutron scattering, absorption and fission, allowing for the boundary conditions on the system. This equation has solutions of the form $\psi(\mathbf{r}, E, \omega, t) = \psi_k(\mathbf{r}, E, \omega)e^{\lambda_k t}$, where the functions ψ_k are eigen-function of the operator $\mathbf{L}$ including boundary conditions, i.e.

$$\mathbf{L}\psi_k(\mathbf{r}, E, \omega) = \lambda_k\psi_k(\mathbf{r}, E, \omega),$$

and λ_k are the corresponding eigen-values. For a given initial distribution

$\psi(\mathbf{r}, E, \omega, t = 0)$ the solution at later times can be expanded in the eigen-functions ψ_k, i.e.

$$\psi(\mathbf{r}, E, \omega, t) = \sum_k c_k \psi_k(\mathbf{r}, E, \omega) e^{\lambda_k t} \,,$$

with the coefficients c_k determined by the initial state. If we denote by λ_l the largest of the eigen-values of $\mathbf{L}$,[10] we can write

$$\psi(\mathbf{r}, E, \omega, t) = c_l \psi_l(\mathbf{r}, E, \omega) e^{\lambda_1 t} \{1 + \sum_{k \neq 1} \frac{c_k \psi_k}{c_1 \psi_1} e^{-(\lambda_l - \lambda_k)t}\}.$$

As t tends to infinity all the terms in the sum tend to zero and

$$\psi(\mathbf{r}, E, \omega, t) \to c_l \psi_l(\mathbf{r}, E, \omega) e^{\lambda_1 t} \,.$$

We see that the long time solution is a distribution in space, energy and direction of motion, modulated by an exponential term which will cause the distribution to explode, maintain itself or decline depending on whether the eigen-value λ_l is positive, zero or negative. Criticality corresponds to $\lambda_l = 0$, and a population of neutrons distributed as $\psi_l(\mathbf{r}, E, \omega)$ will maintain itself — as long as there is no significant burn up of the fissile material, and, of course, subject to statistical fluctuations — and cause energy to be released at a steady rate through fission.

Since the result is independent of the initial neutron distribution a Monte Carlo simulation of the evolution of a population of neutrons, in a particular assembly and composition, using an arbitrary starting distribution can be used to estimate the largest eigen-value. This can be repeated for different configurations to arrive at the one corresponding to criticality. As time progresses the distribution in space and energy will approach a function proportional to $\psi_l(\mathbf{r}, E, \omega)$. When this steady state is established the steady rate of change in population size, or more usually the total sum of weights, can be estimated from their value at two times separated by a time τ, yielding

$$\lambda_l \simeq \frac{1}{\tau} \ln\{\sum_j w_j(t+\tau)/\sum_j w_j(t)\}.$$

Clearly the speed with which this limiting behaviour will be attained will depend on how close the initial distribution is to the steady state one, and the accuracy of the estimation needs to improve the closer the determined value of λ_1 approaches to zero. This method has been used to determine criticality,[11] and we will see its use in chapter 11 in solving the eigen-value problem arising from Schrodinger's equation, but the description here is of greater value as exposition of the underlying theory than as a practical prescription for solving the criticality problem, for which alternative approaches are available.

The most important of these approaches centers round the differentiation of the neutron population by families or generations. The zeroth generation is understood as the initial assumed distribution, an ensemble of n_0 particles in total. All neutrons arising from fissions (not scattering) induced by neutrons from the initial distribution, a total n_1 in number, form the first generation, and so on, the kth generation consisting of $n_k \geq 0$ particles. Although, apart from the zeroth distribution, the full population of any generation does not, in general, coexist simultaneously in the system because of the different neutron velocities and the stochastic nature of the interactions, the idea of generation is conceptually well defined, and can be easily monitored in a Monte Carlo simulation. A *multiplication constant*, k, is defined; this is the ratio of the average number of neutrons in a generation to the number in the previous generation,

$$k = n_{i+1}/n_i, \tag{5.1}$$

and in the steady state it assumes the value unity. Generation number rather than time is taken as the parameter describing evolution, and the growth parameters in the two modes of description are related, $k(\lambda_l)$, such that $k(0) = 1$. As in the time domain simulation, an arbitrary spatial and energy distribution of neutrons, the zeroth generation, is assumed and their transport in the assembly followed to produce a first generation family, having a different spatial and energy distribution. If this is continued, a stage will be reached where the ratio of the population size in successive generations, will reach a steady value. After this settling down has occurred, say at the jth generation, a weighted estimate of the multiplication constant on the basis of s further generations can be made,

$$k^* = \frac{\sum_{i=j}^{j+s-1} n_i k_i}{\sum_{i=j}^{j+s-1} n_i} = \frac{M - n_j}{M - n_{j+s}}, \tag{5.2}$$

where $M = \sum_{i=j}^{j+s} n_i$, and the standard error is $\sqrt{\frac{(M-n_j)(M+n_j-2n_{j+s})}{2(M-n_{j+s})^3}}$.

Although the prescription just described is a very straightforward one, the actual simulation of real reactor assemblies is a formidable project, largely because of the great variation with energy of the fission cross-section. The energy of neutrons produced in fission is typically 10^6 eV, whereas the cross-section for fission is greatest for neutrons with thermal energies, less than a tenth of an electron volt. To bring the neutrons down to these energies a moderating material, typically graphite, is used. The many collisions, of the order of 100, with the graphite nuclei required to thermalise the neutrons must be simulated. The thermal motions of these nuclei also cannot always be ignored. In addition the many resonances in the fission cross-section means that it goes through extremely rapid change over very small energy intervals, and can only be usefully described in tabulated form. We will take for illustration an example shorn of these many complications, the case of a fast assembly containing only uranium.

EXAMPLE 9.8: Enrichment of natural uranium

Natural uranium consists of the two fissionable isotopes, ^{235}U and ^{238}U, with a ratio by mass, $m_{235}/(m_{235}+m_{238}) = 0.0072$. When there is no moderator present most fissions will be produced by energetic neutrons, however in this case no mass of uranium will go critical. Here we investigate the degree of enrichment of ^{235}U necessary in a uranium sphere of radius R for it to go critical. We will work with the number density $f = N_{235}/N_U$. The number density of uranium nuclei is $N_U = 4.810^{22} cm^{-3}$. The average number of neutrons emitted in the fast fission of ^{235}U and ^{238}U are 2.50 and 2.66, respectively. They have an average energy of about 2 MeV, but with a spectrum which extends to beyond 10 MeV — selection from it has been considered in example 5.18. For simplicity, however, we will reduce the problem from three variables to one, the neutron positions, by averaging over energies and ignoring neutron scattering. The cross- section for absorption and fission, σ_a and σ_f, depend on energy, their values, in barns (1 barn = 10^{-28} m^2), for the two isotopes, weighted according to the fission spectrum are

$$\sigma_a(235) = 0.108 \qquad \sigma_f(235) = 1.25$$
$$\sigma_a(238) = 0.080 \qquad \sigma_f(238) = 0.30.$$

More rigorously, this weighting should be done using the actual neutron energy spectrum existing when the steady state distribution obtains, this can be done by iteration if the energy dependences of the cross-sections is available. The number of neutrons, ν, emitted in a fission is a stochastic quantity and can vary from zero to more than six. They can be assumed emitted isotropically and the probability distribution for ν from ^{235}U can be assumed which is as follows,

ν	0	1	2	3	4	5	6
$p(\nu)$	0.027	0.158	0.339	0.302	0.130	0.034	0.010

We begin with an initial distribution of n_0 neutrons — all starting at the origin, or uniformly distributed throughout the sphere with coordinates $r_i = R\xi_i^{1/3}$, are possibilities. When scattering is not included the only angle required is that between the neutron velocity and the radial direction at the point ϑ. Both the initial generation, and all neutrons arising from fission, will be assumed isotropically distributed, this corresponds to $\mu \equiv \cos\vartheta$ uniformly distributed on [-1,1], i.e. $\mu_i = 2\xi_i - 1$. This completes specification of the zeroth generation. Each neutron in the generation is processed as follows. The distance to interaction is exponentially distributed with interaction length $\lambda = (N_U f \sigma_t(235) + N_U(1-f)\sigma_t(238))^{-1}$, where σ_t denotes total cross-section, and can be selected as $s = -\lambda \ln \xi$. The radial coordinate of the interaction point is $r' = (r^2 + s^2 + 2\mu r s)^{1/2}$. If $r' > R$ the neutron is abandoned and the next, if any, from the generation sent for processing. If $r' < R$ and the probability of absorption, $(f\sigma_a(235)+(1-f)\sigma_a(238))/(f\sigma_t(235)+(1-f)\sigma_t(238))$ is greater than a random number the neutron is also treated in this way, otherwise a number of fission neutrons is selected from the distribution $p(\nu)$, each of them assigned the position r' and a randomly selected value of μ, and these quantities stored in the next generation array. When all the neutrons of a generation are processed in this way the number in the new generation is used to calculate the score, either the simple ratio, equation 5.1, or the cumulative ratio, equation 5.2. If sufficient generations have been processed that an equilibrium spatial distribution has evolved this score will be a measure of criticality.

Starting 100 neutrons at the origin of a sphere of radius 15 cm with a value of enrichment $f = 0.77$, the results in Fig. 9.4(a) show three examples, corresponding to different random number sequences, of the number of neutrons as a function of generation. The very different behaviour — the neutrons may all have disappeared or have multiplied uncontrollably before attaining a steady distribution — illustrates the difficulty arising in a calculation which is very simple in principle. The cumulative estimate k^* of criticality for a history which survived for many generations is shown in Fig. 9.4(b). Because the size of successive generations may be very much smaller, or larger than n_0, a *population-stabilising* algorithm must be introduced. We can do this by renormalising each generation's population to n_0 before processing it. This can be done by randomly choosing n_0 neutrons from among

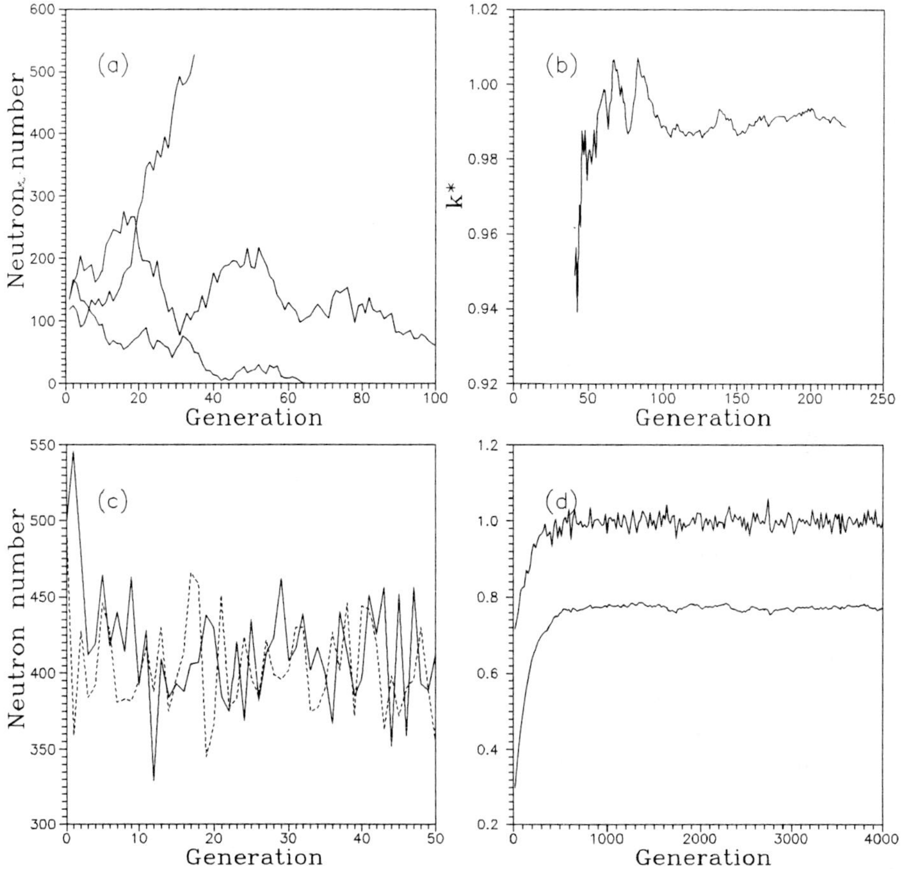

Figure 9.4: (a) Three samples of the simulated time evolution of the neutron population, initially equal to 100 at the origin of a sphere of radius 15 cm, showing the stochastic nature of the phenomenon. (b) The cumulative criticality estimate k^* derived from a history which survived for many generations. (c) The renormalised population as a function of generation number, for the case of enrichment factor $f = 0.5$; the solid curve corresponds to all the particles initially at the origin, and the dashed curve to the case where they were uniformly distributed throughout the sphere at the beginning. (d) Convergence to the critical composition, lower curve, and the criticality factor, upper curve.

those of the new generation if it already contains more than n_0, to form the new generation, or, if it contains less than n_0, duplicating as many as necessary, randomly selected from among the population to make up the deficit. The cumulative estimate, equation 5.2, cannot be used in this case. Fig. 9.4(c) shows, for $f = 0.5$, how the number in successive generations behaves when renormalised in this way to $n_0 = 400$. Also seen there is how the effects of the initial distribution rapidly die out, the solid and dashed curves refer to the initial particles concentrated at the origin, and uniformly distributed throughout the sphere, respectively.

The critical value of concentration can be converged on by renormalising in this way and altering f in a feedback manner during the progress of the simulation; an initial value is guessed and a total of n_0 processed through one generation. If the resulting number of neutrons n_i is greater (less) than n_0, we know that the value of f chosen was too large (small). The value of f is then decreased (increased) by some increment, e.g. $f \to f + 0.1(n_0 - n_i)/n_0$. In practice it is better to do this revision using an accumulated excess/deficit over a number of generations rather than after every one, 20 were used in the present case. Starting with $f = 0.3$ and taking $n_0 = 400$, the progress to criticality, k, and to the critical value of f are shown in the upper and lower curves in Fig. 9.4(d). The value of f obtained, $f = 0.7681 \pm 0.0002$, would not be applicable to a real uranium sphere because of the simplifying assumptions made here, especially the ignoring of neutron scattering, but the methods used illustrate the essentials for a complete investigation.

9.6 Parallel Processing and Particle Transport

As will be clear from the discussion in this chapter so far, the tracking of individual particles from their birth to the end of their contribution are always independent procedures, traditionally done in a serial manner, often with light demands on active memory. For this reason the computing time demands in particle transport simulations is the paramount factor in determining the feasibility of the method. But it is equally clear that the basic structure of the simulated phenomenon, e.g. cascade showers, has an intrinsic parallelism which cries out for an algorithm which will encompass this fact. The increased availability in recent years of parallel computer architecture, in which many processors can operate simultaneously and quasi-independently, has opened up powerful methods for doing these calculations.

Technical problems of coding for such calculations depend on the particular architecture involved. The most straightforward approach is to farm individual primary particles out to a family of distributed processors, each with its own memory bank, which are individually running the same programme, essentially simultaneously, and tally the output scores from each. The management element then only has to handle initialisation and output bookkeeping. The main problems which have to be addressed are:

(i) the availability of good quality random numbers,

(ii) load balance, and

(iii) possible individual memory limitations.

It is very important that correlations do not exist between the random number streams employed by the different processors; we briefly discuss random number generation for parallel computing in section 3.2.5.[12] Since the degree of complexity of the cascade which arises varies from one primary particle to another (even at the same energy, and especially if particles are traversing regions of different geometry), the required processing time is unpredictable. It is a stochastic quantity, and there will be a spread in the times at which their simulation is finished, leading to some of the processors being idle — this is known as load imbalance. If each processor is to sequentially run simulations for several primary particles, by starting new ones off as soon as a processor becomes free we can expect the load imbalance among them to largely cancel out. Otherwise some load balancing will have to be programmed into the management of the processors. This will also be necessary in the case where an individual processor's memory is smaller than the ongoing storage requirements of an individual cascade, e.g. due to a very high primary energy or the indiscriminate accumulation of low weight particles. The complications involved in allowing for these factors will very likely be offset by the gain in efficiency achievable when a multiprocessing facility is available.

Notes

1. In R. Alcouffe, R. Dautray, A. Forster, G. Ledanois and B. Mercier, (Eds.), *Monte-Carlo Methods and Applications in Neutronics Photonics and Statistical Physics*, (Berlin: Springer-Verlag, 1985).

2. The principles of an Ionisation Calorimeter are described in: C. W. Fabjan and T. Ludlam (1982), *Calorimetry in High-Energy Physics*, Ann. Rev. Nucl. Part. Sci., **32**, 335-89; a description of a 13 ton instrument is given in: D. Acosta et al.(1991), *Electron, pion and multiparticle detection with a lead/scintillating-fibre calorimeter*, Nucl. Instrum. Methods Phys. Res., **A308**, 481-508. An introduction to the simulation of cosmic ray showers can be found in chapter 17 of *Cosmic rays and Particle Physics* by T. K. Gaisser, (Cambridge: Cambridge University Press, 1990), and detailed accounts of the simulation of electromagnetic and hadronic cascades are to be found in A. N. Kalinovskii, N. V. Mokhov and Yu. P. Nikitin, *Passage of High-Energy Particles through Matter*, (New York: American Institute of Physics, (1985) 1989).

3. Our discussion of the methods described in this chapter will be mainly heuristic. Formal, mathematical, accounts of the material are to be found in, *Monte Carlo Principles and Neutron Transport Problems*, (Reading, Mass.: Addison Wesley, 1969) by J. Spanier and E. M. Gelbard, or in *Minimization of Computational Costs of Non-analogue Monte Carlo Methods*, (Singapore: World Scientific, 1991), by G. A. Mikhailov. A very comprehensive account of the topic, starting from first principles is to be found in I. Lux and L. Koblinger, *Monte Carlo Particle Transport Methods: Neutron and Photon Calculations*, (Boca Raton, Fla.: CRC Press, 1991).

4. Berger, M. J., in Adler, B., S. Fernbach and M. Rotenberg (Eds.), *Methods in Computational Physics*, Vol 1, pp 135-215, (New York: Academic Press, 1963).

5. Nelson, W.R., H. Hirayama and D.W.O. Rogers (1985), *The EGS4 Code System*, SLAC Report No.265, pp 1-398.

6. The method was introduced in the context of semiconductor carrier simulation by H. D. Rees (1968), *Calculation of Steady State Distribution Functions by Exploiting Stability*, Phys. Lett. **26A**, 416-17.

7. The method was first presented in H. Messel and D. F. Crawford, *Electron-Photon Shower Distribution Functions*, (Oxford: Pergamon, 1970).

8. See for example, C. Moglestue, *Monte Carlo Simulation of Semiconductor Devices*, (London: Chapman & Hall, 1993), or C. Jacoboni and P. Lugli, *The Monte Carlo method for semiconductor Device Simulation*, (Vienna: Springer-Verlag, 1989).

9. See for example Henry, A. F., *Nuclear reactor Analysis*, (Cambridge, MA: MIT Press, 1975).

10. Strictly speaking the eigen-value with the largest real part; unlike the similar Schrodinger equation considered in chapter 11 the operator **L** need not be hermitian.

11. Davis, D. H., *Critical-Size Calculations for Neutron Systems by the Monte Carlo Method*, in Adler, B., S. Fernbach and M. Rotenberg (Eds.), *Methods in Computational Physics*, Vol 1, pp 67-88, (New York: Academic Press, 1963).

12. Discussed in M. J. Gorrod, M. J. Coe and M. Yearworth (1989), *Parallel Processing of Monte Carlo Simulations using a Transputer Array*, Nucl. Instrum. Methods Phys. Res., **A281**, 156-61.

Exercises

9.1 Neutrons interact in a medium according to the exponential distribution, $p(x)dx = e^{-x/\lambda}dx/\lambda$. Taking $\lambda = 0.3$ and a block of thickness 1.0, find the distribution of the first point of interaction of 1000 incident neutrons and make a histogram of them.

Repeat the above calculation, except that in those cases where the interaction occurs at $x > 1/3$ ignore it and restart the neutron at $x = 1/3$ and find its interaction point using a *new* random number. Compare the histograms of interaction points in the two cases.

The two distributions — within statistical error — should be the same. It doesn't matter whether one considers neutrons after the partition as having started there or to have travelled up to that point. Their future pattern of interactions is independent of their previous history.

9.2 Neutrons with an energy spectrum (with ϵ in suitable units)

$$p(\epsilon)d\epsilon = 2\epsilon e^{-\epsilon^2} d\epsilon, \qquad \epsilon \geq 0,$$

are incident on a wall of thickness 6 cm. The characteristic absorption length for such particles is a function of their energy, and is given in centimetres by $\lambda_{\text{abs}} = \sqrt{40/\epsilon}$.

On the basis of a sample of 20 neutrons from the incident beam, estimate

(a) the transmission coefficient of the shield, and

(b) the mean energy per incident particle deposited in the block,

in each case with its associated error. For the same 20 particles evaluate, analytically, the average probability of interacting in the block, and thus the energy deposited. Compare the consistency of this average with the result previously found.

9.3 In a radiology problem it might be necessary to select from the angular distribution of photons emitted as Bremsstrahlung by an electron of energy E and momentum p, which can be approximated as

$$p(\vartheta)d\vartheta = \frac{K \sin\vartheta d\vartheta}{(E - pc\cos\vartheta)^2}, \qquad 0 \leq \vartheta \leq \pi$$

where K is a normalising constant. If $E/pc = 2$, show that the angle can be selected using

$$\vartheta = \cos^{-1}\{(3 - 4\xi)/(3 - 2\xi)\}.$$

9.4 The probability distribution for the number of neutrons emitted in the fission of a uranium nucleus is given in example 9.8. Write a subroutine to select from this distribution, and use it to generate a sample for comparison with the distribution.

9.5 As a first approximation, the energy distribution of neutrons from uranium fission can be written as

$$p(y)dy = \frac{2}{\sqrt{\pi}} y^{1/2} e^{-y} dy$$

with $E = 1.29y$ MeV. Show that the transformation $u = \sqrt{y}$ leads to the distribution in example 5.17 and explain how such an energy spectrum could be selected from.

9.6 A better approximation to the energy spectrum of neutrons than that in exercise 9.5 is the Watt spectrum,

$$p(E)dE = Ce^{-E/T} \sinh \sqrt{bE}\, dE, \qquad E \geq 0,$$

for which a selection algorithm has be arrived at in example 5.18. Verify that the following extension of the rejection technique, due to Kalos, will also select variates from this distribution.

Write $K = 1 + bT/8, L = T(K + \sqrt{K^2 - 1})$ and $M = K - 1 + \sqrt{K^2 - 1}$, and set $E = -L \ln \xi_1$ if $-bL \ln \xi_1 \geq [M(1 - \ln \xi_1) + \ln \xi_2]^2$, where ξ_1 and ξ_2 are from the uniform distribution.

9.7 The two γ-rays in the decay of a π^0 meson are emitted back to back, isotropically in the rest frame of the meson. When such a particle with energy E_π(velocity βc) decays, the energy of a γ-ray in the laboratory is given by

$$E_\gamma = \frac{E_\pi}{2}(1 + \beta\mu^*),$$

where $\mu^* \equiv \cos\vartheta^*$ is uniformly distributed on [0,1]. Show that the energies and emission angles (with respect to the momentum vector of the pion) of the two γ-rays can be selected as,

$$E_i = \frac{E_\pi}{2}(1 \pm \beta\xi), \qquad \vartheta_i = \tan^{-1}\frac{\sqrt{1-\xi^2}}{\gamma(\xi \pm \beta)},$$

where γ is the Lorentz factor of the pion.

9.8 Consider the following one-dimensional toy problem. The absorption coefficient, and total interaction coefficient for a particle in a medium are $\mu_a = 0.3$ and $\mu_T = 1.0$, respectively, and when absorption does not occur the particle continues from the interaction point in the same direction. For a shield of thickness D, in units of μ_T^{-1}, we know the result for the transmission probability: it is simply $T = e^{-\mu_a D}$.

Take $D = 30$ and simulate the passage of $N \geq 20000$ particles to compare with this result. Do this

(a) by taking $T = N_{\text{out}}/N$, with its associated error, i.e. quote $T \pm \sigma_T$, and

(b) using survival biasing, i.e. attaching a weight $(1 - \mu_a/\mu_T)$ at every interaction and averaging the N output weights to find the mean and its standard error, and compare the results of the two methods.

9.9 The differential cross-section for Rutherford scattering by a shielded nucleus (otherwise the total cross-section is infinite) is given by

$$\frac{d\sigma}{d\Omega} = \frac{A}{\sin^4\frac{\vartheta}{2}}, \qquad \eta \leq \vartheta \leq \pi.$$

Show that $A = \frac{\sigma_T}{4\pi}(\text{sech}^2\frac{\eta}{2} - 1)$ where σ_T is the total cross-section, and that scattering angles can be selected according to the prescription,

$$\vartheta = 2\sin^{-1}\left\{\frac{\sin\frac{\eta}{2}}{\sqrt{(1-\xi\cos^2\frac{\eta}{2}}}\right\}.$$

Chapter 10

MONTE CARLO AND THERMAL PHYSICS

> True originality is to be found rather in the use made of a model than in the rejection of all models and masters.
>
> Oscar Wilde

10.1 Introduction

The foundations of Thermal Physics, Statistical Mechanics, provides us with a natural environment for the use of stochastic methods of analysis. It is not surprising that Monte Carlo methods have played a major role in this field over the years. Analytical methods are all but impotent in revealing more than the gross behaviour of multi-particle systems when the inter-particle interactions are of the complexity which is actually observed in most systems. Examples are the inter-molecular potential in a fluid, the forces experienced by a charge in a magnetised plasma, the magnetic cum exchange-force interaction occurring between magnetic dipoles in a crystal, etc. We have seen in earlier chapters how to study the ground state of such systems, i.e their configuration at zero temperature. Here we are interested in the behaviour of a system as the temperature is varied; the experimental behaviour being modeled might be the onset of a phase transition such as to a ferromagnetic or superfluid state. Systems both out of equilibrium and those in equilibrium can be studied by the methods described here, although the former are of much greater technical difficulty and most of our discussion will refer to the equilibrium case. We know that all the behaviour, ultimately, is a consequence of quantum phenomena, but, provided suitable parameterised interaction potentials are included, the formalism of classical statistical mechanics is usually adequate. We will have something to say about quantum statistical mechanics in chapter 11 on quantum Monte Carlo.

The techniques of statistical mechanics calculations have a much wider range of applicability than was suggested at the time of their introduction. Many of the concepts described in this chapter are now of great importance in general Monte Carlo calculations in quantum theory, notably in quantum field theory applied to High Energy physics and Condensed Matter physics. An introduction to these developments is given in chapter 11. The seminal reference for the material in this chapter is the paper of Metropolis *et al.*[1]

10.2 Simulation of Thermal Systems

10.2.1 General

The nature of the physical problem will determine how we specify the system of interest. Important specifications are recognised by particular nomenclature, as follows (we use fluid variables but general thermodynamic variables are implied).

The system is totally isolated; this is known as the *microcanonical ensemble*, typical coordinates for its specification would be the number of particles, the volume and the total energy (N, V, E). Such a system will come to some state where the time average of physical quantities becomes constant — equilibrium. A temperature, T, can be associated with such an equilibrium, in the sense that T is the temperature of the surroundings at which the mean energy per degree of freedom of the system, $k_B T/2$, would equal that in the isolated system, $E/\nu N$, where ν is the number of degrees of freedom per particle.

The system may be in thermal contact with a heat reservoir characterised by a temperature T, allowing continuous energy transfer back and forth between them; this is the *canonical ensemble*, typically specified by the coordinates (N, V, T). Again the system will reach a steady equilibrium state — dictated by the reservoir — with, of course, local fluctuations. This is the most important situation in theoretical formulation in physics, and the one with which we will be most concerned. The configurations of a system in this ensemble are distributed as the Boltzmann distribution. The practically important (N, P, T) ensemble can often be treated by relating it to the canonical ensemble through thermodynamic relations.

If in addition to energy the system of interest can also exchange particles with its environment, i.e. N is no longer constant, we have the *grand canonical ensemble.* Suitable coordinates in this case are (μ, V, T), where μ, the so called chemical potential, is given by

$$\mu = ({}^{\partial F}/_{\partial N})_{V,T}, \tag{2.1}$$

where $F = E - TS$, with S the entropy, is the Helmholtz free energy.

Monte Carlo methods are only of limited applicability in the microcanonical ensemble, which is more easily studied using the rival method of

Molecular Dynamics. As we will see further on, they are especially suited to treat the canonical ensemble, and can be extended to the grand canonical ensemble.

Thermodynamic coordinates are not appropriate for simulating a thermal system. To do that we must specify the configuration of the system in terms of microscopic variables — position, momentum, dipole moments, etc. We will denote the set of such coordinates required to specify the system, generally, by $\{\mathbf{q}\}$. For a gas $\{\mathbf{q}\} = \{\mathbf{r}, \mathbf{p}\}$, the set of position and momentum coordinates of the particles, while for magnetic dipoles confined to sites on a lattice $\{\mathbf{q}\} = \{\sigma\}$, the set of values of the particle spins, etc. We know that the macroscopic coordinates, in general, are not uniquely related to the microscopic configurations of the system. For the case of an isolated system the configuration variables $\{\mathbf{q}\}$ will be confined to a hyper-surface in the joint coordinate-momentum phase space whose equation is

$$H(\mathbf{q}) = E_s,$$

where E_s is the specified energy and $H(\mathbf{q})$ is the Hamiltonian of the system. We can formally write the distribution function of the coordinates as

$$p(\mathbf{q})d\mathbf{q} = \delta(E_s - H(\mathbf{q}))d\mathbf{q}. \tag{2.2}$$

We describe selection from this microcanonical ensemble in section 10.2.2.

In the case of the canonical ensemble the values of the macroscopic variables, e.g. (N, V, T), are a consequence of the equiprobability of the combined microstates of the system plus its environment, the reservoir which determines the temperature. The relative likelihood of the occurrence of particular microstates of the system at T depends on the energy (or Hamiltonian function) of the system in a manner specified by the familiar Boltzmann distribution,

$$p(\mathbf{q})d\mathbf{q} = Z^{-1}e^{-\beta H(\mathbf{q})}d\mathbf{q}, \tag{2.3a}$$

where $\beta = 1/k_B T$, and the normalising constant is given by

$$Z(N,\beta) = \int e^{-\beta H(\mathbf{q})}d\mathbf{q}. \tag{2.3b}$$

We now need to find ways for simulating micro-configurations of systems according to an ensemble; we will start with the microcanonical ensemble.

10.2.2 The Microcanonical Ensemble

From a theoretical viewpoint it is sometimes desirable to model a system of fixed energy, e.g. when studying the coexistence region in a first order phase transition where the energy can change at fixed temperature. This means using the microcanonical ensemble. Markov chain methods can be adopted to this task.[2] The first of two such methods we describe depends on approximately sampling from the distribution in equation 2.2. Focusing on the case of a fluid, to maintain a fixed energy changes in space and momentum coordinates must be coupled. Thus, although an initial configuration of space and momentum coordinates for which the sum of the kinetic and potential energies add up to E_s may be generated, it would be rather difficult to select other configurations for which the sum will also *exactly* equal E_s. For practical sampling of configurations some finite thickness of the hyper-surface in phase space will have to be introduced. Using a truncated Lorentz distribution, for example, as a representation of the delta function (equation 2.4.17), we replace equation 2.2 by

$$\begin{aligned} p(\mathbf{q})d\mathbf{q} &\propto \frac{\epsilon}{(E_s - H(q))^2 + \epsilon^2} d\mathbf{q} \qquad E_s - \Delta \leq H(q) \leq E_s + \Delta, \\ &= 0 \text{ otherwise,} \end{aligned}$$

where for a system of N particles, for example, $d\mathbf{q} = d^{6N}q$. This correspond to phase space points being distributed within a shell of thickness Δ on either side of the energy surface and with a density which falls off more rapidly the smaller ϵ is. The normalisation of this distribution would not be easy, but we can select from in its present form by using the Metropolis algorithm as follows:

(i) choose an initial configuration, $\mathbf{q}^{(0)}$, with energy E_s,

(ii) generate a tentative configurations, $\mathbf{q}^{(t)}$, from the previous configuration $\mathbf{q}^{(i)}$ by changing some or all of the six coordinates of a particle by amounts lying in a 'hypercube' of side Δq (different, of course, for positions and momenta), e.g.

$$q_j \to q_j + (2\xi_j - 1)\Delta q,$$

(iii) determine the new configuration by selecting a uniform random deviate ξ and following the prescription,

$$\begin{aligned} \mathbf{q}^{(i+1)} &= \mathbf{q}^{(t)} \qquad \text{if } \frac{p(\mathbf{q}^{(t)})}{p(\mathbf{q}^{(i)})} > \xi, \\ \mathbf{q}^{(i+1)} &= \mathbf{q}^{(i)} \qquad \text{otherwise,} \end{aligned}$$

where ξ is another uniform unit random deviate.

By simultaneously taking a pair of particles through complimentary steps in displacement and momentum no motion of the centre of mass need be introduced by the random walk.

To develop a second method, the *Thermal Demon*, we remind ourselves that the problem is to get the energy distributed throughout the system, and at the same time to ensure the 'natural' local fluctuations resulting, for example, from local interactions of particles or propagating disturbances in the system. This method depends on the concept of a demon, akin to the demon used by Maxwell in his discussion of reversibility. The method will be best explained by spelling out the algorithm.

We are interested in a system of total energy E, but, as in the previous algorithm, for purposes of applying the algorithm we are prepared to play loose with a part, E_d, of this energy. Provided $E_d << E$ when we are accepting configurations, the energy of the system can be taken to be E, and a temperature can be associated with it in the sense described in the introduction. We start off by choosing an arbitrary initial configuration having an energy $E - E_d$, and assign the remaining energy $E_d > 0$ to a separate, structureless, system, the demon. At this stage E_d can have any value, from zero to E (the method also has some uses for $E < 0$ but we will not consider them). In executing the Markov chain a tentative change is made to the configuration of the system, $\{\mathbf{q}\} \to \{\mathbf{q}'\}$, e.g. a single particle is displaced or a spin flipped, and the associated change in energy, $\Delta E = E(\mathbf{q}') - E(\mathbf{q})$, calculated. The details of the algorithm are as follows:

> if $\Delta \mathrm{E} \leq 0$ the configuration $\{\mathbf{q}'\}$ is accepted, and the energy surrendered thereby is given to the demon, i.e. $E_d \to E_d + \mid \Delta E \mid$; if $\Delta E > 0$ but $\Delta E \leq E_d$ the configuration $\{\mathbf{q}'\}$ is accepted, and the energy of the demon reduced, i.e. $E_d \to E_d - \Delta E$; otherwise the old configuration is retained as the new one.

The demon can be thought of as a particle with one degree of freedom being 'buffeted' by the large system, and in the process distributing the energy of the latter throughout the system. Carrying out the above prescription, equilibrium between the demon and the system will be established, the system acting, in effect, as a reservoir for the demon. The energy of the latter will then be distributed as the Boltzmann distribution, this distribution at the same time defining the temperature of the system. We have

$$p(E_d)\ dE_d = Ce^{-\beta E_d}\ dE_d,$$

with $T = 1/k_B\beta = < E_d > /k_B$. Hence the temperature of the system can be inferred from the mean energy of the demon, provided, of course, that it is much less than the energy of the system. The demon can move systematically through the system, in which case there is no stochastic element in the algorithm, or it may hop randomly from point to point. In both of the algorithms described, as in all Markov chain methods, it is clear that some number of configurations in the early stages of the sampling must be discarded as not representing samples from the steady state. For fastest equilibration the demon should hop with large steps through the system and not move systematically from particle to neighbouring particle.

More than one demon can be introduced, and in the limit where their number is very large the selection is from a canonical ensemble, with the disadvantage that the temperature cannot be specified *a priori*; it evolves from equilibration. A major advantage of this algorithm over the Metropolis algorithm, described in section 10.2.4, is that there is no need to calculate the exponential term in the Boltzmann factor, and all programming can be done in integer arithmetic making it much faster — this is important, for example, in quantum field theory simulations.

A slight modification of the algorithm does allow for selection from the canonical ensemble at a specified temperature T. En route between every pair of points the demon can visit a heat reservoir at T. Its energy, independent of its existing value, is then thermalised, i.e. reassigned by selection from the Boltzmann distribution, e^{-E_d/k_BT}. A more direct application of this philosophy is the *Heat Bath* method which we describe next.

10.2.3 The Heat Bath Method

The most straightforward method for selecting from the canonical ensemble is the Heat Bath method. This algorithm depends on the fact that when a system is in equilibrium at T, every part of it is in equilibrium, and the distribution of the coordinates of any constituent (particle, degree of freedom) separately follows the Boltzmann distribution, i.e.

$$p(q_i)dq_i \propto e^{-\beta E(\ldots,q_i,\ldots)}\,dq_i. \tag{2.4}$$

We could bring an individual particle to equilibrium at T by bringing a heat reservoir in contact with that particle, i.e. by assigning it a coordinate based on this distribution at T — without reference to the pre-existing value of that coordinate. This, in general, is a false equilibrium. It will not correspond to true equilibrium because the coordinates of the neighbouring particles (to an extent determined by the range of the interactions in the system), which enter into the selection through their occurrence in E_i, do not have values characteristic of equilibrium at T. However, it is intuitively obvious that physical iteration of this exercise, or its mathematical simulation, i.e. repeatedly sweeping through the particles and bringing each to ostensible equilibrium at T, will bring the system eventually to an overall state of equilibrium at T. Continuing the process will then be equivalent to selecting configurations of the system at equilibrium. The selected configurations are elements in a Markov chain; in the notation of chapter 5 we have $W_{ij} = p(q_j)$, independent of q_i. As with all such methods the chosen configurations will be correlated since successive samples will differ only in the coordinates of one, or at most a few, particles, but in so far as the change in an individual coordinate is not *a priori* restricted, it may be quite large and the correlations should be less than in the more restricted Metropolis method, in which the allowed coordinate changes are usually limited in some way. For this reason the method has some advantages. However, selection of new coordinates for a particle according to equation 2.4, and therefore the method, will only be practical if the dependence of the energy on a coordinate is rather explicit, or of a particular kind. In such cases any of the methods we described in chapter 5 can be used to select the coordinate value from equation 2.4. Where it can be implemented, it is an important method because of the possibility of speeding up the selection of equilibrium configurations by incorporating the technique of *successive over relaxation.*

EXAMPLE 10.1

For the spin 1/2 Ising model the coordinate of a particle, which we now denote by σ, assumes only two values, $\sigma_i = \pm 1$. In an external field (see section 10.4) the contribution of a particle to the energy E is $-\epsilon_i\sigma_i$, with $\epsilon_i = (H + \frac{J}{2}\sum_{j(i)}\sigma_j)$, where the summation is over the nearest neighbours of the particle — in two dimensions, for example, there are only five possible values for this sum. Thus the probability distribution to be selected from is,

$$p(\sigma_i = \pm 1) = e^{\pm\beta\epsilon_i} / (e^{-\beta\epsilon_i} + e^{\beta\epsilon_i}).$$

EXAMPLE 10.2

If the contribution to the energy of a system depends on the coordinates of a particular particle, denoted by x_i, in a quadratic manner, i.e. we can write $E = E_i +$ terms which do not depend on x_i, with $E_i = a_i(x_i - b_i)^2$, where b_i might be zero, and a_i and b_i, in general, would depend on the coordinates of other particles. The distribution equation 2.4 then becomes a Gaussian with mean b_i and variance $1/(2a_i)$. Several methods for selecting from a Gaussian exist (see examples 5.15, 5.16, some of which can be very efficient).

10.2.4 The Metropolis Algorithm

In most cases the dependence of interaction energy on the coordinates involved will be too complicated to allow use of the Heat Bath algorithm. Selection from the canonical ensemble is then invariably done using the Metropolis algorithm, or some version of it. We have already, when discussing Markov chain methods of variate selection in section 5.5.4, described these algorithms. We have seen its general application in chapter 6, and its suitability to the task of interest here, the selection of configurations from the Boltzmann distribution, when describing simulated annealing. We repeat here the steps in its implementation.

(i) Choose an initial phase-space coordinate configuration, $\{\mathbf{q}^{(0)}\}$, for the system.

(ii) Generate a tentative configuration, $\{\mathbf{q}^{(t)}\}$, from the previous configuration $\{\mathbf{q}^{(i)}\}$; usually this is done by altering the coordinates of just one particle.

(iii) Determine the associated change in energy, $\Delta E = E(\{\mathbf{q}^{(t)}\}) - E(\{\mathbf{q}^{(i)}\})$; $p(\{\mathbf{q}^{(t)}\})$ is greater or less than $p(\{\mathbf{q}^{(i)}\})$ according as this quantity is negative or positive.

(iv) If $\Delta E \leq 0$ set $\{\mathbf{q}^{(i+1)}\} = \{\mathbf{q}^{(t)}\}$.

(v) If $\Delta E > 0$ select a uniform random deviate ξ and set $\{\mathbf{q}^{(i+1)}\} = \{\mathbf{q}^{(t)}\}$ if $\xi < e^{-\beta\Delta E}$, otherwise set $\{\mathbf{q}^{(i+1)}\} = \{\mathbf{q}^{(i)}\}$.

In some applications, e.g. where the effort in calculating the change in energy in step (iii) is very demanding, a number, n_h, of repeated efforts to select new configurations by selecting values for the same single coordinate in step (ii) are made. If n_h is large the value of this coordinate will reach a state of equilibrium with the remainder of the configuration, so in the limit $n_h \to \infty$ the algorithm becomes the same as the Heat Bath method.

EXAMPLE 10.3

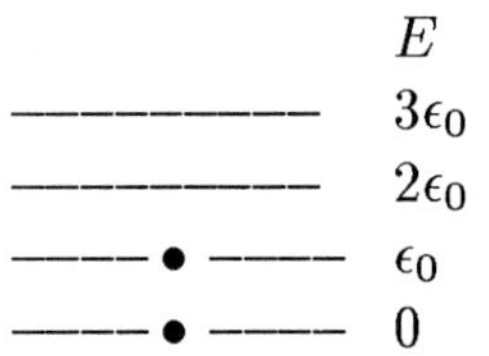

We consider here a toy problem of selecting configurations of two particles distributed among four energy levels as shown in the figure, and constrained such that not more than one particle can occupy each level, at a temperature $T = 1$, in units of (ϵ_0/k_B). We are asked to find the transition matrix for the case where only one particle can be moved at a time, and then only to an immediately adjacent level.

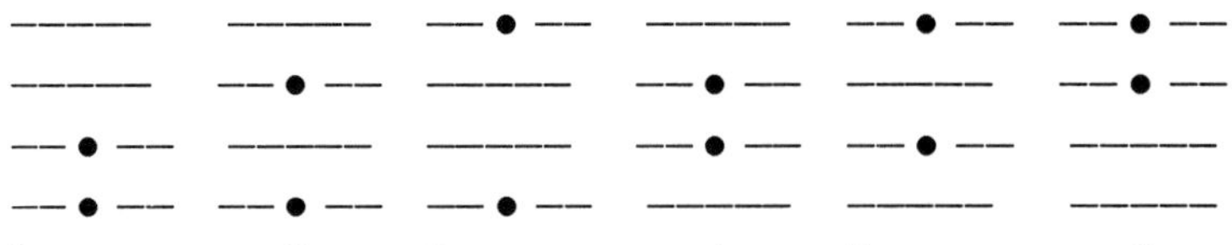

i	1	2	3	4	5	6
E/ϵ_0	1	2	3	3	4	5
p_i	e^{-1}/Z	e^{-2}/Z	e^{-3}/Z	e^{-3}/Z	e^{-4}/Z	e^{-5}/Z
ν_i	1	3	2	2	3	1

We have labelled the six possible configurations, and shown their energies and probabilities of occurrence according to the Boltzmann distribution. Z is the partition function, which here, of course, can easily be calculated by normalising $p(i)$, and the exact probabilities can be used to confirm the validity of our algorithm. Also shown are the number of configurations, ν_i, other than the configuration itself, accessible in a single move as described. Because these values are not all the same we cannot set the elements of $\mathbf{G}$ (in section 5.5.4) equal to $1/(\nu+1)$. A symmetric matrix for making the moves which can be made stochastic is

$$G_{ij} = 1/\sqrt{(\nu_i+1)(\nu_j+1)},$$

i.e.

$$\mathbf{G} = \frac{1}{2}\begin{pmatrix} \bullet & 1/\sqrt{2} & 0 & 0 & 0 & 0 \\ 1/\sqrt{2} & \bullet & 1/\sqrt{3} & 1/\sqrt{3} & 0 & 0 \\ 0 & 1/\sqrt{3} & \bullet & 0 & 1/\sqrt{3} & 0 \\ 0 & 1/\sqrt{3} & 0 & \bullet & 1/\sqrt{3} & 0 \\ 0 & 0 & 1/\sqrt{3} & 1/\sqrt{3} & \bullet & 1/\sqrt{2} \\ 0 & 0 & 0 & 0 & 1/\sqrt{2} & \bullet \end{pmatrix}$$

where the diagonal elements are not filled in. Combining this with $\min[1, p_j/p_i]$ we find

$$\mathbf{W} = \begin{pmatrix} .870 & .130 & 0 & 0 & 0 & 0 \\ .354 & .434 & .106 & .106 & 0 & 0 \\ 0 & .289 & .605 & 0 & .106 & 0 \\ 0 & .289 & 0 & .605 & .106 & 0 \\ 0 & 0 & .289 & .289 & .292 & .130 \\ 0 & 0 & 0 & 0 & .354 & .646 \end{pmatrix}.$$

In Fig. 10.1 we compare the exact distribution, solid line, with the results of applying this transition matrix, starting from state No. 1 ($i = 1$) above, for a sample of 200, and for a sample of 10000.

In practice it is usually not necessary to build up the explicit transition matrix as was done in this toy example, its elements can be continuously generated in a simulation, but it is important that the algorithm be properly constructed. The most transparent way of ensuring this is to impose the detailed balance condition as we did in the above example.

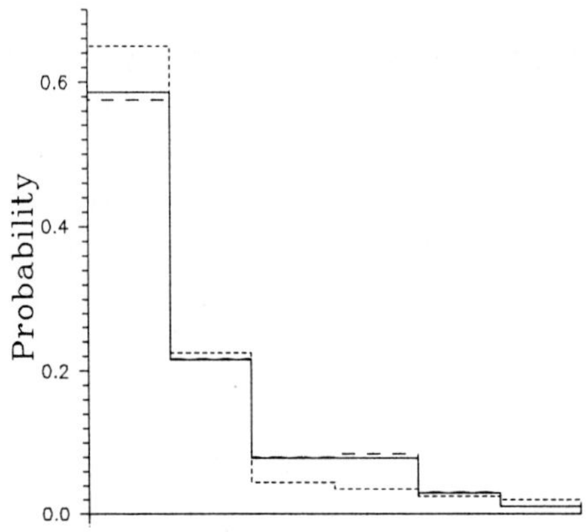

Figure 10.1: Comparison of the exact Boltzmann distribution for the six state system in example 10.3, solid curve, with that produced by the Markov chain generating matrix based on a sample of 200 configurations - - small dashed curve, and a sample of 10000 configurations — long dashed curve.

10.2.5 Correspondence with a Real Physical System

Real physical systems will evolve in time through sequences of configurations which display the thermal stochastic behaviour, but will do so in a more purposeful way than our sample configurations. Its profile will be determined by the dynamics underlying the interparticle interactions. The Markov chain of configurations produced by any of the earlier described algorithms must not be mistaken for a simulated dynamical history in this sense. The sequence may bear some resemblance to the true dynamical development, e.g. when flipping magnetic dipoles on fixed lattice sites. Or it may not at all resemble such development, e.g. when independently stepping an individual molecule by arbitrary steps through a gas. This failure to reflect the dynamics in a problem is one of the weaknesses of Monte Carlo methods.

10.3 Simulation in Classical Statistical Mechanics

10.3.1 Fundamentals

So far in this chapter we have spent a lot of time on the technical aspects of thermal physics, without any clear discussion of the nature of the fundamental problems which are of interest. Let's concentrate on a system in equilibrium at a temperature T. For the important case of a system of particles in which the mutual interactions do not depend on the momenta of the particles, the energy E in equation 2.2 will be separable into a sum of kinetic and interaction energy terms, the former only depending in a simple way on the momenta coordinates, $K = (\sum p_i^2/2m_m)$, and the latter on the remaining coordinates, which we continue to denote by $\mathbf{q}$, $U(\mathbf{q})$. Restricting ourselves to the probability distribution for these coordinates, the momentum coordinates in the Boltzmann distribution can be integrated over to yield

$$p(\mathbf{q}, T)d\mathbf{q} = \frac{1}{Q}e^{-\beta U(\mathbf{q})}d\mathbf{q}.$$

Here, all the manipulations glossed over earlier have been subsumed by a normalising constant, $Q(N, \beta)$, known as the *configurational* (or *internal*) *partition function*, where, obviously

$$Q(N, \beta) = \int dq_1 dq_2 ... dq_N e^{-\beta U(q_1, q_2, \ldots, q_N)} .$$

For a gas of indistinguishable particles it is related to the partition function equation 2.3*b*,

$$\begin{aligned} Z(N, \beta) &= Z_{\mathrm{id}} Q(N, \beta)/V^N \\ &= Q(N, \beta)/N!\Lambda^{3N}, \end{aligned}$$

where Z_{id} is the partition function for an ideal gas and $\Lambda = h/\sqrt{2\pi m}$. Ideally, we would like to be able to evaluate this important function, since most thermodynamic quantities can be derived from it — for example for a system of particles the Helmholtz free energy F, compare equation 2.1, is given by,

$$F(N,\beta) = -\beta^{-1} \ln Z(N,\beta). \tag{3.1}$$

In most cases, however, this proves to be an insuperable difficulty and we have to use more primitive relations. The mean value of any function of the coordinates of a configuration, e.g. the function $U(\mathbf{q})$ itself, can be estimated from the integral

$$< U(T) > \; = \int U(\mathbf{q}) p(\mathbf{q}, T) d\mathbf{q}, \tag{3.2a}$$

$$= \int U(\mathbf{q}) \frac{1}{Q} e^{-\beta U(\mathbf{q})} d\mathbf{q}. \tag{3.2b}$$

In the same way the value of $< U^2 >$, and thus the variance $\sigma_U^2 =< U^2 > - < U >^2$ can be obtained. From this an estimate of the specific heat $C(T) \equiv d < U > /dT$ can be found by differentiating equation 3.2b with respect to β and noting that $\partial Q/\partial\beta = -Q < U >$ giving the relation[3]

$$C = k_B \beta^2 \sigma_U^2. \tag{3.3}$$

The fluctuations in the magnetisation, $\mathbf{M}$, of a system in a magnetic field $\mathbf{B}$ can be related to the susceptibility, $\chi_B \equiv \mu_0 d < M > /dB$, by a similar relation

$$\chi_B = \mu_0 \beta \sigma_M^2. \tag{3.4}$$

Multidimensional integrals of the form of equation 3.2 have frequently been encountered in earlier chapters, where we have seen that samples of the integrand based on uniform selection of phase space points would give very poor estimates of the integral because of the exponential term $e^{-\beta U(q_1, q_2, \ldots q_N)}$ which would have appreciable values only in very restricted, but not easily determined, regions of the space. From this arises the necessity of using importance sampling in the evaluation. The method is also familiar— sample points can be selected from the Boltzmann distribution by using one of the methods described at the beginning of the chapter.

10.3.2 Finite Size Effects

The purpose of our calculations is to make comparison with experimental results on macroscopic systems, yet it is clear that the number of particles involved in such a system is enormously greater than the number which could be carried along in a practical simulation using the most extravagant computing resources. One way of attempting to mimic a larger system using a base system of N particles is to surround this system by images of itself — this is the method of *periodic boundary conditions* (or *toroidal boundary conditions*). The simplest case of such conditions is shown for a system of two kinds of particles in Fig. 10.2 where the base system is surrounded on four sides by identical systems. Depending on the the range of interactions involved, a greater degree of surrounding may be sometimes desirable, possibly extending to an infinite array of images. Sample configurations refer only to the base system, but the interactions of a particle in it both with the other particles in the base system and with the surrounding particles will be taken into account — the latter possibly by an averaging process — in the selection process. Such methods go some way to overcoming the limitations of small systems. For example, they remove the artificial surface

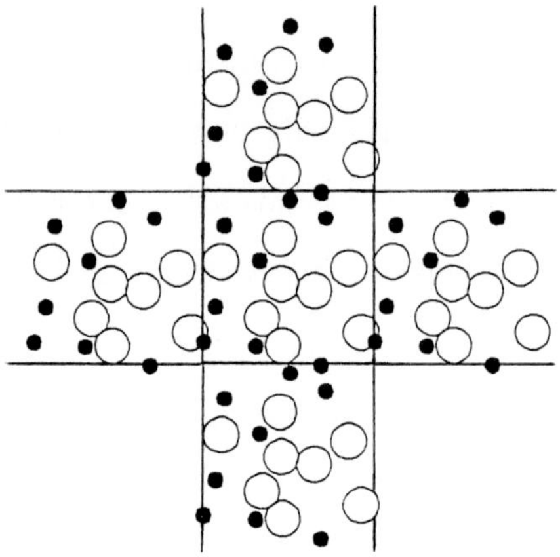

Figure 10.2: A system of two types of particles surrounded by four images of itself, the simplest implementation of periodic boundary conditions.

effects of an isolated system, but still do not completely remove the limitations, e.g. density fluctuations or other correlations on the scale of the base system — as might occur in a phase separation.

An obvious method of arriving at results for a large system which can sometimes be employed is to carry out simulations for several practical values of N and attempt to extrapolate to infinite size, e.g. by extrapolation as a function of $1/N$.

10.3.3 Time in Simulations

As noted in section 10.2.5, the frequency of configuration sampling in a simulation is not in general isomorphic to the time progression in the physical evolution of a real system. It is however convenient to talk about a time in a calculation; for a system of N particles we will consider one unit of time to have elapsed after N attempts to sample a configuration, or equivalently, one sweep through the system. This unit is known as one Monte Carlo step per site (MCs/s). In particular the time required for a system to be brought to equilibrium, which will determine the feasibility of a calculation, will be measured in these units. Modifications of the basic algorithms already described will become necessary when such equilibration times become too long — the phenomenon of *critical slowing down.*

10.4 Application to a Physical System

The simplest statistical mechanical system which comes to mind is one composed of a number of gas molecules; however, this turns out to be too complicated to use as an introduction to the method, and the subtleties which may arise, and we leave it until later in the chapter. For an example we will use a system which can be described in terms of a finite number of discrete coordinates. A system whose configurations can be described in a concise manner is a set of electric or magnetic moments located on lattice sites — as might occur in a crystalline solid — and whose orientations are discrete as a result of space quantisation. Models based on these ideas can be developed for a surprising variety of physical problems, and only a few exact results have been determined for them, so that great scope exists for numerical modeling. We discuss such moments associated with particles in general terms by using 'spin' variables, σ, with components σ_i which can assume various values, and the interaction energy is expressed in terms of these components. For example, for a three component spin, the exchange interaction — the *Heisenberg interaction* — between a pair of spin-1/2 particles can be written as

$$u = -J\sigma^{(1)}.\sigma^{(2)},$$

where the magnitude of each spin vector is conventionally taken as unity,

$$\sigma_x^{(i)^2} + \sigma_y^{(i)^2} + \sigma_z^{(i)^2} = 1.$$

The so called *X-Y* interaction corresponds to the spins being confined to a plane, i.e. $\sigma_z = 0$, so we have

$$u = -J(\sigma_x^{(1)}\sigma_x^{(2)} + \sigma_y^{(1)}\sigma_y^{(2)}).$$

The simplest case, where only one spin component is allowed for, conventionally $\sigma_z^{(i)}$, which then assumes the discrete values ± 1, gives us the *Ising interaction*, for which we have

$$u = -J\sigma_z^{(1)}\sigma_z^{(2)}.$$

With the assumption that such interaction only occurs between nearest neighbour dipoles in the system, it forms the basis of the *Ising model*, which we will investigate in the next example. Hereafter we will only be concerned with one-component spins and will change the notation so that the energy of interaction of particles i and j is, ($\sigma_z^{(i)} \to \sigma_i$, etc.),

$$u = -J\sigma_i\sigma_j.$$

A series of models where this spin component can assume q distinct values, and the interaction energy of a pair is non-zero only when their spins are parallel,

$$u = -J\delta(\sigma_i, \sigma_j),$$

where δ is the Kronecker delta symbol, can be constructed; they are known as q-state *Potts* interactions. The case of $q = 2$ corresponds to the Ising interaction with strength $J/2$.

EXAMPLE 10.4: A magnetic dipole system

We consider a two-dimensional system of spin 1/2 magnetic dipoles arrayed on a square lattice. The projections of a spin on any axis, s_i, can assume only two values, which we write in terms of a 'coordinate' σ_i as $s_i = \frac{1}{2}\sigma_i$, where the two possible values of the coordinate are $\sigma_i = \pm 1$. The reference axis can be thought of as in the plane, or perpendicular to it. A configuration $\{\sigma\}$ of the system is then specified in terms of the set of values of σ_i at each site, n_+ of which will have values $\sigma_i = +1$ and the remaining $(N - n_+)$ with $\sigma_i = -1$. The magnetic moment of a dipole, $\mu\sigma_i$, $(\mu = g\mu_B/2)$, is then proportional to its coordinate. Such magnetic moments interact among themselves, and with any externally applied magnetic field $\mathbf{B}$. Taking the direction of such a field as the reference direction, this latter interaction is of the form $-\mu\sigma_i B = -H\sigma_i$, where H is conventionally used for the combination of constants. In the case where these magnetic dipoles are the result of electron spins their interaction with each other will be strongly determined by quantum exchange effects which dominate over the longer range dipole-dipole interaction. In this case the interaction of a pair is proportional to $\sigma_i\sigma_j$ — the Ising model. It is assumed that a dipole interacts only with its nearest neighbours, of which there are only four in our case, in this way. In terms of our coordinates, the interaction energy of dipole i is then $-J\sigma_i \sum_{j(i)} \sigma_j$, where J is a combination of constants (the minus sign is conventional) and the index on the summation indicates that it should run over all sites j which are neighbours of site i. The total energy, summing over the N dipoles, is then given by

$$U(\{\sigma\}) = -H\sum_{i=1}^{N}\sigma_i - \frac{J}{2}\sum_{i=1}^{N}\sigma_i\sum_{j(i)}\sigma_j, \tag{4.1}$$

where the factor of one-half in the second term arises from the requirement that the interaction energy of a dipole pair not be double counted. The expression 4.1 is now best thought of as a spin-based Hamiltonian, with parameters H and J. It could, as our derivation suggests, represent a model for a ferromagnetic (or, if $J < 0$, an antiferromagnetic) substance, but its form is also isomorphic to the energy

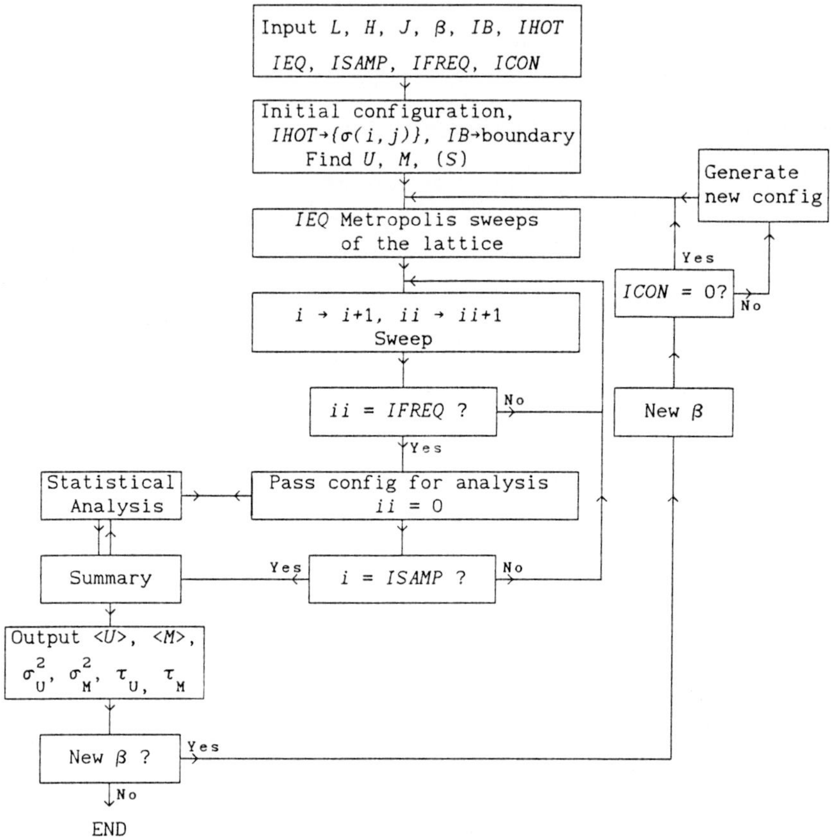

Figure 10.3: A flowchart for example 10.4.

of a binary alloy, or lattice gas (where the two values of σ denote presence or absence of a particle at a site). If the coupling constant J itself is allowed to be a function of site position, $J \rightarrow J_{ij}$, and each such parameter is randomly distributed according to some probability distribution we are dealing with so called *spin glasses.*

The type of quantities we can compute for comparison with experiment are the energy as a function of temperature, the specific heat using equation 3.3, the magnetisation — or long range order — of the system, and, via equation 3.4, the susceptibility. The magnetisation per dipole of a particular configuration, in scaled units, is $M(\{\sigma\}) = \frac{1}{N}\sum_{i=1}^{N} \sigma_i$. For discrete states the integral in equation 3.2 becomes a sum and we have

$$< M(T) >= \sum_{\{\sigma\}} M(\sigma) p(\sigma, T),$$

where the sum is over the 2^N distinct configurations of the dipoles, and the probability is given by $Z^{-1}e^{-\beta U(\{\sigma\})}$. For a system as small as a 20×20 array of dipoles the number of configurations is already 10^{12}, and even allowing for the appreciable degree of degeneracy of these states, direct enumeration of the terms in the sum becomes unrealistic. We select instead a total of K configurations ($K << 2^N$) from the Boltzmann distribution and use the estimate

$$< M >= \frac{1}{K} \sum_{k=1}^{K} M(\sigma^{(k)}), \tag{4.2}$$

with a similar expression for $< M^2 >$.

A flowchart for a Monte Carlo programme to determine the thermal properties of such a model is shown in Fig. 10.3. The two-dimensional system of $N = L \times L$ dipoles on a square array is isolated or surrounded by a periodic boundary according as *IB* equals zero or one. The starting configuration is random or completely aligned depending on whether *IHOT* equals one or zero. An example of the former, with periodic boundary layers, is shown in Fig. 10.4. Starting from a high temperature, a sequence of temperatures is chosen, at each of which an *equilibration* followed by a *sampling* pass is carried out. Successive configurations in a pass are obtained by attempting to flip the spin at one site, $\sigma_n \to \sigma'_n = -\sigma_n$, the change in energy calculated (noting that $\sigma'_n - \sigma_n = -2\sigma_n$ we have $\Delta U = 2H\sigma_n + 2J\sigma_n \sum_{j(n)} \sigma_j$), and the Metropolis algorithm employed to decide on the next configuration. The last sampled configuration at one temperature is conveniently used as the starting configuration for the equilibration pass at the next temperature, as in annealing (ICON = 0 generates a new starting configuration at each temperature if desired). These passes consist of IEQ and ISAMP sweeps, respectively, where a sweep corresponds to N attempts to select a new configuration (flip a spin) by the Metropolis method. Such a sweep can be carried out systematically through the array but is best carried by the, slower, method of selecting the N sites at random. Each of the NSAMP configurations at equilibrium considered for evaluating quantities of interest, e.g. equation 4.1, although separated by a stretch of N configurations (one sweep) may

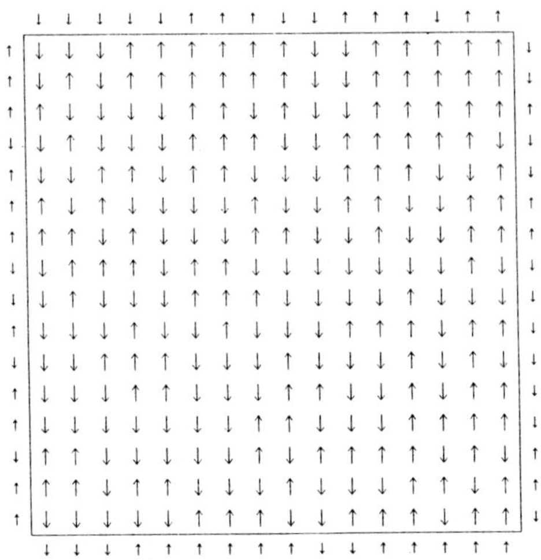

Figure 10.4: A 'hot' starting configuration for an 16×16 dipole system with periodic boundary conditions.

still be too strongly correlated to allow meaningful averages to be calculated. Configurations separated by IFREQ, (≥ 1), instead can be used, i.e. in equation 4.1 we have $K = \text{NSAMP/IFREQ}$. Even then, in certain temperature regions, the correlations among the sample configurations may be such that the simple estimate of the variance, e.g. on the magnetisation, $\hat{\sigma}_M^2 = < M^2 > - < M >^2$, underestimates it and it must be corrected according to the procedures described in section 2.3.3,

$$\sigma_M^2 = \frac{K}{K-1}\hat{\sigma}_M^2 + \frac{K}{K-2J}\hat{\rho}^{(J)}$$

where J is the order of the highest significant correlation occurring in the sequence of samples, and

$$\hat{\rho}^{(J)} = \sum_{j=1}^{J} \hat{\rho}_j'$$

with $\hat{\rho}_j'$ given by equation 2.3.4. The correlation length τ, calculated using equation 2.3.16, is also a useful measure of the degree of independence of the samples included in the calculations, the effective number of independent data points is $K/(1+2\tau)$. In the following, because of the large sample sizes, the subroutine CORREL in the Appendix for calculating the running mean and variance, as well as the correlation

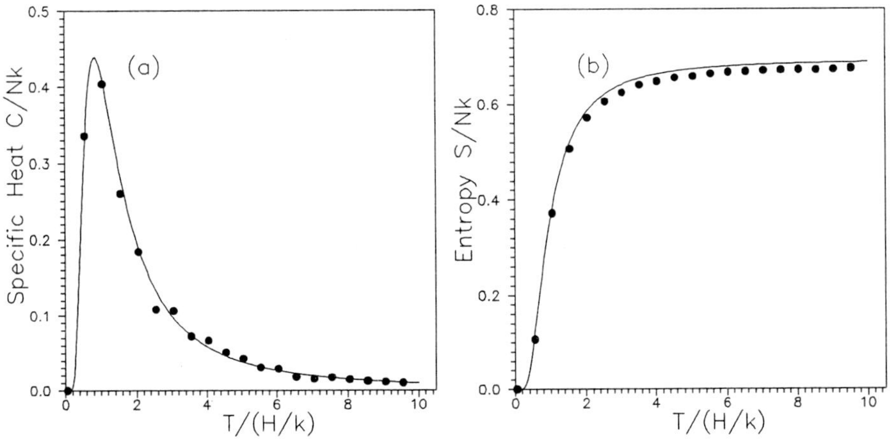

Figure 10.5: Temperature dependence of the properties of a non-interacting dipole system in an external magnetic field (a) the specific heat, and (b) the entropy per particle in units of k_B. The solid curve in each case is the known analytical result.

coefficients, has been used, and the corrected standard deviation on the mean, equation 2.3.15, evaluated using CORVAR. When correlations in a sequence were significant at separation $K/4$ the analysis of the data was abandoned.

A good check on such a programme is to consider the simple case of non-interacting dipoles in an external field, i.e. $J = 0$, which is easily solved for analytically.[4] No extra-boundary dipoles are involved in such a calculation. As an example, we show in Fig. 10.5(a) for a 16×16 array, taking IEQ = 50, NSAMP = 200 and IFREQ = 1, the temperature dependence of the specific heat. Units are chosen such that the temperature $\vartheta = T/(\mu B/k_B)$ and specific heat $c = C/Nk_B$. Comparison is made with the exact analytical result, $c = \vartheta^{-2}\text{sech}^2 1/\vartheta$.

As the comparison shows, fairly modest effort in this case reproduces the exact results. The entropy, <S>, can also be easily calculated in this case — in a configuration with n_+'up' spins it is $k_B \ln[N!/n_+!(N - n_+)!]$ — in units of Nk_B the result for large N is

$$s = \ln(2\cosh 1/\vartheta) - \vartheta^{-1}\tanh 1/\vartheta.$$

The discrepancy with the simulated results seen in Fig. 10.5(b) is a consequence of the 'large N' qualification on the theoretical expression.

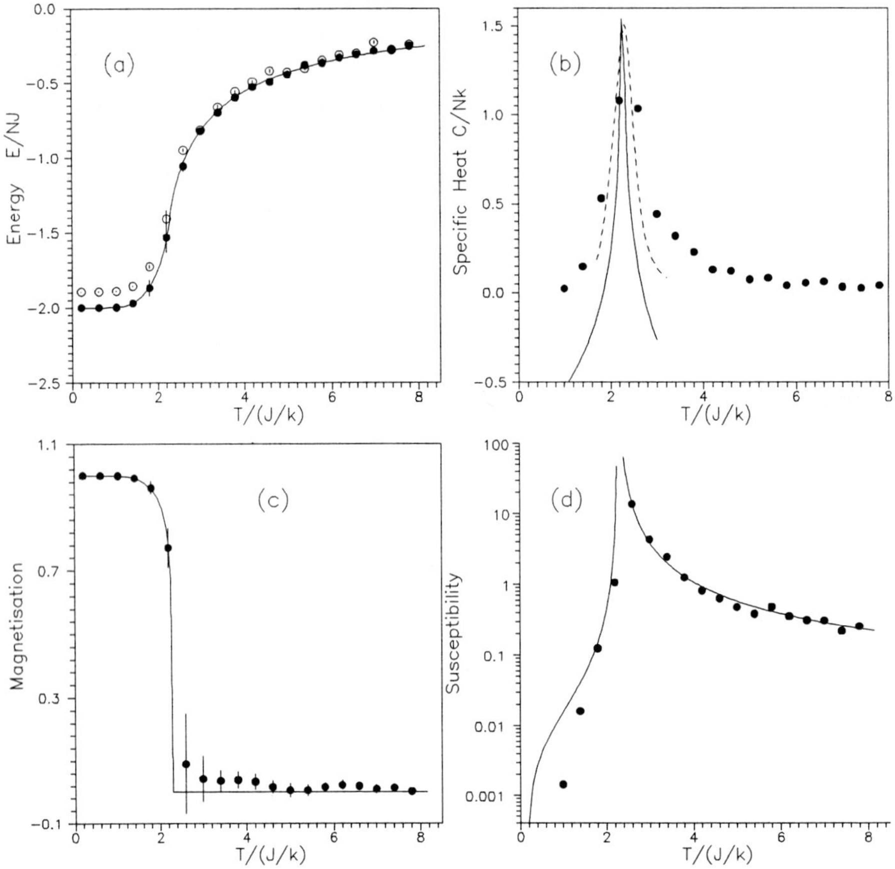

Figure 10.6: Results for the Ising model in zero magnetic field with periodic boundary conditions, and their comparison with analytical results — solid curves. The quantities plotted are (a) the energy — closed and open circles correspond to the case of periodic boundary conditions and an isolated system, respectively, (b) the the specific heat — the dashed curve here is the result of simulations using histogram analysis described later in the text, (c) the magnetisation and, (d) the susceptibility per particle.

The case with $B = 0, J \neq 0$ is more interesting. Although this model has also been solved in two dimensions the solution is a complicated exercise, a computer study of it will reveal more directly the underlying behaviour. At the same time it will show up some of the difficulties and subtleties which attend such a calculation, and prepare one for the three-dimensional problem, for which no analytical solution is known. The results for a 16×16 array, with IEQ $= 500$, NSAMP $= 200$, IFREQ $= 4$ are shown in Figs 10.6(a) to 10.6(d); the units are such that the magnetisation per dipole $m =< M > /N\mu$, and susceptibility $\chi = \chi_B/(N\mu_0\mu B)$. Compared to the smooth behaviour seen in the paramagnetic case, a sharp transition in the specific heat, and magnetisation, in the neighbourhood of $T = 2$ (in units of J/k_B), is observed, with rather large errors in the magnetisation in the transition region. The latter are a reflection of the large variance in the magnetisation, and in the energy, as reflected in the plots of the susceptibility and the specific heat shown. Such behaviour is characteristic of a phase change; in the present case a transition from a paramagnetic to a ferromagnetic state. In Fig. 10.6(a) we also compare the results for the 16×16 array when considered in isolation (denoted by open circles) with our results using periodic boundary conditions.

We can compare some of these results with the analytical expressions calculated for a system of infinite size.[5] In units J/k_B where the temperature $\vartheta = k_BT/J$, the transition occurs at a value

$$\vartheta_c = -2/\ln(\sqrt{2} - 1) = 2.2692.$$

The theoretical functions are compared with our simulation results in Fig. 10.6. Although the comparison in these figures for a relatively small system is encouraging, we note the rather large errors in the transition region. Without the theoretical curves superimposed, it would not be easy to accurately locate the transition temperature, or to determine the temperature dependence of the various quantities in its neighbourhood.

10.5 Improved Algorithms

We now look at methods for improving the efficiency of Monte Carlo calculations of the type described in the previous example. Studies in the region of the phase transition are of the greatest interest, both for thermal systems and in analogous lattice gauge theory calculations. The study of *critical exponents*, and the phenomenon of *finite size scaling* as described in chapter 4 on percolation, which are concepts central to an understanding of the results of these models for macroscopic size systems, depend on the behaviour of the specific heat and susceptibility in the region of the transition. For this reason the transition region must be studied in fine detail, a task which is made even more difficult by the greater errors which occur in this region (see Fig. 10.6) due to the increased fluctuations occurring there as well as to the appreciable correlations which arise in the sampling. To improve the algorithm, we will look in turn at methods which maximise the information extracted from the programme as described so far, and, later, describe improvements to the programme which will accelerate the rate at which independent samples of configurations are generated.

10.5.1 Histogram Analysis

In carrying out the simulation at two successive temperatures in the previous example almost all the information generated in the simulation at the previous temperature has been cast aside, except, possibly, for taking the final configuration as the new starting configuration. The two simulations have been treated almost as independent calculations. However, it is clear that there must be considerable overlap between the properties of the systems simulated when the temperature difference between them is small. We could hope that a more comprehensive understanding of the temperature dependence of quantities might be extracted. This is especially important when the fluctuations of some quantities in the temperature region of interest become large. These considerations are the basis for the *multiple histogram*, or *spectral density* method.[6] It is possible, on the basis of this method, to use data from a simulation at a single temperature, or at a small number of temperatures, to map out the temperature dependence over a finite region, e.g. we might hope to locate the position of the peak in the specific heat accurately and economically. We will start with a description of the single histogram method.

At a given temperature, the distribution in energy of the system, at equilibrium, is given by

$$p(U,\beta_r) = g(U)e^{-\beta_r U}/Q(\beta_r),$$

where we specify a reference temperature, β_r, at which a simulation is carried out. Here $g(U)$ is the number of configurations having an energy U — the density of states — and the configuration sum is,

$$Q(\beta_r) = \sum_U g(U)e^{-\beta_r U} .$$

If, in the simulation, the K sampled configurations at 'equilibrium', are grouped in terms of U to form a histogram, $H(U,\beta_r)$ — in the Ising model calculation these can be just the discrete U values — an approximation to the energy distribution is

$$p(U,\beta_r) \simeq h(U,\beta_r) = H(U,\beta_r)/K.$$

We can thus approximate the density of states by a scaled histogram,

$$g(U) \simeq Q(\beta_r)h(U,\beta_r)e^{\beta_r U} .$$

We can therefore write at any other temperature, β,

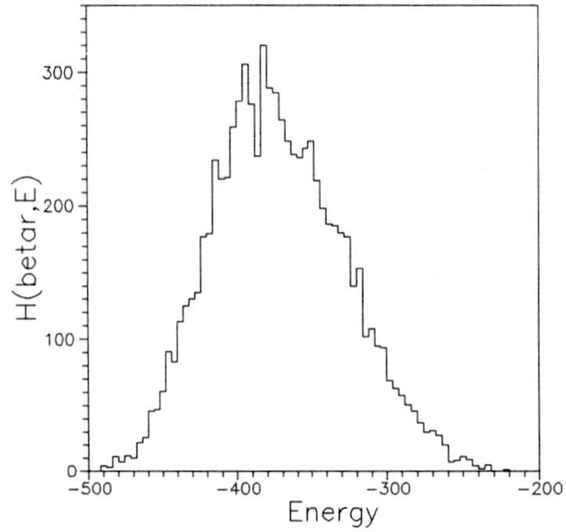

Figure 10.7: The energy histogram for the Ising model calculation in example 10.4 at a reference value of $\beta = 0.4407$.

$$p(U,\beta) = Ch(U,\beta_r)e^{-(\beta-\beta_r)U} , \tag{5.1}$$

where the, undetermined, normalising constant, $C = Q(\beta_r)/Q(\beta)$, is independent of U. By selecting energies from this distribution rather than selecting individual configurations, the thermal average of any function of the energy could be calculated at the temperature β. We will now apply these ideas to the Ising model example.

EXAMPLE 10.4 (continued)

Returning to our Ising model example, with the same parameters as before, we have taken $\beta_r = 1/\vartheta_c = 0.4407$, equilibrated from a hot start with IEQ = 1000 sweeps, and taken the subsequent sample of NSAMP = 8000 sweeps to form the energy histogram $H(U,\beta_r)$. This is shown in Fig. 10.7. At any value of β, on the basis of this histogram, the average of any function of U can be evaluated using expression 5.1, e.g.

$$< U^n >= \frac{\sum_i U_i^n H(U_i, \beta_r) e^{-(\beta - \beta_r) U_i}}{\sum_i H(U_i, \beta_r) e^{-(\beta - \beta_r) U_i}},$$

where the sums are over the histogram cells — in practice an offset of the energy will have to be introduced to avoid overflow in the calculation. This has been done to estimate $<U>$ and $<U^2>$, and thus the variance on the energy, for a range of temperatures around β_r. The corresponding values of specific heat are shown in Fig. 10.6(b) by the dashed curve.

When compared with the exact result it is seen that a very good determination of the behaviour of the specific heat in the neighbourhood of the reference temperature can be obtained on the basis of a simulation at just one temperature. It is clear, however, that there are limitations on the usefulness of a phase space estimator derived at a single temperature in this way. Because of the limited statistical accuracy with which the histogram can be determined, the accuracy with which different regions of U will be sampled by configurations at the reference temperature will vary, $\propto \sqrt{H(U, \beta_r)}$, while the role of these different regions in determining the thermal average at other temperatures will be different, potentially introducing a bias into the calculations. Over a limited region of temperature around the reference value, however, we can see that a statistically well determined histogram, $H(U, \beta_r)$, can be very useful in determining temperature dependences. The more β differs from β_r in equation 5.1, the more important the value of the histogram in the tails becomes in determining the value of the integrals, but this is just the region where, because of statistics, its determination is least accurate. The method of multiple histograms uses bridging between density of states histograms generated at a few separated temperatures; it provides a valuable method for estimating the partition function, and thus the Helmholtz free energy — the closely related technique of umbrella sampling is discussed in section 10.6.3.

10.5.2 Critical Slowing Down

As the temperature approaches the transition temperature in example 10.4 the extent to which sampled configurations are independent of each other rapidly decreases. This is seen in the increasing errors in the magnetisation based on the same number of samples in Figs 10.6(a) and (c). This unwelcome feature becomes more obvious with the increasing size of the system, and calculations become more and more demanding of computer time as the transition temperature is approached. Quantitatively we can see this from the behaviour of the correlation length, where the correlation length for the energy, τ_U can be estimated for a sample of size N by equations 2.3.3d and 2.3.16; more explicitly

$$\begin{aligned}\tau_U &= \sum\left(1-\frac{k}{N}\right)\frac{\mathrm{cov}(U(t),U(t+k)]}{\sigma_U^2}\\ &\simeq \frac{N}{2s^2}r^{(J)},\end{aligned}$$

with $r^{(J)}$ and s^2 determined using equations 2.3.11 and 2.3.12, respectively. The quantity τ_U+1 is shown as a function of the array side L at the critical temperature of the infinite lattice, corresponding to a value of $\beta = 0.441$, by the solid points in Fig. 10.8. Accurate estimation of these correlations requires a large amount of computation; the data in the figure are based on samples of only 2000 sweeps at each array size, and are subject to increasing fluctuations as L increases. At large values of L a dependence, $\tau_U \approx L^2$, is expected. Remembering that to maintain the same level of statistical

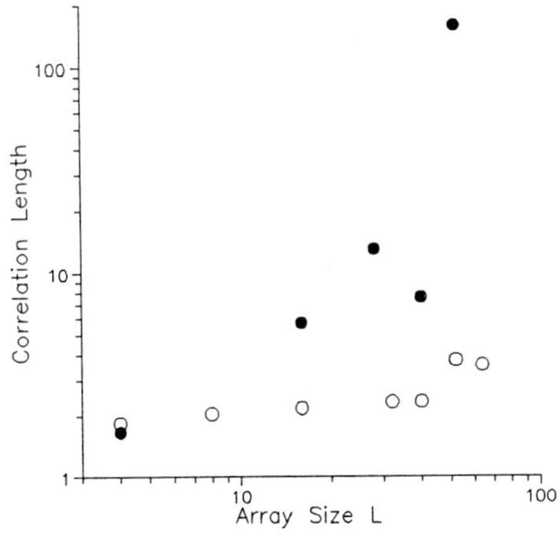

Figure 10.8: The energy correlation length for the Ising model calculation at the infinite lattice critical temperature, $\beta = 0.441$, for different array sizes — solid points. The open points are the results obtained using the cluster algorithm described in section 10.5.3.

accuracy, i.e. a constant number of equivalent *independent* samples, the actual sample size must be increased by a factor of about 2τ (section 2.3.3), and that the actual work done per sweep itself is $\propto L^2$. We see that the factor by which the computation effort scales with array size in this case is $\approx L^4$. It is clear that some new techniques will be required to extend this type of calculation to 'large' L. The phenomenon described is known as *critical slowing down*, and its origins may be seen by referring to Fig. 10.9. Fig. 10.9(a) shows, for a 64×64 array at $\vartheta = 2.3$, the spin configuration resulting after 700 sweeps, starting from a wholly random arrangement. The nett magnetisation is 8.1%. In Fig. 10.9(b) a configuration with the same 8% magnetisation but generated at random is shown. The much greater presence of extended clusters in the former case is very obvious. When the array evolves systematically to such a configuration it is very difficult, using the single spin flip algorithm, to access configurations which are markedly different from it, in a short number of steps, and so the correlations in a

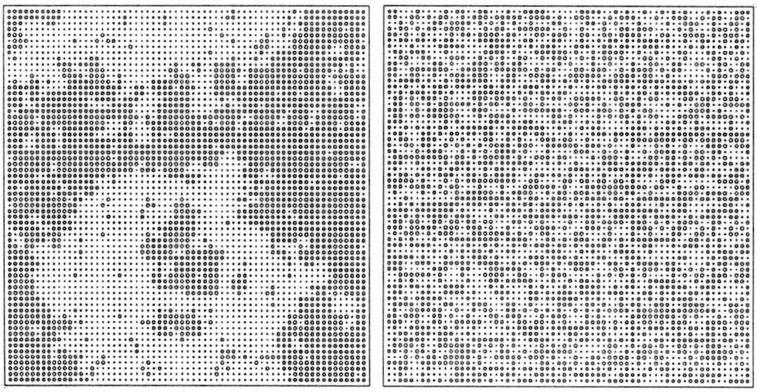

Figure 10.9: Two configurations with the same magnetisation (a) the result after 700 sweeps near the critical temperature through the array starting from a zero magnetisation random configuration, and (b) a randomly generated array with the same magnetisation.

sequence of generated configurations is very great. Since parallel spins are energetically favoured (for $J > 0$), there is a relatively low probability for a spin in the interior of a cluster to be flipped, and if it is there is a strong probability that it will be restored on the next sweep. Only spins on the interface between clusters play a significant role in bringing about new configurations, gradually causing one cluster to grow at the expense of the other. This boundary movement is plausibly of a diffusive nature, with motion in computer time t proportional to $\sqrt{t}/L^2$, and so the rate of access to new configurations can be expected to be correspondingly slow.

This phenomenon of critical slowing down is well known in algorithms which relax a configuration of coordinates to an 'equilibrium' state, e.g. in iterative solutions of finite difference method equations or in sampling paths in a path-integral calculation. In all of these cases relaxation to the asymptotic solution becomes ever more slow, and essentially for the same reason as here — the slow diffusion of information from boundaries. Among the methods of improving on such algorithms we may mention *over-relaxation* and the *Multigrid* method in which the array is redefined through a sequence of grids of various sizes. Here we shall describe another, elegant, approach for expanding the range of configurations sampled, motivated by the obvious similarity of Fig. 10.9(a) to the clusters occurring in a percolating system, chapter 4, which can be applied to the Ising model.

10.5.3 Cluster Algorithms: The Swendsen-Wang Algorithm

As the temperature is decreased in an Ising model calculation, in the absence of an applied field, clusters of spins of both signs will tend to expand as we approach the transition. This gives rise to a near trapping of the system in a pocket of phase space which can only be overcome by a collective rearrangement of large blocks of spins, rather than individual ones. The Swendsen-Wang algorithm exploits a mapping between a configuration of spins and a bond configuration in a percolating system to carry out such collective rearrangements.[7] A recourse here to methods used in studying percolation will not come as a surprise, especially if one compares the analogous behaviour of magnetic moment and susceptibility in Figs. 10.6(c) and (d) with percolation probability and non-spanning cluster size shown in Figs. 4.3(a) and (b). In the case of the percolating system, long range order — percolation — sets in when the occupation probability exceeds the critical value p_c. In the Ising case an analogous onset of coherence occurs when $\beta > \beta_c$ ($T < T_c$). We note, however, that critical slowing-down is absent in the percolating problem; globally new configurations can be generated in sequence because we don't have to worry about any equilibrium condition. For the case of the Ising model, or more generally a q-state Potts model, configurations at equilibrium can be directly related to bond configurations of a percolating system (and the thermodynamic properties of the magnetic system inferred from the analogous properties of the percolating system in the neighbourhood of the transition[8]). Here we will just describe how the percolation analogy can be used to design an algorithm which selects configurations which have small correlation, and hence see how critical slowing down can be, to a large extent, overcome.

The algorithm depends on identifying clusters of adjacent like-spins and reversing such clusters rather than individual spins in a single operation. For this purpose we can make use of the cluster handling techniques described in chapter 4. The prescription for defining a spin cluster, and for flipping a spin configuration — the transition probability of going from a configuration $\{\sigma\}$ to one $\{\sigma'\}, W(\sigma, \sigma')$ — must be such that the occurrence probability of any spin configuration, $p(\{\sigma\}, \beta)$, is the Boltzmann distribution at β. The Markov chain defined in the algorithm satisfies detailed balance,

$$p(\{\sigma\}, \beta)W(\sigma, \sigma') = p(\{\sigma'\}, \beta)W(\sigma', \sigma),$$

and defines spin clusters in such a way to ensure this. At any value of β not all immediately adjacent like-spins adhere to a cluster, rather *bonds* are established between two such spins with the temperature dependent probability,

$$p = 1 - e^{-2/\vartheta}, \qquad \vartheta = 1/\beta J.$$

Clusters of such bonds (along with isolated spin sites) are the objects that are flipped. All the spins in each cluster are flipped with the probability described below.

The algorithm, for the Ising model in a magnetic field for which we write the temperature $\vartheta = T/(J/k_B)$, consists of four steps.

(i) Convert the spin configuration into a bond configuration by creating a bond between neighbouring pairs of spins with probability $p = 1 - e^{-2\delta(\sigma_i, \sigma_j)/\vartheta}$. Nearest neighbours with parallel spins, $\sigma_i = \sigma_j$, are linked with a probability $p = 1 - e^{-2/\vartheta}$, with no bonds occurring between unlike spins.

(ii) Determine all the distinct clusters of sites connected by bonds — including single site clusters — and label them. If $H \neq 0$, find the number of sites in each such cluster, its size $S(\lambda), S \geq 1$, where λ is the cluster label — this latter step will not be necessary in the absence of an applied field.

(iii) Reassign the spin value at each site in cluster λ independent of its previous value, all the sites in the cluster assuming a value σ_λ as follows:

▷ $\sigma_\lambda = -1$ with probability $p = (1 + e^{2rS(\lambda)/\vartheta})^{-1}$,

▷ $\sigma_\lambda = +1$ with probability 1-p,

where $r = H/J$. For the case of $H = 0$, these probabilities are each equal to 1/2. Do this for all clusters.

(iv) Ignore the bonds; the resulting spin configuration is the new sample configuration.

When there are sizable spin clusters present in the starting configuration it is easy to see that its transformed spin configuration may be one which is very different from it, and which, using sequential spin flips, could only be accessed from it after a very long time.

The difficult stage in the algorithm is the cluster monitoring one, but we have seen in the section on percolation, chapter 4.4, how this may be done. In Fig. 10.10 we see the stages of the Swendsen-Wang algorithm, incorporating the Hoshen-Kopelman cluster counting algorithm for a 6×6 array at $\vartheta = 2.27$ with $H = 0$. Fig. 10.10(a) shows a typical spin configuration at this temperature (solid and open circles indicate the two spin directions). The establishment of bonds between like spins according to the above prescription — the probability is 59% — resulted in the bond configuration shown in Fig. 10.10(b). Five clusters of bonds can be seen, two with a single spin each, one containing two spins, one containing five spins and the remainder forming a cluster of 27 spins. When the spins associated with these clusters were reversed with probability 0.5, the cluster of five retained its spin direction while the remaining four clusters each reversed their direction. The resulting spin configuration of the array, with periodic boundary conditions restored, is seen in Fig. 10.10(c). Because the algorithm is known to generate equilibrium configurations of the system, this is a *bona fide* configuration of the spins at this temperature, yet one could imagine how much effort would be required using single spin flip steps to progress from the configuration in Fig. 10.9(a) to that in Fig. 10.10(c). Such an algorithm greatly reduces the correlation between successive configurations and in many simulated models greatly reduces the effect of critical slowing down.

EXAMPLE 10.4 (continued)

For our Ising model in zero applied field, as in the case of the Metropolis algorithm at the infinite-lattice critical value, $\beta = 0.441$, the cor-

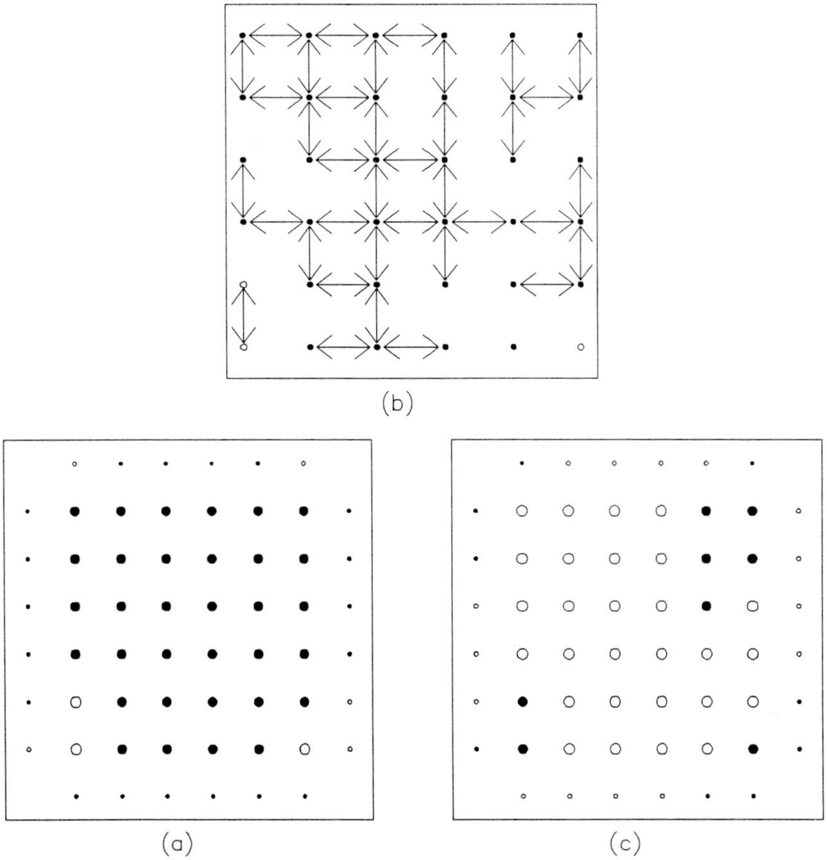

Figure 10.10: The stages in the Swendsen- Wang algorithm for a 6×6 array with periodic boundary conditions, (a) a typical configuration with the two spin directions indicated by open and closed circles, (b) the resulting bond configuration, generated according to the prescription given, forming five clusters, and (c) the final configuration as a result of randomly reversing clusters of spins.

relation length as a function of array size L has been found, based on samples of 2000 sweeps following 2000 equilibrating sweeps from a 'cold' start. The quantity $1 + \tau_U$ is shown in Fig. 10.8 by the open circles. A comparison with the corresponding quantity obtained using the local updating algorithm, as described in section 10.4, and shown by the solid points in the figure shows that critical slowing down is almost completely absent when the global — cluster — algorithm is used.

10.6 Simulation of Fluids

10.6.1 General

The study of fluids and solids of atomic, molecular and ionic species is a major field for the application of Monte Carlo methods to Statistical Mechanics problems. This is especially so at high densities where traditional methods like virial expansions become inadequate. We bypassed it earlier in the chapter, not because of its unimportance, but rather because, in addition to the features we described earlier — periodic boundary conditions, etc. — additional complications arise which make it unsuitable for developing the methods of Monte Carlo application to thermal physics problems.[9]

At its simplest a study will typically investigate the occurrence of a phase transition in a system of N particles of a single species interacting via a specified interparticle potential energy function $u(r)$. The coordinates involved in specifying a configuration will simply be the spatial coordinates of the N particles, $\mathbf{r}^{(N)} \equiv (\mathbf{r}_1, \mathbf{r}_2, \ldots, \mathbf{r}_N)$, confined to a volume V. Simulations can readily be carried out at a fixed temperature or fixed volume. Quantities of interest will be the internal energy, equation 3.2 and the specific heat, equation 3.3; more precisely these quantities augmented by the contribution of kinetic energy terms, $3N/2\beta$ and $3Nk_B/2$, respectively, the ideal gas contribution for structureless particles. If the potential function $u(r)$ has continuous derivatives we define $v(r) = r du/dr$ and write the *Virial* function,

$$\mathcal{V}(r) = \sum_{i<j}\sum v(r_{ij}) = \sum_{i<j}\sum r_{ij} du(r_{ij})/dr_{ij}.$$

The thermal average of this quantity is related to the pressure through the equation[10]

$$\frac{pV}{N} = k_B T - \frac{<\mathcal{V}>}{3N}. \tag{6.1}$$

The pair correlation function or radial distribution function $g(r)$, which is a measure of the structure of a fluid, is also important — it can be compared with its corresponding experimental values determined from the angular distribution of X-rays scattered by the fluid. This function measures the probability of finding a particle at a distance r from any given particle, and so if $\rho = N/V$ is the average density of particles, the average number of particles in the system whose separation lies in $(r, r + dr)$ is

$$4\pi r^2 \rho g(r) dr,$$

with the normalisation,

$$\int 4\pi r^2 \rho g(r) dr = N - 1 \simeq N.$$

For a given value of r, a sphere of radius r surrounding the ith particle in a particular configuration will contain $n_i(r)$ particles. The average of this over the N particles,

$$G(r) = \frac{1}{N}\sum_{1}^{N} n_i(r),$$

when averaged over configurations yields an integral distribution from which $g(r)$ can be obtained by differentiation,

$$g(r) = \frac{1}{4\pi r^2 \rho} \frac{d < G(r) >}{dr}.$$

When non-spherical particles are involved, e.g. dipole-dipole interactions, pair correlation functions in angular variables can also be defined. The correlation function offers an alternative method for obtaining the pressure by using the relationship[11]

$$p = \frac{\rho}{\beta} - \frac{\rho^2}{6} \int_0^\infty 4\pi r^3 \frac{du(r)}{dr} g(r) dr.$$

The demarcation of phase transitions in these problems simply on the basis of energy derivatives, e.g. specific heats, becomes difficult in small systems because of the relatively large contribution of inter-phase surface energies and the tendency to separate into one phase which often results in trapping of the system near the transition temperature. Less ambiguous results may be obtained from the temperature dependence of the free energy F. Calculation of this may be attempted using thermodynamic integration of the energy and pressure in the canonical ensemble, since

$$({}^{\partial \beta F}/_{\partial \beta})_{N,V} =< E >, \qquad ({}^{\partial F}/_{\partial V})_{N,T} = -p, \tag{6.2}$$

where $< E >=< U > +$ the mean kinetic energy <K>, $3N/2\beta$ for structureless particles. A less time consuming and cumbersome method is to use the partition function and the fundamental relation, equation 3.1, and some elegant techniques have been developed for this purpose, one of which we describe in section 6.3 of this chapter.

The potential energy term,

$$U(\mathbf{r}^{(N)}) = \sum_{i=1} \sum_{j>i} u(r_{ij}), \tag{6.3}$$

involves the interparticle interactions. Different levels of complexity in these interparticle potentials are possible. A simple example which has been widely studied is the hard sphere potential

$$\begin{aligned} u(r) &= \infty \text{ for } r < r_0 \\ &= 0 \text{ for } r > r_0. \end{aligned}$$

The implementation of the Metropolis algorithm is very straightforward in this case (although the virial function is not defined), and it serves as a suitable starting point in developing a programme to work in this field. One of the most studied interparticle potentials for a system of molecules is the Lennard-Jones potential,

$$u(r) = 4\epsilon[(\frac{\sigma}{r})^{12} - (\frac{\sigma}{r})^{6}],$$

which represents a severe short range repulsion plus a milder long range attraction, with a well depth $-\epsilon$ at an equilibrium separation of $r_0 = \sigma 2^{1/6}$. With this potential it will be convenient to take units of σ^{-3} for number density and ϵ/k_B for temperature. When considering charged particles or polar molecules, longer ranged, e.g. Coulomb, potentials must be used.

With a potential specified for the problem, temperature dependences can be studied by averaging over sampled configurations, e.g. using the Metropolis algorithm as described earlier, and the various thermodynamic quantities estimated. The procedure, however, is usually less straightforward than in our discussion of the fixed dipole system. In general, unless core storage is at a premium, the array of energy values, $u_{ij} \equiv u(r_{ij}) = u_{ji}$, and a parallel array, v_{ij}, will be stored rather than recalculated at every step of the Markov chain. After a tentative move of a molecule,

$$\mathbf{r}_k \rightarrow \mathbf{r}_k + \delta, \qquad r_{jk} \rightarrow r'_{jk} = \mid \mathbf{r}_k - \mathbf{r}_j + \delta \mid,$$

the values $u(r'_{jk})$ must be calculated and the change in energy,

$$\Delta U = \sum_{j \neq k} u(r'_{jk}) - u(r_{jk}), \tag{6.4}$$

found. The values of u_{jk}, and v_{jk}, for $j \neq k$ are updated when a new configuration is accepted. Without going into excessive technical detail we note some of the complications which arise in the next paragraph.

10.6.2 Pitfalls

A problem may occur in sampling configurations at relatively high density, and ergodicity may not be as obvious as in the case of the Ising model simulation. Interchange of particles may be very difficult (this will be a function of the potential employed, and might reflect 'long' relaxation times in a real physical system), and the necessarily finite Markov chain may become effectively trapped in a limited region of configuration space — it is *quasi-ergodic.* One way to minimise the possible occurrence of this effect is to use more than one chain in the simulation, with the initial state of each of them sensibly different. Another method, which we describe in the next section, is that of non-Boltzmann or umbrella sampling.

In almost all cases periodic boundary conditions will be used to reduce the contribution of surface effects. The extent to which the system is surrounded by periodic image cells will depend on the range of the forces (or their truncated approximation). Whenever a tentative movement brings a particle outside the basic cell it is brought into the cell, and all the images, through the opposite wall, so that only the motions of the N particles in the core cell come into the calculations directly. The larger the basic system the more realistic are these periodic conditions as an approximation to a macroscopic system; as long as the scale of significant fluctuations is not much greater than the basic cell size the simulation should closely mimic the properties of a large system.

The convenient feature of nearest neighbour only interactions will no longer be viable, and a scheme allowing for a truncation of the range of interaction becomes necessary, if only to have a manageable number of terms in the expression for the energies which have to be calculated at every attempted step in the Markov chain, see equation 6.4. A common prescription is the *minimum-image distance* method, in which only N particles are included in the energy calculation at all times, i.e. a total of $N(N-1)/2$ interactions and distinct values in the arrays u_{ij} and v_{ij} above, and a total of N-1 terms in equations 6.3 and 6.4. If the distance of a particle j from particle i in the core cell is greater than the distance from i to the image nearest to it of particle j in the surrounding cells, it is this latter interaction alone which is included. Sometimes it is adequate to exclude the interaction of any such pair when its separation is greater than a value r_c, where r_c has a value smaller than half the edge side of the core cell (typically $2\sigma - 3\sigma$ in the case of a L–J fluid). Corrections may be necessary to compensate for such a cut-off, e.g. if interested in the pair correlation function at distances larger than half the edge of the core cell.

When using the Metropolis algorithm a decision must be made on the displacement step size to be used in generating tentative new configurations. Considerations along the lines described in section 6.2.3 on simulated annealing, reflecting a compromise between the exploration of the configuration space not being too slow and frequent acceptance of tentative configurations will be relevant. In particular the step size may need to be changed at different stages of a simulation. Acceptance of tentative states in more than 50% of moves is desirable although the actual value is a matter of trial and error, guided by the requirement of a thorough exploration of the configuration space.

The existence of a phase transition might be revealed in a plot of an isotherm, e.g. βp versus V, but the trapping, superheating and supercooling, alluded to earlier will make it very difficult to accurately locate the transition. In addition, high accuracy studies of such phenomena are much more time consuming than for the lattice spin system described earlier; acceptance of a trial flip of a spin may be of the order of 100 times faster for the latter system than a corresponding move of a particle.

10.6.3 Umbrella Sampling

It has so far seemed very natural to use the Boltzmann function as the importance function in finding the equilibrium values of any quantity, call it $A(\mathbf{r})$, i.e.

$$< A >= \int d\mathbf{r}\pi(\mathbf{r})A(\mathbf{r}),$$

with $\pi(\mathbf{r}) = e^{-\beta U(\mathbf{r})}/Q_\beta$. The important regions are just those where the energy $U(\mathbf{r})$ is smallest. It may happen that in the configuration space two regions exist in which small values of U occur separated by a 'barrier' region where the energy is relatively large. In this case it is unlikely, at the temperature in question, that exploration will ever move from one of the regions to the other when sampling with the Boltzmann distribution; the trapping that we saw when discussing simulated annealing, in section 6.2.3, occurs. We are at liberty, however, to take a different importance function $\pi^*(\mathbf{r})$, i.e. we write

$$< A >= \int d\mathbf{r}\pi^*(\mathbf{r})\{\frac{\pi(\mathbf{r})}{\pi^*(\mathbf{r})}A(\mathbf{r})\},$$

where $\pi^*(\mathbf{r})$ allows for selection which avoids trapping in one of the regions.[12] We need a sampling scheme which extends the region in configuration space which is explored. Such a method of sampling the canonical ensemble from a non-physical distribution which is more spread over the configuration space than results from the Boltzmann distribution is known as *umbrella sampling.* Sometimes the Boltzmann distribution is also a convenient form for $\pi^*(\mathbf{r})$ with, for example, a different (higher) temperature,

$$\pi^*(\mathbf{r}) = e^{-\beta' U(\mathbf{r})}/Q_{\beta'}.$$

The equilibrium value of A is then $Q_{\beta'}/Q_\beta$ times the average of the quantity

$$A(\mathbf{r})e^{-(\beta-\beta')U(\mathbf{r})}.$$

The ratio $Q_\beta/Q_{\beta'}$ is readily seen to be the average using π^* of $e^{-(\beta-\beta')U(\mathbf{r})}$. Although this may improve the ergodicity of the sampling it is also likely to negate much of the advantage of employing importance sampling. We can be more bold in a choice of importance function, instead of changing the temperature we can change the interaction potential — no bias will be introduced since it is compensated for by the quantity averaged. Thus we can take

$$\pi^*(\mathbf{r}) = e^{-\beta U'(\mathbf{r})}/Q'_\beta,$$

with the result being the average of the quantity

$$A(\mathbf{r})e^{-\beta(U(\mathbf{r})-U'(\mathbf{r}))}$$

divided by

$$Q_\beta/Q'_\beta = < e^{-\beta(U(\mathbf{r})-U'(\mathbf{r}))} >_{\pi*} .$$

For example if we know the details of the barrier region we could take $U'(\mathbf{r}) = U(\mathbf{r}) - W(\mathbf{r})$ with

$$\begin{aligned} W(\mathbf{r}) &= B, \text{ for } \mathbf{r} \text{ in the barrier region,} \\ &= 0 \text{ elsewhere,} \end{aligned}$$

where B is the approximate relative height of the barrier.

The above example is not the only possible use of umbrella sampling. Obviously an arbitrary interaction potential or temperature, or both, can be employed to explore, more or less efficiently, the configuration space of interest.

Notes

1. N. Metropolis, A. W. Rosenbluth, M. N. Rosenbluth, A. H. Teller and E. Teller (1953), *Equation of State Calculations by Fast Computing Machines*, J. Chem. Phys., **21**, 1087-92. This paper laid the foundations for simulation studies in statistical mechanics, introducing both importance sampling using Markov chains and periodic boundary conditions.

2. Brady, J. W., J. D. Doll and D. L. Thompson (1981), *Cluster dynamics: A classical trajectory study of $A_n^* \to A_{n-1}^* + A$*, J. Chem. Phys., **74**, 1026- 28. The thermal demon is described in Creutz, M. (1983), *Microcanonical Monte Carlo Simulation*, Phys. Rev. Lett., **50**, 1411-14, and in chapter 15 of part 2 of H. Gould and J. Tobochnik, *An Introduction to Computer Simulation Methods* (Reading, Mass.: Addison Wesley, 1988).

3. See for example chapter 3.3 of D. A. McQuarrie, *Statistical Mechanics* (New York: Harper & Row, 1976).

4. See for example chapter 3 of F. Mandl, *Statistical Physics* (London: John Wiley & Sons, 1971).

5. These 'exact' results for an infinite array are; for the energy per dipole:

$$< E > /NJ = -\coth(2/\vartheta)\{1 + \frac{2}{\pi}K_1(\zeta)[2\tanh^2(2/\vartheta) - 1]\}$$

where

$$\zeta = 2\sinh(2/\vartheta)/\cosh^2(2/\vartheta),$$

and K_1 is the complete elliptic integral of the first kind; for the magnetisation per dipole (C. N. Yang(1985), Phys. Rev., **87**, 404):

$$\begin{aligned} < M > /N &= \{1 - \sinh^{-4}(2/\vartheta)\}^{1/8} \quad \vartheta \leq \vartheta_c, \\ &= 0 \quad \vartheta \geq \vartheta_c, \end{aligned}$$

while in the neighbourhood of the transition, $\vartheta \approx \vartheta_c$, the specific heat is given by,

$$C/Nk_B \approx \frac{2}{\pi}(\frac{2}{\vartheta_c})^2\{-\ln \mid 1 - \frac{\vartheta}{\vartheta_c} \mid + \ln({}^{\vartheta}c/2) - 1 - \frac{\pi}{4}\}$$

The exact behaviour of the specific heat for a lattice with finite L can be found in A. E. Ferdinand and M. Fischer (1969), *Bounded and Inhomogeneous Ising Models. I. Specific-Heat Anomaly of a Finite Lattice*, Phys. Rev. **185**, 832-846. The susceptibility is given by, (E. Barouch, B. M. McCoy and T. T. Wu (1973), Phys. Rev. Lett., **31**, 1409),

$$\chi \simeq \beta[c_{0\pm} \mid 1 - \vartheta_c/\vartheta \mid^{-7/4} + c_{1\pm} \mid 1 - \vartheta_c/\vartheta \mid^{-3/4}]$$

where the subscript +(-) refers to $\vartheta > (<)\vartheta_c$ and the coefficients are

$$c_{0-} = .0255369719, c_{0+} = .9625817322,$$

$$c_{1-} = -.0019894107, c_{1+} = .0749881538.$$

6. The method of histogram analysis is described by A. M. Ferrenberg and R. H. Swendsen (1988), *New Monte Carlo Technique for Studying Phase Transitions*, Phys. Rev. Lett., **61**, 2635-38. Its long historical background is given by the same authors in *Optimized Monte Carlo Data Analysis*, Computers in Physics, **3**, 101-104, (1989).

7. The algorithm was introduced in R. H. Swendsen and J-S. Wang (1987), *Nonuniversal Critical Dynamics in Monte Carlo Simulations*, Phys. Rev. Lett., **58**, 86-88. A clear justification for it, along with the listing of a Fortran programme for its implementation, is given in J-S. Wang and R. H. Swendsen (1990), *Cluster Monte Carlo Algorithms*, Physica A, **167**, 565-79.

8. The determination of energy, specific heat, magnetisation and susceptibility via the average number of occupied bonds, the fluctuations on this number, the percolation probability and the non-spanning cluster size in a percolating array is described by C-K Hu and K-S Mak (1989), *Monte Carlo study of the Potts model on the square and simple cubic lattices*, Phys. Rev., B**40**, 5007-14.

9. For a practitioner's account see J. P. Valleau and S. G. Whittington, *A Guide to Monte Carlo for Statistical Mechanics: 1. Highways,* in Bruce

J. Berne (Ed.), *Statistical Mechanics, Part A: Equilibrium Techniques,* (New York: Plenum, 1977), pp 137-68. Much of the formalism of the theory of classical fluids, and elucidation of many of the difficulties we only point out can be found in an unsurpassable account by Allen, M. P. and D. J. Tildesley, *Computer Simulation of Liquids* (Oxford: Clarendon Press, 1987); they also provide many valuable programme segments suitable for original work in this field. A valuable collection of reprints, with commentaries, is Ciccotti, G., D. Frenkel and I. R. McDonald (Eds.), *Simulation of Liquids and Solids* (Amsterdam: North-Holland, 1987).

10. For example, section 13.3 of reference 3.

11. See section 13.23 of reference 3.

12. Umbrella sampling is described and applied to a Lennard Jones fluid in Torrie, G. M. and J. P. Valleau (1977), *Nonphysical Sampling Distributions in Monte Carlo Free Energy Estimation: Umbrella Sampling,* Jour. Comput. Phys., **23**, 187-99.

Exercises

10.1 Write a routine which uses the Metropolis algorithm to generate variates from the distribution

$$p(x)dx = Ce^{-\beta V(x)}dx,$$

with

$$V(x) = 6x^4 - 2x^3 - 9x^2 + 5,$$

using a targeting step $x_i \to x_j = x_i + (2\xi - 1)\delta$, where ξ is a uniform random deviate, and the acceptance probability is $\min[1, p(x_j)/p(x_i)]$.

Make a plot of $V(x)$.

For values of $\beta = 0.1, 0.2$ and 1.0 generate samples of 1000 variates, in each case starting with $x_0 = 1$, and a value of δ which results in about 50% acceptance of attempted moves.

Compare histograms of your samples with distribution $p(x)$ above.

This example illustrates quasi-ergodicity, although in principle all the parameter space will be explored, in some cases — corresponding to low temperatures — the walker may become trapped in some region of the space.

10.2 For the four-state dipole system shown ($s = 1/2$),

i		
1	$\Uparrow$	$\Uparrow$
2	$\Downarrow$	$\Uparrow$
3	$\Uparrow$	$\Downarrow$
4	$\Downarrow$	$\Downarrow$

for which the energy of a configuration, in units of k_BT, is given by

$$\Phi_i = 1 + s_{1z} + s_{2z},$$

use the Metropolis algorithm to construct a transition matrix which would select configurations distributed as the Boltzmann distribution. Only one spin flip is to be allowed in a step.

Chapter 11

QUANTUM MONTE CARLO

> ...we could imagine and be perfectly happy, I think, with a probabilistic simulator of a probabilistic Nature ...
>
> Richard P.Feynman

11.1 Introduction

We may not be in a position to do what would have pleased Feynman,[1] but we will go half way; we will see how deterministic simulation of probabilistic problems in nature can be carried out. In this chapter we look at the application of Monte Carlo methods in problems in quantum mechanics and quantum statistical mechanics. The topics broadly fall under the heading of Diffusion Monte Carlo, Green Function Monte Carlo and Path Integral Monte Carlo. One of the welcomed, and unexpected, features of the solution of quantum problems by stochastic methods is that the solutions can, in principle be exact. This contrasts with other commonly used methods for solving such problems, such as variational or perturbation methods, in which a residual uncertainty in the result is unavoidable.

11.1.1 Background

Rather than starting with the Hamiltonian formalism of classical mechanics and requiring the substitutions $\mathbf{p} \to -i\hbar$ grad, $E \to i\hbar\partial/\partial t$, to arrive at his equation for the probability amplitude, Schrodinger was aware that an alternative approach was possible starting instead from the formalism of stochastic processes. This approach is also valuable in itself in giving a different insight into the equations underlying quantum theory, and it has led in recent years to practical (stochastic) methods for solving these equations. Understanding the intimate link which exists between the formalism of quantum theory and that of stochastic processes is also a great asset when approaching an understanding of Feynman's path integral formalism of quantum mechanics — the sum over paths can be rather easily appreciated from a consideration of the solution of problems in classical statistical mechanics.

In what follows we expand on our brief observation on the relation of Schrodinger's equation to random walks (see section 8.3.4) and look into the practical utility of random walks for solving Schrodinger's equation for a multiparticle system. Towards the end of the chapter we will briefly consider how the path integral formalism follows from these considerations, and how Monte Carlo methods may be used in evaluating path integrals.

A revealing insight into the relation between random walks and Schrodinger's equation can be obtained from a 'hand waving' argument along the following lines. A fundamental starting point for quantum theory is Heisenberg's Uncertainty Principle. At any time t the position and momentum of a particle are uncertain by amounts δx and δp, respectively — we can think of these quantities as standard deviation uncertainties. They are related via the inequality

$$\delta x \delta p \geq \hbar/2.$$

For non-relativistic motion we can write

$$\begin{aligned}\delta p &= m\delta v \\ &= m\delta(dx/dt) \\ &= md(\delta x/dt).\end{aligned}$$

Thus, taking the minimum uncertainty case, we have

$$\delta x d(\delta x) = \frac{\hbar}{2m} dt.$$

If we integrate this over a time interval δt (not representing an uncertainty in the time) we have the relation

$$\frac{(\delta x)^2}{2\delta t} = \frac{\hbar}{2m}.$$

The uncertain change in x is related to the time step in the same way as the space and time steps are related in a diffusive random walk with a constant diffusion coefficient, the latter given by

$$D = \hbar/2m.$$

We can arrive at this physically motivated result also in a purely mathematical way.

11.1.2 Schrodinger's Equation

The relationship of random walks to quantum mechanical equations has already appeared in section 8.3.4, where we noted the parallel between Schrodinger's time dependent equation for a particle moving in a potential $V(r)$,

$$\frac{\partial \Psi}{\partial it} = \frac{\hbar}{2m}\nabla^2\Psi - \frac{V(r)}{\hbar}\Psi, \tag{1.1}$$

and the equation for the distribution of particles diffusing subject to decay with a mean lifetime, μ^{-1}, which depends on position,

$$\frac{\partial p}{\partial t} = D\nabla^2 p - \mu(r)p.$$

The diffusion coefficient, $\hbar/2m$, is just the expression we arrived at on physical grounds earlier, while the decay probability is directly proportional to the value of the potential. The essential difference between the two equations is the necessity in the quantum mechanical case to consider imaginary values of the time variable. The Schrodinger equation and the diffusion equation are related by analytic continuation in the time variable. This same feature relates Feynman's path integrals to the Weiner integrals of statistical mechanics. A second difference relates to interpretation. While the function $p(r,t)$ in the diffusion equation represents a probability, or concentration, it is the square of the function $\Psi(r,t)$ which we usually associate with probability in quantum mechanics. These apparent difficulties notwithstanding, this parallel with random walks is a practical starting point for solving Schrodinger's equation which we now describe.

11.2 Diffusion Monte Carlo

11.2.1 The Method

As we have noted the relationship between the diffusion equation for unstable particles and his own equation was realised by Schrodinger himself,[2] while the possibility of solving it by simulated random walks was suggested by Fermi.[3] The equation, which for ease of presentation we will deal with in one dimension is

$$i\hbar\frac{\partial\Psi}{\partial t} = \frac{-\hbar^2}{2m}\frac{\partial^2\Psi}{\partial x^2} + V(x)\Psi. \tag{2.1}$$

As is shown in introductory Quantum Mechanics texts the general solution consistent with a reference state $\Psi(x,0)$ at $t = 0$ may be written as an expansion in the eigenfunctions of the Hamiltonian, i.e.

$$\Psi(x,t) = \sum c_n\psi_n(x)e^{-iE_nt/\hbar}, \tag{2.2}$$

where the constants c_n are the coefficients of the expansion of the reference state at $t = 0$.[4] For atomic problems, it is convenient to choose $\hbar/\alpha^2mc^2$ as the unit of time and $\hbar/\alpha mc$ as the unit of length, where $\alpha \equiv e^2/4\pi\epsilon_0c\hbar \simeq 1/137$. With this rescaling of the time and space variables, i.e. writing

$$\tau = t/(\hbar/\alpha^2mc^2),$$

$$z = x/(\hbar/\alpha mc),$$

we have the numerically manageable unit of energy $\alpha^2mc^2 \simeq 27eV$. Writing $\mu(z) \equiv V(z)/\alpha^2mc^2$, the equation takes the form

$$i\frac{\partial\Psi}{\partial\tau} = -\frac{1}{2}\frac{\partial^2\Psi}{\partial z^2} + \mu(z)\Psi. \tag{2.3}$$

In practical problems some energy other than the rest mass may be found more convenient in defining the time unit.

Although derived for real values of the variable τ, mathematically this equation is equally valid for complex τ. Treating this more general case will, we shall see, enables us to find some solutions in a simple way. We now consider equation 2.3 for complex values of τ, in particular for pure imaginary values of τ, which we write as $\tau = -i\vartheta$, ϑ real — in the literature this transformation, a clockwise rotation of $\pi/2$ in the complex plane, is known as a Euclidean rotation. Equation 2.3 then takes the form

$$\frac{\partial \Psi}{\partial \vartheta} = D\frac{\partial^2 \Psi}{\partial z^2} - \mu(z)\Psi, \tag{2.4}$$

with $D = 1/2$. This is just the diffusion equation with capture (multiplication when $\mu < 0$). The function $\Psi(z, \vartheta)$ mimics the time dependent distribution of a 'gas' of particles diffusing in space. We can thus try to solve for Ψ by random walk methods using the formalism for this purpose developed for the diffusion equation in chapter 8 — the algorithm is known as Diffusion Monte Carlo (DMC). We know that its solution can be written

$$\begin{aligned} \Psi(z,\vartheta) &= \sum c_n \psi_n(z) e^{-\epsilon_n \vartheta} \\ &= c_0 \psi_o(z) e^{-\epsilon_0 \vartheta} \{1 + \approx \sum_{n=1} \psi_n e^{-(\epsilon_n - \epsilon_0)\vartheta}\} \end{aligned} \tag{2.5a}$$

where $\epsilon_n \equiv E_n/\alpha^2 mc^2$, and ϵ_0 is the smallest of the eigenvalues, i.e. the ground state energy in the potential $V(x)$. At sufficiently large values of ϑ all the terms in the sum will become negligible compared with unity giving

$$\Psi(z,\vartheta) \xrightarrow[\vartheta \to \infty]{} \psi_0(z) e^{-\epsilon_0 \vartheta}, \tag{2.5b}$$

i.e. the ground state wavefunction modulated by the exponential factor. We have assumed $c_0 \neq 0$, i.e. the initial reference state is not orthogonal to this ground state. In general the solution would tend to the state of lowest energy which is not orthogonal to $\Psi(x,0)$. In a random walk simulation of this diffusion equation 2.4, $\Psi(z,\vartheta)$ would be represented by the density of walkers at z at time ϑ, so equation 2.5*b* tells us that, starting with walkers distributed as $\Psi(z,0)$, if we carry out a walk for a sufficiently long time the positions of walkers should be distributed according to the function ψ_0. This is true for *any* initial distribution $\Psi(z,0)$ — provided only that it is not orthogonal to ψ_0, i.e. $c_0 \equiv < \Psi(z,0) \mid \psi_0 > \neq 0$. In practice, however, depending on whether E_0 is greater than or less than zero, the total number of walkers at ϑ may all have disappeared or have multiplied uncontrollably due to the exponential factor in equation 2.5*b*. A *population-stabilising* algorithm must be introduced. We can do this by considering equation 2.1 for a related potential,

$$\mathcal{V}(x) = V(x) - E^*,$$

where E* is a constant — at the moment unknown. The eigenvalues of this new problem will just be $(E_n - E^*)$, where E_n are the eigenvalues of the original problem. For our new problem equation 2.5*b* then takes the form, with $\epsilon^* \equiv E^*/\alpha^2 mc^2$,

$$\Psi(z,\vartheta) = \psi_0(z) e^{-(\epsilon_0 - \epsilon^*)\vartheta} . \tag{2.6}$$

We see that if, in selecting E^*, one had the luck to choose it equal to E_0 the total number of walkers would, in the long time limit, remain constant. In other words an estimator for the ground state energy E_0 is just that value of the constant E^* which stabilises the number of walkers in the random walk simulation dictated by

$$\begin{aligned} D(z) &= 1/2, \\ \mu(z) &= \mathcal{V}(z)/\alpha^2 mc^2, \end{aligned}$$

and that stable distribution is the ground state wavefunction.

The eigenvalue E_0 can be converged on by altering E^* in a feedback manner during the progress of the simulation; an initial value is guessed and a total of N walkers processed through one step of the walk. If the resulting number of walkers is greater (less) than N, we know from equation 2.6 that the value of E^* chosen was too large (small). The value of E^* should then be decreased (increased) by some increment, and the number of walkers rescaled to N, e.g. by deleting (replicating) at random from among the total number. This new set of N is then processed through another step, and the process repeated. This can be done for 'many' steps — large ϑ — and terminated when a value of E^* which stabilises the population is found. This value of E^* is the *growth* estimator of the ground state energy E_0,

$$\epsilon_0 = \lim_{\vartheta \to \infty} \epsilon^*, \tag{2.7}$$

and the distribution of the walkers follows the ground state wavefunction ψ_0. In practice, the revision of E^* and population renormalisation should not be done after every step, but after every fixed number of steps, to overcome the effects of the fluctuations. Of course, for a finite sample N, there will be fluctuations in the stabilising value E^*. E^* does not settle to a unique value, its mean value over a sequence of samples in the asymptotic region can be calculated to estimate E_0.

We can arrive at an alternative estimator for the ground state energy by integrating equation 2.4, including E^*, over all space,

$$\int_{-\infty}^{\infty} dz \frac{\partial \Psi}{\partial \vartheta} = D \int_{-\infty}^{\infty} dz \frac{\partial^2 \Psi}{\partial z^2} - \int_{-\infty}^{\infty} dz [\mu(z) - \epsilon *]\Psi,$$

and using the boundary conditions that grad(Ψ) must vanish at infinity to eliminate the first integral on the right-hand side. If we consider the $\vartheta \to \infty$ limit, we can use equation 2.6 to rewrite the left-hand side to get

$$-\epsilon_0 \int_{-\infty}^{\infty} dz \Psi(z, \vartheta) = - \int_{-\infty}^{\infty} dz \mu(z) \Psi(z, \vartheta),$$

or

$$\epsilon_0 = \overline{\epsilon} \equiv \lim_{\vartheta \to \infty} \int_{-\infty}^{\infty} dz \mu(z) \Psi(z, \vartheta) / \int_{-\infty}^{\infty} dz \Psi(z, \vartheta). \qquad (2.8)$$

The quantity $\overline{\epsilon}$ is known as the *local* estimator of the energy. In a practical calculation where the wavefunction is characterised by a sample of N coordinates drawn from it we have

$$\overline{\epsilon} \simeq \frac{1}{N} \sum_{1}^{N} V(z_i) / \alpha^2 mc^2. \qquad (2.9)$$

Practical aspects of the DMC algorithm are best described in the context of an example.

EXAMPLE 11.1: The Morse potential

As in all Monte Carlo applications the value of the method is only obvious when multivariable problems are encountered, however we shall use a simple example in one dimension, the motion of a particle of mass m in a Morse potential, $V(x)$, to illustrate the method, where

$$V(x) = V_0 (e^{-2a(x - x_0)} - 2e^{-a(x - x_0)}).$$

It is now convenient to scale the time and space variables as $\tau \equiv t/\omega^{-1}$ and $z \equiv (x - x_0)/\sqrt{\hbar/m\omega}$, with $\omega \equiv a\sqrt{2V_0/m}$, and as before $\vartheta \equiv i\tau$. The unit of energy is now $\hbar\omega$. Equation 2.4 then assumes the form

$$\frac{\partial\Psi}{\partial\vartheta} = D\frac{\partial^2\Psi}{\partial z^2} - \mu(z)\Psi,$$

with $D = 1/2, \mu(z) = \mathcal{V}(z)/\hbar\omega = 0.5\zeta^{-2}(e^{-2\zeta z} - 2e^{-\zeta z}) - \epsilon^*$ where $\zeta^2 \equiv \hbar\omega/2V_0$ and $\epsilon^* \equiv E^*/\hbar\omega$. We consider the case of $\zeta = 1$. To develop an algorithm we note that, since no convective term occurs in the equation 2.4, we take $T_+(z) = T_-(z) = T(z)$. One of the step sizes $\Delta\vartheta$ and Δz is arbitrary; we will fix $\Delta\vartheta$. We can consider the two mechanisms, decay and stepping, separately and write $2T = 1$, then $\Delta^2/2\Delta\vartheta \equiv D = 1/2$, i.e. the step size $\Delta = \sqrt{\Delta\vartheta}$. The value of the potential at some points, and therefore the magnitude of μ, may be very large and a better, but slower, expression for the decay probability is $1 - \exp(-\mu\Delta\vartheta)$. An initial distribution for the walkers must be chosen; the final result does not depend on this distribution but a good choice (in a serious calculation it might be based on an earlier variational estimate of the wavefunction) will lead to faster convergence. Starting them all at the origin is an obvious choice; in the present case selecting them from a uniform distribution will lead to faster convergence.

Figure 11.1 shows a flowchart for a basic program to carry out this algorithm. The array Z, contains NWALK coordinates at input, i.e. a sample from the trial wavefunction. With the units used, a typical scale for the problem would be unity; a uniform distribution on [-1,1] has been adopted. The array Z should be declared larger than NWALK to allow for birth of new walkers. A starting value for ϵ^* must be guessed; a simple method is to use the initial distribution of walkers to find $\bar{\epsilon}$ from equation 2.9. The method, as we know, is exact in the limit of vanishing $\Delta\vartheta$ so we should use the smallest possible value. On the other hand a very small value initially will require very many steps for the walkers to redistribute themselves from the initial arbitrary distribution. A relatively large value may be used for the early stages of the walks, which is later reduced to a small value before any results are extracted.

A walker at z is processed as follows:

- it is stepped to left or right by an amount STEP. Rather than

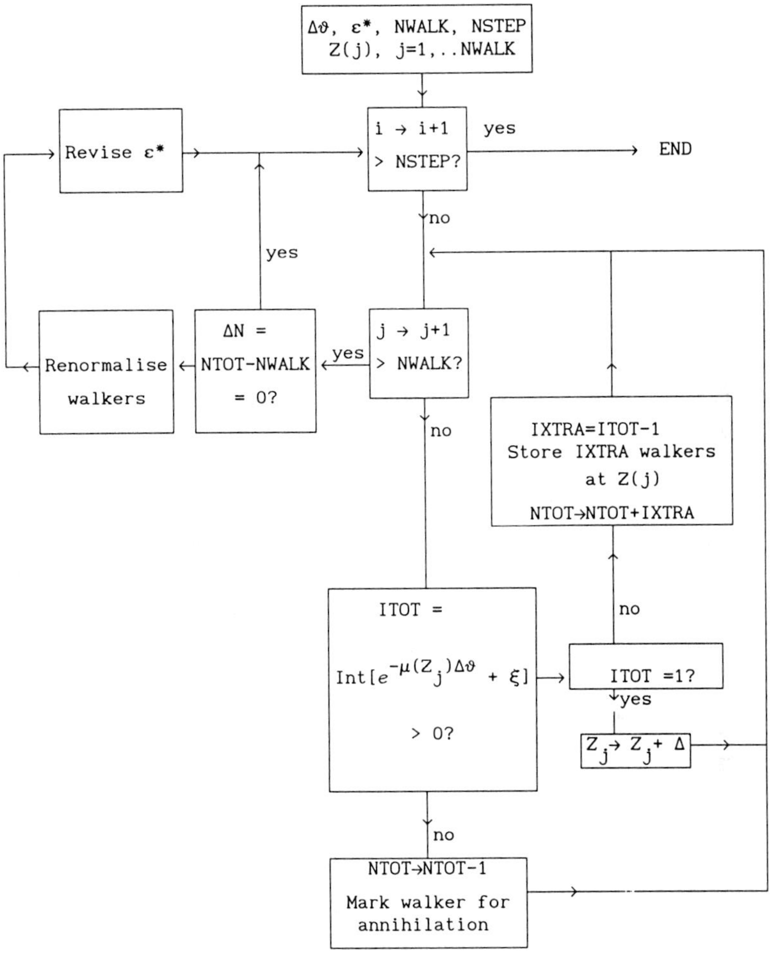

Figure 11.1: A flowchart for a random walk solution of Schrodinger's equation — the method of Diffusion Monte Carlo (DMC).

a fixed step size, Δ, it is better to select STEP from a continuous distribution having a standard deviation Δ; a variate from the uniform distribution on $(-\Delta\sqrt{3}, \Delta\sqrt{3})$ is such a quantity and is used here.

- If ITOT= $[e^{-\mu\Delta\vartheta} + \xi]$ equals zero, here ξ is a unit random variate and the square brackets indicate the integer part, the walker is marked for annihilation at next housekeeping, otherwise it is replaced by ITOT walkers at z.

- When all NWALK walkers have been taken through one step a total of NTOT walkers will exist. The excess, or deficit, $\Delta N =$ NTOT-NWALK is calculated and the number of walkers renormalised as follows: if there is an excess these are eliminated at random from among the walkers while in the case of a deficit the coordinates of $\mid \Delta N \mid$ of the walkers, chosen at random, are duplicated. The energy ϵ^* is rescaled, e.g. if the sign of ϵ^* is already correct, $\epsilon^* \to \epsilon^* - \alpha \mid \epsilon^* \mid \Delta N/\text{NWALK}$, where α is a numerical factor. Except in the initial stages of the walk, this updating of ϵ^* is not done after every time step but the accumulated excess/deficit over a number of steps, 10–50, can be used. A value of $\bar{\epsilon}$ using equation 2.9 based on the sample can then be calculated, i.e. $(2N)^{-1}\sum_i(e^{-2z_i} - 2e^{-z_i})$.

- This is continued until all walkers have been taken through NSTEP time steps, after which the array Z will correspond to a sample from the ground state wavefunction. Averages over many samplings of ϵ^* and $\bar{\epsilon}$ will yield two estimates of the ground state energy.

The top curve in Fig. 11.2 (based on a population of 300 walkers for 10000 steps with a time step $\Delta\vartheta = 0.02$) shows how ϵ^*(its first estimate was obtained using the initial distribution and equation 2.9, and updated every step until $\vartheta = 10$ and every 25th step thereafter using a value $\alpha = 25$) is influenced by the initial guess and how, when this influence has died out, it fluctuates thereafter about a steady value at large times. The correlations between successive values of ϵ^* means that estimation of the uncertainty in this value needs to take correlations into account, following section 2.3.3; The lower curve shows $\bar{\epsilon}$ (displaced by -0.5) interrogated every 25th step and only based on the single sample of 300 walkers at that time. The effect of finite step size $\Delta\vartheta$ on the estimate of E_0, based on walks of 5000 steps, can be seen

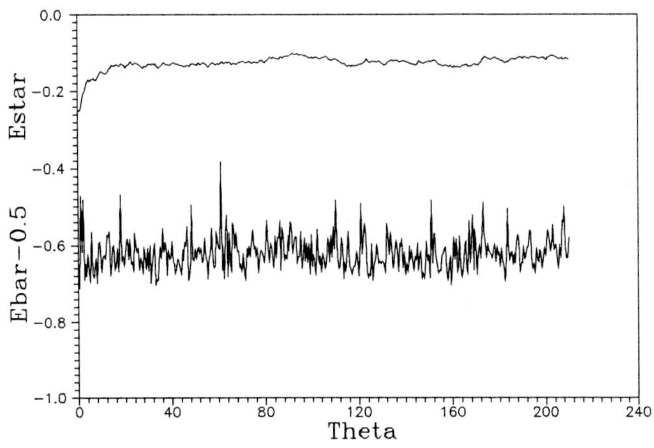

Figure 11.2: The growth estimator ϵ^* of the ground state energy for the Morse potential as a function of simulation time — upper curve. The lower curve shows the corresponding dependence of the local estimator $\bar{\epsilon}$, displaced by 0.5 for clarity of presentation.

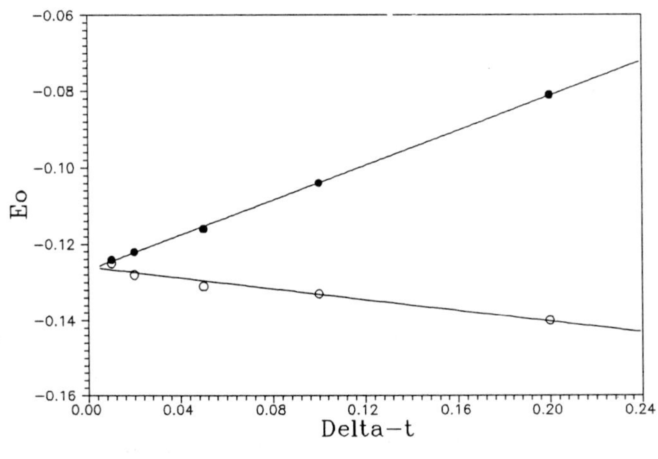

Figure 11.3: The dependence of the ground state energy on the simulation time step $\Delta\vartheta$. The open symbols refer to the growth estimator ϵ^*, and the solid ones to the local estimator $\bar{\epsilon}$.

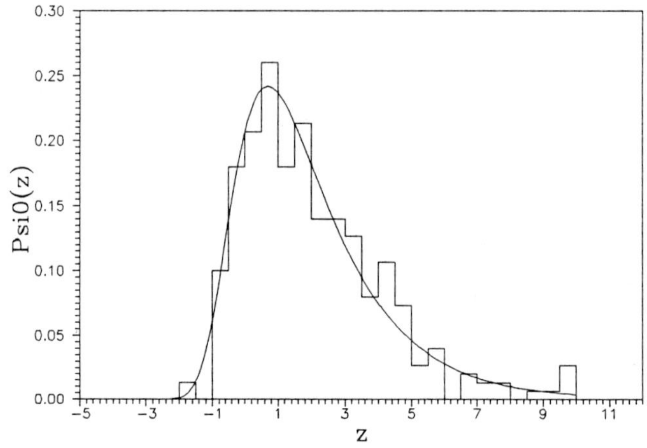

Figure 11.4: A histogram of the final distribution of walkers compared with the known ground state wavefunction $\psi_0(z)$.

in Fig. 11.3 where the open (solid) circles refer to $\epsilon^*(\overline{\epsilon})$. The trend toward the known value of the ground state energy, in these units $\epsilon_0 = 1/2 - \zeta^2/8 - 1/2\zeta^2 = -0.125$, as $\Delta\vartheta \to 0$ is clearly seen. We also see that whereas this limiting value is approached from below in the case of the growth estimator, the local estimator approaches it from above — as in a variational calculation.
The distribution of the walkers at the end is shown as a histogram in Fig. 11.4, where it is compared with the known ground state wavefunction

$$\psi_0(z) = \frac{1}{\sqrt{\pi}} e^{-(z/2 + e^{-z})} , \tag{2.10}$$

the solid curve in the figure. This function is normalised such that

$$\int_{-\infty}^{\infty} \psi_0(z) dz = 1.$$

The solution of the Morse potential can be achieved with a moderate sized walk (a relatively large step size, and tolerable fluctuations in the estimators). Other potentials which show more rapid variation or exhibit singular behaviour, such as the Coulomb potential, will show much greater fluctuations, due to rapid multiplication induced by large negative values of the potential, and less certain convergence to a solution, and the method is only practical in a refined version which suppresses these fluctuations.

11.2.2 Variance Reduction

In the example just discussed, relatively large fluctuations in the stabilising energy occur — these can be seen for the data in Fig. 11.2 when presented on an enlarged scale as the top curve in Fig. 11.5; because of these, convergence to a value for E_0 is slow, and it is difficult to be confident of when a reliable value is reached. This occurs mainly because of the great variability of the branching probability μ, which depends on $V(x)$. In addition we can expect some error of truncation origin in the best estimate because of the finite time step used in the random walk. To get a substantive improvement using DMC some form of variance reduction technique must be incorporated in the algorithm. Importance sampling on the random walk can be devised in which configurations from a distribution different from Ψ are generated but which converge to the same energy. We now describe such a scheme.[5] We shall continue to seek a solution of equation 2.4 corresponding to a Hamiltonian

$$\begin{aligned} H &= -\frac{\alpha^2 mc^2}{2}\frac{\partial^2}{\partial z^2} + \mathcal{V} \\ &= H_0 - E^* \end{aligned} \tag{2.11}$$

where again we have replaced V by $\mathcal{V} = V - E^*$, with E^* to be determined from a solution of

$$\frac{\partial \Psi(z,\vartheta)}{\partial \vartheta} = D\frac{\partial^2 \Psi(z,\vartheta)}{\partial z^2} - \frac{\mathcal{V}(z)}{\alpha^2 mc^2}\Psi(z,\vartheta), \tag{2.12}$$

with $D = 1/2$. We introduce an importance function, not depending on time but about which we know nothing else at the moment, $\phi_T(z)$, and multiply it into equation 2.12 on both sides. Then, defining

$$\rho(z,\vartheta) \equiv \phi_T(z)\Psi(z,\vartheta)$$

the function ρ satisfies an equation of the following form,

$$\frac{\partial\rho}{\partial\vartheta} = D\frac{\partial^2\rho}{\partial z^2} - \frac{\partial}{\partial z}(\frac{2D}{\phi_T}\frac{d\phi_T}{dz}\rho) - (\frac{\mathcal{V}(z)}{\alpha^2mc^2} - \frac{D}{\phi_T}\frac{d^2\phi_T}{dz^2})\rho.$$

Using equation 2.11 we can replace d^2/dz^2 in the last bracket by

$$\frac{d^2}{dz^2} = \frac{-H+\mathcal{V}}{D\alpha^2mc^2}.$$

The equation then takes the form,

$$\frac{\partial\rho}{\partial\vartheta} = D\frac{\partial^2\rho}{\partial z^2} - \frac{\partial}{\partial z}(v(z)\rho) - \mu(z)\rho, \tag{2.13}$$

with $D = 1/2$,

$$v(z) = \frac{2D}{\phi_T(z)}\frac{d\phi_T}{dz}, \tag{2.14a}$$

$$\mu(z) = \epsilon_{\text{loc}}(z) - \epsilon^*, \tag{2.14b}$$

$$\epsilon_{\text{loc}}(z) = \frac{H_0\phi_T(z)}{\alpha^2mc^2\phi_T(z)}. \tag{2.14c}$$

From this latter quantity we immediately recognise the relationship of the, so far ill defined, importance function ϕ_T to the trial wavefunction occurring in Variational Monte Carlo in chapter 7. This suggests that ϕ_T should mimic, as far as possible, the solution we are seeking $\psi_0(z)$. We are now left with equation 2.13 to solve, i.e. to find the value of ϵ^* which will result in a steady state $\rho(z)$, in which case $E_0 = \alpha^2mc^2\epsilon^*$, and in principle,

$$\psi_0(z) = \rho(z)/\phi_T(z).$$

The stochastic interpretation of equation 2.12 or 2.13 is clear; from our discussion in section 8.3.5 we recognise the first and third terms in this Fokker-Planck equation representing diffusion and branching/annihilation, respectively; in addition we have a term representing drift of the walkers with a local drift velocity $v(z)$. This velocity, $\propto \mathrm{grad}(\ln \phi_T)$, is directed towards regions where the importance function is large, and will thus encourage walkers to spend time in the 'important' regions of the space. There are two attractive features of this formalism, first the drift has the effect at any point of preferentially diverting 'walkers' so that regions where the trial function is large are sampled, without biasing the final outcome of the walk. Second, if ϕ_T is a reasonable approximation to the wavefunction, the multiplication term, ϵ_{loc}, involving the 'local' energy will show much smaller fluctuations (none at all in the case where ϕ_T coincides with ψ_0) than occur using $V(z)$. This can be so even if $V(z)$ contains singularities, which give rise to great fluctuations in the simple approach.

The distribution from which we are asymptotically sampling using this method, even though we don't know it explicitly, is $\phi_T\psi_0$ and the mean of the local energy over a sample is an estimate of the integral

$$\frac{\int dz\rho(z)\epsilon_{\mathrm{loc}}(z)}{\int dz\rho(z)}, \tag{2.15}$$

which, using equation 2.14*c*, gives

$$\frac{1}{\alpha^2 mc^2}\frac{< \psi_0 \mid H \mid \phi_T >}{< \psi_0 \mid \phi_T >} \tag{2.16}$$

$$= \frac{E_0}{\alpha^2 mc^2}. \tag{2.17}$$

The result is exact, allowing for truncation and statistical errors, and *independent of the importance function* ϕ_T. This is to be contrasted with

Variational Monte Carlo using a similar trial function, where the configurations sampled are from $|\phi_T|^2$, resulting in the estimation of an upper bound

$$\frac{1}{\alpha^2 mc^2}\frac{<\phi_T \mid H \mid \phi_T>}{<\phi_T \mid \phi_T>},$$

rather than the exact value implied by equation 2.16. As before, we have both a growth estimator ϵ^*, equation 2.7, which stabilises the population of the ensemble, and a local estimator equation 2.15, estimated by

$$\bar{\epsilon} \simeq \frac{1}{N}\sum_1^N \epsilon_{\text{loc}}(z_i). \tag{2.18}$$

A practical algorithm is only slightly different from that described in example 11.1. The starting points of the walkers $\{z_i\}$ should be sampled from the distribution $|\phi_T|^2$, which for configurations would typically be done using the Metropolis algorithm, the local energy calculated using equations 2.18 and 2.11 and used as a starting value of ϵ^*. A walker at z_i is stepped by an amount $\pm\Delta + v(z_i)\Delta\vartheta$ (where, as before, Δ is a variate from a distribution with variance $2D\Delta\vartheta$) and replaced by ITOT walkers, where ITOT= $[e^{-\mu(z_i)\Delta\vartheta}+\xi]$ and μ is given by equation 2.14*b*. Renormalisation of the walker population and energy rescaling is done after all the walkers have gone through a fixed number of steps, ≥ 1, as before.

In three or more dimensions the corresponding relations are, using the original unscaled variables,

$$\frac{\partial\rho}{\partial it} = D\nabla^2\rho - D\ div(\rho\mathbf{v}) - \mu\rho, \tag{2.19}$$

with $D = \hbar/2m, \mathbf{v} = \frac{\hbar}{2m}\,\text{grad}(\ln|\phi_T|^2)$ and $\hbar\mu = (H\phi_T/\phi_T - E^*)$. For a system of N particles, a 'walker' will now correspond to a configuration $\{\mathbf{R}\}$

of $3N$ coordinates and a 'step' to $3N$ steps for these particles (the walkers can be moved one at a time successively or together simultaneously) to arrive at a new configuration. Members of an ensemble of NWALK of such configurations must each be processed through one step, with new members being added or old ones removed as determined by the multiplication term.

Before returning to our example to illustrate this discussion, we enlarge a bit on the relationship between DMC with importance sampling and the Variational Method, mentioned earlier.

11.2.3 Variational-Diffusion Monte Carlo

It is interesting to see that a special case of the method described in the previous section corresponds to the Variational Monte Carlo method discussed in chapter 7. Instead of the Hamiltonian, H, in equation 2.11 let us consider H', where

$$H' = H - \phi_T^{-1} H \phi_T.$$

Since $H'\phi_T = 0$ we know that the ground state energy and and eigenfunction of this Hamiltonian are $\epsilon_0 = 0$ and $\psi_0 = \phi_T$, respectively. Repeating the analysis of section 11.2.2 will give rise to an equation corresponding to equation 2.13 for $\rho(z)$ with D and $\mathrm{v}(z)$ as before, and $\mu(z) \propto H'\phi_T = 0$. In other words, simulating the Fokker-Planck equation with the same diffusion and drift but without decay or multiplication will, in the long run, lead to the walkers being distributed as $|\phi_T|^2$. We will have no growth estimator, but by equation 2.15 the local energy will converge to

$$\epsilon_{\mathrm{var}} = \frac{1}{\alpha^2 mc^2} \frac{< \phi_T \mid H \mid \phi_T >}{< \phi_T \mid \phi_T >}.$$

As a prelude to a DMC calculation it is possible, since no growth estimator needs to be guessed at, to start with a parameterised form $\phi_T(\{\alpha\})$ and an initial distribution $\rho(z,0)$ selected from $|\ \phi_T\ |^2$, and propagate the walkers with diffusion and drift but no annihilation. After some time the local energy will converge to $\epsilon_{\text{var}}(\{\alpha\})$. Of course this could also be done using the methods described in chapter 7, but will be very simple to programme in this way in advance of a full DMC, for which the optimum variational trial function and its associated energy can be used as the starting point.

EXAMPLE 11.2: The Morse potential with importance sampling

We return to the Morse potential considered in example 11.1, and introduce an importance sampling function,

$$\phi_T(z) = \frac{2^\gamma}{\sqrt{\Gamma(\gamma)}} e^{-(\gamma z + e^{-z})} \ .$$

In fact, for $\gamma = 1/2$ this is just the exact ground state wavefunction of the Morse potential; by considering other values of the parameter γ we will be able to see how the 'distance' between the trial function and the true wavefunction affects the calculation.
Using equations 2.14a and 2.14b we have

$$v(z) = e^{-z} - \gamma,$$

$$\epsilon_{\text{loc}}(z) = \frac{1}{2}[e^{-z} - (\gamma - e^{-z})^2] + V(z).$$

The initial sample is selected from $|\ \phi_T(\text{z})|^2$, which using the substitution $u = 2e^{-z}$ gives a probability distribution

$$p(u)du = \frac{1}{\Gamma(2\gamma)} u^{2\gamma-1} e^{-u} du.$$

For values of $\gamma \geq 1/2$ variates from this distribution can be selected using the method in example 5.9.
The walk has been executed under the same conditions as in Example 11.1, i.e. 300 walkers stepped for 10000 steps with $\Delta\vartheta = 0.02$ with

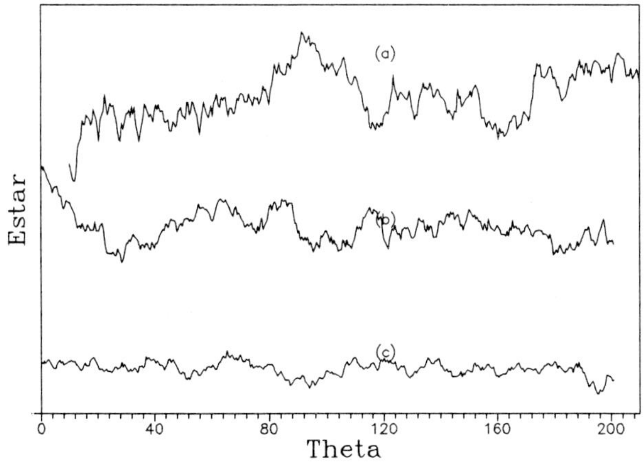

Figure 11.5: The evolution of the growth estimator for the ground state energy determined using importance sampling with the parameter $\gamma = 0.75$ and 0.6 for curves (b) and (c), respectively. Curve (a) correspond to the upper curve in Fig. 11.2, with no variance reduction. The three curves have been displaced for clarity of presentation.

updating every 25 steps. For $\gamma = 0.75$ and 0.6 the resulting ϵ^* as a function of simulation time is shown as curves (b) and (c), respectively in Fig. 11.5, where the results are compared with the results of the simple diffusion algorithm previously shown in Fig. 11.2 — the scale for all three curves is the same. The decrease in fluctuations as the trial function approaches the true wavefunction is clearly seen, for that case, $\gamma = 0.5$, all fluctuations would vanish.

Before concluding this section, we briefly consider two limitations of the methods so far described. All our discussion has concentrated on the ground state energy. If the initial distribution $\Psi(z,0)$ (or ϕ_T when using variance reduction) is orthogonal to the ground state ψ_0, c_0 in equation 2.5a will be zero and the long time distribution will behave as

$$\Psi(z,\vartheta) = \psi_1(z) e^{-(\epsilon_1 - \epsilon^*)\vartheta} .$$

The stabilising energy will tend to the energy of the first excited state which is not orthogonal to $\Psi(z,0)$. In practice, it often will not be easy to arrange such orthogonality for the trial function and the method has had little application to excited states so far.

The second limitation of DMC is that it requires that the wavefunction, obtained as a probability density, is everywhere positive. If the problem only involves bosons the ground state can always be chosen positive everywhere, however, if the wavefunction alternates in polarity in different regions, e.g. due to antisymmetry requirements of the Pauli exclusion principle for identical fermions, problems arise. The method can still be applied separately in each such region of definite sign (the negative of the function being considered in regions where the wavefunction is negative). To do this presupposes knowledge of the boundaries, nodal surfaces, separating these regions, something which we normally don't have. Having a trial function offers a way, albeit approximate, of overcoming this problem. One way of proceeding is to simply guess these nodes and and impose them on the trial function ϕ_T, which thus must have the same polarity as the wavefunction at every point so that the sampled function $\phi_T\psi_0$ is always positive. If in the course of the walk a walker would cross a node, this walker (configuration) is annihilated. This is known as the *fixed node approximation.* When the nodes imposed differ from those in the true wavefunction it can be shown that the energy determined is an upper bound to the true eigenvalue. The problem of applying DMC to fermion problems cannot be said to be fully solved. More elaborate variants of the method described can be developed but they are very demanding of computer time. They are discussed in the research literature.

11.3 Green Function Monte Carlo

11.3.1 The Short Time Green Function

The method of DMC described in the last section can be presented more formally in terms of Green functions, and this formulation can lead us to more accurate algorithms for solving Schrodinger's equation. Schrodinger's time dependent equation in the variables z and ϑ defined in section 11.2.1 becomes

$$\frac{\partial \Psi(z,\vartheta)}{\partial \vartheta} = -H(z)\Psi(z,\vartheta), \tag{3.1}$$

where $H(z) = -D\frac{\partial^2}{\partial z^2} + \mu(z)$ — the diffusion equation. We define the Green function, $G(z,z',\vartheta)$, corresponding to this equation such that

$$\Psi(z,\vartheta) = \int dz' G(z,z',\vartheta)\Psi_0(z'), \tag{3.2}$$

where $\Psi_0(z') \equiv \Psi(z',0)$. This Green function is a solution of

$$\frac{\partial G(z,z',\vartheta)}{\partial \vartheta} = -H(z)G(z,z',\vartheta), \tag{3.3a}$$

$$G(z,z',0) = \delta(z-z'). \tag{3.3b}$$

The form of this function for the free particle case, μ a constant, we have seen before, equation 8.3.8,

$$G_0(z,z',\vartheta) = \frac{1}{(4\pi D\vartheta)^{1/2}} e^{-\frac{(z-z')^2}{4D\vartheta} - \mu\vartheta}. \tag{3.4}$$

We will immediately consider the case with importance sampling included, i.e. the evolution of the function $\rho(z,\vartheta) = \phi_T(z)\Psi(z,\vartheta)$. Equations for this quantity, analogous to equations 3.1, 3.2 and 3.3, take the form

$$\frac{\partial \rho}{\partial \vartheta} = -\hat{\mathrm{H}}\,\rho,$$

where

$$\hat{\mathrm{H}}(z) = -D\frac{\partial^2}{\partial z^2} + \frac{\partial}{\partial z}v(z) + \hat{\mu}(z),$$

and

$$\hat{\mu} = \frac{1}{\alpha^2 mc^2}\frac{H\phi_T}{\phi_T}.$$

The corresponding Green function, defined by

$$\rho(z,\vartheta) = \int dz'\ \hat{\mathrm{G}}(z,z',\vartheta)\rho_0(z'), \tag{3.5}$$

satisfies the equations

$$\frac{\partial\hat{\mathrm{G}}(z,z',\vartheta)}{\partial\vartheta} = -\hat{\mathrm{H}}(z)\hat{\mathrm{G}}(z,z',\vartheta), \tag{3.6a}$$

$$\hat{\mathrm{G}}(z,z',0) = \delta(z-z'). \tag{3.6b}$$

We know that at large values of ϑ the function ρ evolves via equation 3.5 to $\phi_T(z)\psi_0(z)e^{-\epsilon\vartheta}$ with $\epsilon \equiv (E_0 - V^*)/\alpha^2 mc^2$ where V^* is an arbitrary constant added to the Hamiltonian and E_0 is the energy of the ground state. This evolution can be simulated, i.e. samples from $\rho(z,\vartheta)$ can be generated, starting with an arbitrary distribution $\rho_0(z)$ if the function $\hat{\mathrm{G}}(z,z',\vartheta)$ is known. The diffusion algorithm described above is essentially an iteration, or propagation of equation 3.5 through successive short times, sampling from an approximation to the unknown Green function, valid for sufficiently short time intervals, i.e.

$$\rho(z,\vartheta+\Delta\vartheta) = \int dz'\ \hat{\mathrm{G}}(z,z',\Delta\vartheta)\rho(z',\vartheta).$$

The short time Green function for this we have already seen; it is

$$\hat{G}(z, z', \Delta\vartheta) = \frac{1}{(4\pi D\Delta\vartheta)^{1/2}} e^{-\frac{(z-z'-v(z')\Delta\vartheta)^2}{4D\Delta\vartheta} - \mu(z')\Delta\vartheta},$$

with v and μ defined in equations 2.14*a* and 2.14*b*, and is exact when these quantities are constant.

Thus an improvement on the diffusion algorithm described would be to select steps from a Gaussian distribution of mean $z' + v\Delta\vartheta$ and variance $2D\Delta\vartheta$, i.e.

$$z' \to z = z' + \eta\sqrt{2D\Delta\vartheta} + v(z')\Delta\vartheta,$$

where η is a standardised Gaussian variate. A further obvious improvement to lessen the effect of finite $\Delta\vartheta$ is to take the mean value, $[\epsilon_{\text{loc}}(z)+\epsilon_{\text{loc}}(z')]/2$ in the multiplication term.

Rather ingenious Monte Carlo methods can also be used to generate approximations to the exact Green function,[6] but they are too technically sophisticated for description here.

11.4 Some Applications

The methods described in this chapter have been applied in studies of many multiparticle systems, ranging from single atoms through clusters of atoms to samples large enough to show the behaviour of liquids or solids. The relevance of the work depends on how well the interactions between the particles in the system are known. For example, the properties of a gas of electrons distributed throughout a uniform neutralising positive charged fluid have been determined using these methods. Some quite exotic systems can be studied, for example, the binding energy of the molecule formed by an atom — Cl with 18 electrons, and positronium, a hydrogen like atom formed of a positron and an electron — can be estimated.[7]

An interesting three body system features a deuteron, a tritium nucleus and a negative muon, which can form a bound state. Unlike previous cases described, where it is the ground state energy of the system which is of greatest interest, here the interest centres on determination of the wavefunction. The muon behaves as a heavy electron ($m_\mu \simeq 207m_e$) and its correspondingly smaller molecular orbit reduces the separation of the D-T nuclei such that they overlap to an appreciable extent and fusion can occur, i.e.

$$(DT)_{\mu^-} \to n + \ He^4 + \mu^- + \Delta E,$$

with $\Delta E \simeq 17.6$ MeV. This is known as *cold fusion*, it can occur at room temperature, compared with the $\approx 10^8 K$ temperature required to overcome the Coulomb barrier for the case of the isolated nuclei.[8] As we see, the muon is released in the process — it has effectively catalysed the fusion of the nuclei — and may continue to catalyse more fusions. However, the number of fusions a single muon can mediate is restricted by its finite lifetime, ($\tau_\mu \simeq 2.2 10^{-6}$ sec) which limits the number of fusions in which it can partake. More importantly, its role within that time can be frustrated by capture into an atomic orbit about the α-particle to form a muonic He ion, i.e.

$$(DT)_{\mu^-} \to n + (He^4)^+_{\mu^-} + \Delta E.$$

The muon will stay captured until it decays and thus cannot catalyse any more fusions. To decide on the economic feasibility of using muons to assist energy generation by fusion it is necessary to calculate the probability of the muon being taken out of circulation in this way — the so called 'sticking' probability. Fusion only occurs when the D-T nuclear separation is less than about 10^{-12}cm, the range of nuclear forces. The simplest way of viewing the process when the nuclei are so close is an initial state which is a muonic He^5 ion which expels a neutron — this occurring on a time scale much shorter than characteristic atomic rearrangement times. One of the possible final states of the atom is a muonic He^4 ion. The amplitude of a muon in this state $\psi^{(\mathrm{He})}(r_\mu)$ which occurs in the initial wavefunction $\psi^{(i)}(r_\mu, \mathbf{r}_D - \mathbf{r}_T \simeq 0)$ will be the weight of such a final state in the initial state, i.e. the overlap

$$a = \int \psi^{(i)}(r_\mu)\psi^{(\mathrm{He})}(r_\mu)d^3r_\mu.$$

The probability of this final state, the probability of sticking, is given by $|a|^2$. The approximation of the initial state by a He^5 ion is too crude, a more realistic calculation of the wavefunction of the muon by calculating the three particle system wavefunction must be done. Clearly one of the ways of doing this is using the methods developed in this chapter.

Even though the sticking probability calculated is small, $\simeq 0.9\%$, it means that about one in a hundred fusions ties up the muon so that instead of the approximately 1000 synthesis of D-T mu-molecules followed by fusion which it might achieve in its lifetime it can only give rise to about 100. This further reduces the viability of energy generation in this way.

11.5 Quantum Statistical Mechanics

11.5.1 Introduction

We have implicitly assumed that the multiparticle systems referred to throughout the chapter so far are isolated; in practice the properties of such systems will often be of interest when they are in equilibrium with a heat bath. Methods of studying such systems in the context of classical statistical mechanics, i.e. where quantum effects are small, have been described in detail in chapter 10. We now want to describe how Monte Carlo methods can be employed when this approximation cannot be made. The condition for ignoring quantum effects is that the energy associated with classical thermal fluctuations $\approx k_BT$ be much larger than that associated with appreciable virtual transitions between quantum states. When this condition no longer holds the quantum formalism, in particular the impossibility because of the uncertainty principle of specifying a phase space trajectory, the possibility of barrier penetration, and the spin symmetry requirements on the wavefunction must be allowed for. Fortunately the techniques employed for classical trajectory statistical mechanical calculations need not be entirely abandoned. As we will see, a system can be described by an equation very similar to the Schrodinger time dependent equation which served as the basis for the methods described earlier in the chapter, but with the reciprocal temperature replacing the time, and which leads naturally to Feynman's development of quantum evolution as a sum over paths. Because finite temperatures (times) are of interest, the method of Diffusion Monte Carlo, which depends on the large time limit, is no longer very useful. However, it can be generalised to evaluate path integrals. This formalism not only enables thermal systems in the quantum domain to be studied, but provides the basis for an invaluable generalisation to a wide class of quantum problems.

11.5.2 The Density Matrix Formalism

As a starting point, classical expressions are assumed valid with the conventional replacement of dynamical quantities by Hermitian operators. In particular, the energy is replaced by the Hamiltonian $H = T + V$, and integrals over continuous variables are replaced by sums over quantum states (here always assumed to form complete, orthonormal sets) whenever the states are discrete. The probability of a system being in a quantum state $\mid k >$ at a temperature T (we shall use the quantity $\beta = 1/k_B T$ throughout) is assumed to be the generalisation of the Boltzmann distribution, the expectation value of an operator proportional to $e^{-\beta H}$, i.e.

$$p_k \propto < k \mid e^{-\beta H} \mid k > .$$

The normalising quantity, the partition function Z, is obtained by requiring $\sum p_k = 1$, i.e.

$$p_k(\beta) = \frac{1}{Z} < k \mid e^{-\beta H} \mid k >, \tag{5.1}$$

where the partition function is given by

$$\begin{aligned} Z &= \sum_k < k \mid e^{-\beta H} \mid k > \\ &= Tr(e^{-\beta H}), \end{aligned} \tag{5.2}$$

and Tr denotes the trace of the operator. The trace is the sum of the diagonal elements in a matrix. It is straightforward to show that the value of this trace does not depend on the basis set $\mid k >$ used for the matrix representation, and that the trace of the commutator of two operators vanishes. In particular, if the states $| k >$ are the eigenstates of the Hamiltonian — these states we denote by $| h >$ — we have

$$Z = \sum_h < h | h > e^{-\beta E_h} = \sum_h e^{-\beta E_h} ,$$

the latter expression being just the classical expression (equation 10.2.3*b* in chapter 10 above).

The mean value of any property Q of the system at a temperature β, the analogue of equation 10.3.2, is the sum over the quantum states of the probability of the system being in a state $| k >$ at β times the expectation value of the operator corresponding to Q on the state $| k >$, i.e.

$$\begin{aligned} < Q(\beta) > &= \sum_k p_k(\beta) < k | Q | k > \\ &= \frac{1}{Z(\beta)} \sum_k < k | e^{-\beta H} | k >< k | Q | k > . \end{aligned}$$

Using the rules of matrix multiplication, this equals

$$\frac{1}{Z(\beta)} \sum_h < k | e^{-\beta H} Q | k >,$$

i.e.

$$< Q(\beta) > = Tr(e^{-\beta H} Q)/Tr(e^{-\beta H}). \tag{5.3}$$

Since, as we noted earlier, the trace of a commutator vanishes so the order of the two operators inside the bracket is immaterial. This is the fundamental formula of quantum statistical mechanics. The traces can be written as path integrals, as we shall shortly see, and these provide the starting formalism for path integral Monte Carlo methods, important in condensed matter physics.

The logical next step is to write the probability of finding the system in a volume element $d\mathbf{r}$ of phase space, which is just the sum over the distinct quantum states of finding it in this volume element — we will concentrate on the spatial dimensions of the phase space and use one dimension for ease of presentation, but all of our presentation can be taken over to momentum or spin coordinates and to the many dimensions of interest in practice. We have

$$p(x,\beta)dx = \sum_k \frac{\partial p_k(x,\beta)}{\partial x} dx.$$

By taking the x-representation for the states in equation 5.1 this becomes

$$\frac{dx}{Z(\beta)} \sum_k \frac{\partial}{\partial x'} \int \psi_k^*(x') e^{-\beta H} \psi_k(x') dx',$$

i.e.

$$p(x,\beta)dx = \frac{dx}{Z(\beta)} \sum_k \psi_k^*(x) e^{-\beta H} \psi_k(x).$$

If the eigenstates of the Hamiltonian are taken in the sum this probability (particle density if multiplied by the number of particles in the system) can be written as

$$p(x,\beta)dx = \frac{\rho(x,\beta)}{Z(\beta)} dx, \tag{5.4}$$

where the function $\rho(x,\beta)$ is defined as

$$\rho(x,\beta) \equiv \sum_h \psi_h(x) e^{-\beta E_h} \psi_h^*(x),$$

with the irrelevant order of the factors in each term retained for clarity later.[9] The general element of a *density matrix* $\rho(x,x';\beta)$, which turns out to be a very useful construct, is defined by

$$\rho(x, x'; \beta) \equiv \sum_h \psi_h(x) e^{-\beta E_h} \psi_h^*(x'). \tag{5.5}$$

By analogy, it is usual to write the function $\rho(\text{x},\beta)$ above as $\rho(x, x; \beta)$, the 'diagonal' element. We note that for $\beta = 0$ we have,

$$\rho(x, x'; 0) = \sum_h \psi_h(x) \psi_h^*(x') = \delta(x - x'),$$

the latter equality arising from the ortho-normality property of the eigen-functions. The partition function can obviously be written as

$$Z(\beta) = \int dx \rho(x, x; \beta). \tag{5.6}$$

This is a very valuable formula since if we can find the density matrix it will enable us to numerically evaluate Z, from which most of the thermal properties can be determined. For a system of two particles we have the definition

$$\rho(x_1, x_2; x_1', x_2'; \beta) \equiv \sum_h \psi_h(x_1, x_2) e^{-\beta E_h} \psi_h^*(x_1', x_2'),$$

with obvious extension to larger particle numbers, and to more than one dimension, e.g. $x_1 \to \mathbf{r}_1$, etc.

To determine the density matrix we first look for an equation which it satisfies. Continuing with the single particle case, we find, on differentiating ρ in equation 5.5 with respect to the parameter β the relation

$$\partial \rho(x, x'; \beta)/\partial \beta = -\sum_h E_h \psi_h(x) e^{-\beta E_h} \psi_h^*(x'),$$

and since the $\psi_h(\text{x})$ are the eigenstates of H this can be written as

$$= -\sum_h H \psi_h(x) e^{-\beta E_h} \psi_h^*(x').$$

For a single particle, $H = \frac{-\hbar^2}{2m}\frac{\partial^2}{\partial x^2} + V(x)$. As far as this operator is concerned, x' is a constant and so we can move the operator H to the left, giving

$$= -H \sum_h \psi_h(x) e^{-\beta E_h} \psi_h^*(x'),$$

i.e.

$$\frac{\partial \rho}{\partial \beta} = -H\rho, \tag{5.7}$$

a relation known as the *Bloch equation.* Writing this equation out in one dimension,

$$\frac{\partial \rho}{\partial \beta} = \frac{\hbar^2}{2m}\frac{\partial^2 \rho}{\partial x^2} - V(x)\rho, \tag{5.8}$$

we see that its form is that of the familiar diffusion equation with capture (section 8.3.9) where the parameter β replaces the time, the diffusion coefficient is a constant, and the capture/annihilation is determined by the potential energy function $V(x)$. In principle this then gives us a method of solving for ρ, and thus, using equation 5.6, of obtaining the partition function. Since $\rho(x, x'; 0) = \delta(x - x')$ the solution is just the Green function of this diffusion equation. Diffusion Monte Carlo can be used to solve this equation, but since we are interested in the solution at finite values of β, we cannot fall back on the convergence as $\beta \to \infty$ (zero temperature case) which proved so convenient in the earlier sections of this chapter and the method is more problematic. An alternative formulation however can lead us to a method suitable for Monte Carlo solution.

11.5.3 The Path Integral

The solution to equation 5.8 is the density matrix, which we have written in the form

$$\rho(x,x';\beta)=\sum_h \psi_h(x)e^{-\beta E_h}\psi_h^*(x'). \tag{5.9}$$

This is purely formal, however, since it requires solution for the eigenfunctions of the time independent problem, hardly less demanding than the solution we seek. We can, however, find a formalism — the path integral formalism — which bypasses knowledge of these functions. By inserting a Kronecker δ_{jh}, rewriting some of the terms and summing over j, it is easy to see that an equivalent, if redundant, expression is

$$\sum_h \psi_h(x)(\sum_j \delta_{jh}e^{-\{\beta' E_h+(\beta-\beta')E_j\}}\psi_j^*(x')),$$

where the only non-zero term in the bracket is the same as the term in the previous expression. We now insert a representation for the Kronecker δ,

$$\delta_{jh}=\int dx''\psi_h^*(x'')\psi_j(x''),$$

giving

$\rho(x,x';\beta)=$

$$\int dx''\sum_h\sum_j \psi_h(x)e^{-\beta' E_h}\psi_h^*(x'')\psi_j(x'')e^{-(\beta-\beta')E_j}\psi_j^*(x')$$
$$=\int dx''\{\sum_h \psi_h(x)e^{-\beta' E_h}\psi_h^*(x'')\}\{\sum_j \psi_j(x'')e^{-(\beta-\beta')E_j}\psi_j^*(x')\},$$

i.e.

$$\rho(x,x';\beta)=\int dx''\rho(x,x'';\beta')\rho(x'',x';\beta-\beta'). \tag{5.10}$$

We see that the density matrix at temperature β can be written as a convolution of two such density matrices at β' and β-β'. The expression can obviously be decomposed further to an arbitrary extent. By writing $\beta' = \beta/(M+1)$, $M \geq 0$ an integer, it can be written as M convolutions of density matrices each with temperature β', i.e. (with a slight change in notation),

$$\rho(x, x_0; \beta) = \int dx_1 .....dx_M \rho(x, x_1; \beta')\rho(x_1, x_2; \beta')....\rho(x_M, x_0; \beta'). \quad (5.11)$$

Summing over the diagonal elements, i.e. setting $x = x_0$ and summing, we find the partition function

$$\begin{aligned} Z(\beta) &= \int dx_0 \rho(x_0, x_0; \beta) \\ &= \int dx_0 ...dx_M \prod_{k=0}^{M} \rho(x_k, x_{k+1}; \beta'), \end{aligned} \quad (5.12)$$

where it is understood that $x_{M+1} \equiv x_0$. It seems as though we have made very heavy weather of the partition function for a one-dimensional system (expressing it as an M+1 fold integral) but, as in all Monte Carlo based calculations, the value of our approach only manifests itself when it is applied to many particle/dimensional problems.

As we have noted, the functions involved in the integral are the Green function for the diffusion equation. For the case that $V(x)$ is a constant, equal to V_0, we have (equation 8.3.8)

$$\begin{aligned} \rho(x, x'; \beta') &= p(x, \beta' \mid x', 0) \\ &= \frac{1}{\sigma\sqrt{2\pi}} e^{-(x-x')^2/2\sigma^2 - V_0\beta'} , \end{aligned}$$

with $\sigma^2 = \beta'\hbar^2/m$. For sufficiently small β', i.e. small σ, this function will only be appreciable for $x' \approx x$. This suggests that if $V(x)$ is not too rapidly varying a function of x we can replace V_0 by $V(x)$, or perhaps better by $[V(x) + V(x')]/2$, i.e.

$$\rho(x, x'; \beta') \simeq \sqrt{\frac{m}{2\pi\hbar^2\beta'}} e^{-m(x - x')^2/2\hbar^2\beta' - \beta'[V(x') + V(x)]/2} \;. \quad (5.13)$$

This is the 'short time' approximation to the density matrix. Using this form in equation 5.12 we have

$$Z(\beta) = \left(\frac{m(M+1)}{2\pi\hbar^2\beta}\right)^{(M+1)/2} \int dx_0 ... dx_M \exp[-\beta' \sum_{k=0}^{M} \left(\frac{(x_{k+1} - x_k)^2}{2\hbar^2\beta'^2/m} + \overline{V_k}\right)]. \quad (5.14)$$

In the limit of $M \to \infty$ integrals of this kind are known as Weiner integrals, or path integrals, and the expression is the so called path integral representation for the partition function.

Path integrals play an important role in practical quantum mechanical calculations, in the next section we expand on this, and describe how Monte Carlo methods can be used to evaluate them.

11.6 Path Integrals and Schrodinger's Equation

11.6.1 Introduction

We have seen in section 11.5 how the partition function, the fundamental relation for determining the thermal properties of a system, can be formulated as a Weiner, or path, integral. We will see how these methods can be taken over to solve more general quantum problems, and in particular how solutions of Schrodinger's equation can be obtained using similar integrals, and how Monte Carlo methods can be used in their estimation. Much of the importance of Monte Carlo methods applied to path integrals arises from the fact that, in recent years, they have become an essential technique in modern field theory, especially in quantum chromodynamic QCD calculations. Starting from the quarks and gluons and the gauge invariant theory of strong interactions which describes their interactions, QCD, it is hoped to be able to predict the particle states (eigenstates), mesons, baryons, etc. which result. In particular it is hoped to predict the mass spectrum of these 'elementary particles'. In addition, when the energy density is high — high temperatures — it is suspected that a phase transition will occur in the quark gluon plasma which may have played an important cosmological role at early stages of the evolution of the universe. The properties of this high temperature state and any phase transition — quark thermodynamics — can only be investigated by first principles numerical methods because of the singular property of QCD that the interaction strength increases with separation of the quarks, which invalidates any perturbation type approach. As we have pointed out Monte Carlo based methods of solution of quantum problems can, in principle, lead to exact solutions, which explains the importance of this topic. Path integral calculations are also important in several other branches of physics, notably in polymer physics, nuclear physics, etc.

In mathematical terms what we present in this section is very similar to that in the previous section, however, the motivation, and the insights following from it are on a more fundamental level.

11.6.2 Feynman Path Integrals

Although the methods are usually applied to quantum field theory with an infinite number of degrees of freedom, we will start off, as usual, with the time dependent Schrodinger equation in a single space variable. This is a differential equation which determines the evolution with time of the probability amplitude $\Psi(x,t)$ for a system; we write it as

$$\frac{\partial \Psi}{\partial t} = -\frac{i}{\hbar} H\Psi, \tag{6.1}$$

where H is the Hamiltonian of the system. With the initial state $\Psi(x,0)$ specified, the solution of the equation determines the amplitude, and thus the probability density, $\mid \Psi(x,t) \mid^2$, at later times. If we think of a particle initially with coordinate x_i it can be thought of as being in an eigenstate $\mid x_i >$ of the position operator. We can then formally write the solution of equation 6.1 at a time $t, \mid x(t) >$ as

$$\mid x(t) >= e^{-\frac{itH}{\hbar}} \mid x_i >,$$

where the exponential term is an operator known as the time evolution operator. The probability that by a time t the state evolves to correspond to the particle having a coordinate x_f, which we will denote by $G(x_f, t; x_i, 0)$, is just the projection of this state $\mid x_f >$ on the state $\mid x(t) >$, i.e.

$$G(x_f, t; x_i, 0) =< x_f \mid e^{-\frac{itH}{\hbar}} \mid x_i > . \tag{6.2}$$

We shall henceforth write it as $G(x_f, t; x_i)$; strictly speaking we should allow for an arbitrary initial time $t_i \neq 0$, but if the Hamiltonian does not contain explicit time dependence we will always be able to take $t_i = 0$ and economise on notation. We can view $G(x_f, t; x_i)$ as the probability density for the particle propagating from the location with coordinate x_i to x_f, dx_f in a time t. For the more likely case that the initial state of the particle is not a position eigenstate but lies in an interval dx with probability $\mid \Psi_0(x) \mid^2 dx$, Feynman built in the essentials of quantum theory by applying the superposition principle to propagate the amplitude $\Psi_0(x)$ rather than the probability as would be done in classical statistics, i.e. he wrote the state at the time t in the form

$$\Psi(x,t) = \int dx_0 G(x,t;x_0)\Psi_0(x_0). \tag{6.3}$$

The propagator $G(x,t;x_0)$ represents the strength with which the value of the initial wavefunction at x_0 influences its value at x at the later time t. We also recognise that the function G is the Green function of the equation which Ψ satisfies. This latter we know is Schrodinger's equation,

$$\frac{\partial\Psi}{\partial t} - \frac{i\hbar}{2m}\frac{\partial^2\Psi}{\partial x^2} = -\, iV\Psi/\hbar.$$

With a diffusion constant D and decay probability μ defined by

$$D = \frac{i\hbar}{2m}, \qquad \mu = iV/\hbar,$$

this is the same as equation 8.3.7. For the case of a constant potential V_0, by analogy with equation 8.3.8 we can write its Green function, that for the diffusion equation

$$G_0(x,t;x_0,0) = (\frac{m}{2\pi i\hbar t})^{1/2} e^{\;+\frac{i}{\hbar}\frac{m}{2}\frac{(x-x_0)^2}{t} - \frac{i}{\hbar}V_0 t}.$$

The index of the exponential here can be written as

$$\frac{i}{\hbar}\{\frac{m}{2}[\frac{x-x_0}{t}]^2 - V_0\}t.$$

The velocity of a free particle is constant, here given by $(x-x_0)/t$, so the first term in the bracket is the kinetic energy, T, of the particle, and the whole term can be written as $(i/\hbar)$ times the integral of the particle's Lagrangian,

$L(x,t)$, over the time, $\int L(x,t)dt$. Thus we end up with the final form for the Green function,

$$G_0(x,t;x_0,0) = (\frac{m}{2\pi i\hbar t})^{1/2} e^{+\frac{i}{\hbar}\int L(x,t)dt} . \qquad (6.4)$$

Although the classical notion of paths is ordinarily absent in quantum mechanics because of the Uncertainty Principle, Feynman argued that this form for the propagator can be used provided we sum over the various paths (all 'possible' world lines) linking the initial and final points, i.e.

$$G(x_f,t;x_i) = \sum_\lambda \text{contribution from path}_\lambda(x_i,x_f).^{10}$$

Each such path, and there are infinitely many of them, can be represented by a trajectory $X_\lambda(t)$, and Feynman assumed its contribution to be of the form

$$\text{contribution from path}_\lambda(x_i,x_f) \propto e^{iS[X_\lambda(t)]/\hbar} ,$$

where the functional $S[X_\lambda(t)]$ is the action calculated along the trajectory $X_\lambda(t)$. This action is given by

$$S[X_\lambda(t)] = \int_0^t dt L(x, dx/dt), \qquad (6.5)$$

where L is the Lagrangian for the particle and the integration is carried out along the trajectory X_λ. Thus we can write

$$G(x_f,t;x_i) = \int_{(x_i,0)}^{(x_f,t)} \mathcal{D}[X] e^{iS[X]/\hbar}, \qquad (6.6)$$

where the notation $\mathcal{D}[X]$ (any constant of normalisation can be included) is simply a bald instruction to sum over all the paths. Formally, it denotes integration over all *functions* $X(t)$ which satisfy $X(0) = x_i, X(t) = x_f$. How

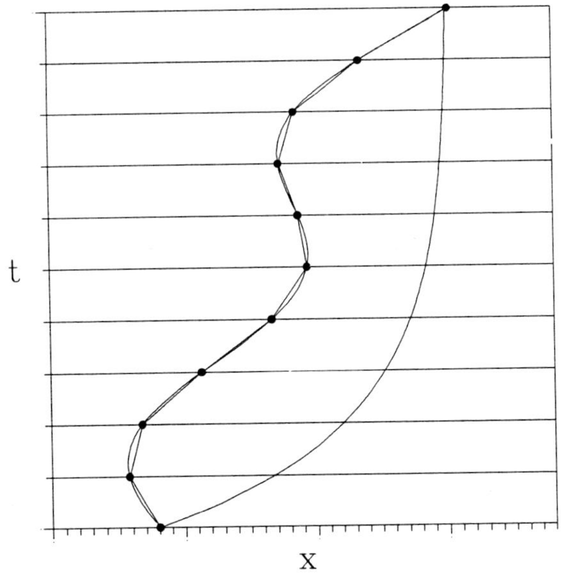

Figure 11.6: Two of the possible paths linking events separated in space and time. An approximation to one of them by a segmented polygonal path is also shown.

to do so needs to be investigated; in all but the few cases where analytical methods have been found, numerical methods working from first principles of path structure must be used. In Fig. 11.6 we show two of the non- denumerable set of paths linking x_i with x_f — a polygonal path which closely approximates one of them is also shown.

Using equation 6.3 we can also find an entirely different expression for the propagator which we will find of value. It is shown in all elementary Quantum Mechanics texts that the solution to equation 6.1 can be written as an expansion in the time independent eigenstates $\psi_n(x)$ of the Hamiltonian (itself not an explicit function of time)

$$\Psi(x,t) = \sum_n c_n e^{-iE_n t/\hbar} \psi_n(x), \tag{6.7}$$

with the coefficients c_n given by

$$c_n = \int dx_0 \Psi(x_0, 0) \psi_n^*(x_0).$$

When this is inserted into the left-hand side of equation 6.3 we find the equivalence

$$G(x_f, t; x_i) = \sum_n \psi_n^*(x_i) \psi_n(x_f) e^{-iE_n t/\hbar} . \tag{6.8}$$

Apart from differences in notation, this expression is mathematically identical to equation 5.9 for the density matrix. It follows that the results derived there can be taken over directly, in particular the convolution formula equation 5.10 takes the form,

$$G(x_f, t; x_i) = \int_{-\infty}^{\infty} dx' G(x_f, t - t'; x') G(x', t'; x_i). \tag{6.9}$$

To evaluate $G(x_f, t; x_i)$, defined by equation 6.6 we make use of the convolution result equation 6.9 to build up the propagator at t from its values at earlier times. We superimpose a time lattice which divides t into $M+1$ subintervals of duration $\epsilon = t/M + 1$ (see Fig. 11.6). Any path is then defined by its intersections with this lattice, $X = \{x_i, x(t_1), \ldots, x(t_M), x_f\}$, where $t_k = k\epsilon$. Using equation 6.9 we have

$$\begin{aligned} G(x_f, t; x_i) &= \int dx_1 G(x_f, t - \epsilon; x_1) G(x_1, \epsilon; x_i) \\ &= \int dx_1 \int dx_2 G(x_f, t - 2\epsilon; x_2) G(x_2, \epsilon; x_1) G(x_1, \epsilon; x_i), \end{aligned}$$

which, noting $t - M\epsilon = \epsilon$, can be expanded to an M-fold integral. If we write $x_0 \equiv x_i, x_{M+1} \equiv x_f$, we have

$$G(x_f, t; x_i) = \int dx_1 ... dx_M \prod_{k=0}^{M} G(x_{k+1}, \epsilon; x_k). \tag{6.10}$$

In principle, we can increase the complexity of this integral to an arbitrary degree by making ϵ ever smaller. For any finite ϵ there is still, clearly, an infinity of paths linking x_{k+1} with x_k, but their 'number' might be expected to decrease with ϵ, e.g. excursions cannot exceed $\approx c\epsilon$. Most of the contribution to $G(x_{k+1}, \epsilon; x_k)$ is given by the tangent to the classical path linking x_k with x_{k+1}. Using equation 6.6 this gives us for the small ϵ case

$$G(x_{k+1}, \epsilon; x_k) = Ae^{\frac{i}{\hbar}\int_0^\epsilon dt\overline{L}(X)} \quad ,$$

where, from equation 6.4, $A = (\frac{m}{2\pi i\hbar\epsilon})^{1/2}$ is some measure of the number of contributing paths, and $\overline{L}$ is the mean value of the Lagrangian on these paths. If we take for $\overline{L}$ its value along the tangent path linking x_k with x_{k+1}, and make use of the Mean Value theorem to evaluate the integral we have in the exponent

$$i\frac{\epsilon}{\hbar}\overline{L}(x_{k+1}, x_k) = i\frac{\epsilon}{\hbar}\{\frac{m}{2}[\frac{x_{k+1} - x_k}{\epsilon}]^2 - \overline{V}_k\},$$

and so

$$G(x_{k+1}, \epsilon; x_k) \simeq (\frac{m}{2\pi i\hbar\epsilon})^{1/2} \exp -\{\frac{(x_{k+1} - x_k)^2}{2i\epsilon\hbar/m} + \frac{i\epsilon\overline{V_k}}{\hbar}\}. \qquad (6.11)$$

11.6.3 Evaluation of Path Integrals

We have yet to attach a meaning to the sum over paths; the fact that the integrand in equation 6.6 is a complex quantity will also complicate any numerical evaluation. To treat a simpler case we note that, mathematically, the expression will also be meaningful when the time is treated as a complex variable, so we revert to a ploy which proved useful in the previous chapter of converting to imaginary values of the time. Many problems cannot be solved in this way, but some important problems, especially in quantum field theory, are amenable to solution using complex time. In particular for pure imaginary time we write $t = -i\tau$ with τ real. Further, for a non-relativistic particle interacting via a potential $V(x)$ which does not depend on time we can write the Lagrangian as the difference between the kinetic and potential energy,

$$\begin{aligned} L(x, dx/dt) &= \frac{m}{2}(\frac{dx}{dt})^2 - V(x) \\ &= -\{\frac{m}{2}(\frac{dx}{d\tau})^2 + V(x)\} \\ &= -H(x,\tau), \end{aligned} \tag{6.12}$$

i.e. the Lagrangian at imaginary values of t can be replaced by the negative of the Hamiltonian evaluated at real values of τ. The action, equation 6.5, then takes the form

$$S[X(t)] = iS_R[X(\tau)],$$

where

$$S_R[X(\tau)] = \int_0^T d\tau H[X(\tau)]. \tag{6.13}$$

The propagator, equation 6.6, then only involves real quantities,

$$G(x_f, -i\tau; x_i) = \int \mathcal{D}[X] e^{-S_R[X(\tau)]/\hbar} . \tag{6.14}$$

With the same notation we can write equation 6.7 as

$$G(x_f, -i\tau; x_i) = \sum_n \psi_n^*(x_i)\psi_n(x_f) e^{-E_n\tau/\hbar} . \tag{6.15}$$

If we set $x_f = x_i \equiv x$ and integrate over all starting points we have, using the orthonormality of the ψs,

$$\int_{-\infty}^{\infty} dx G(x, -i\tau; x) = \sum_n e^{-\tau E_n/\hbar} .$$

This is just the partition function for the system at a temperature $k_B T = \hbar/\tau$, i.e.

$$Z(\hbar/\tau) = \int_{-\infty}^{\infty} dx G(x, -i\tau; x). \tag{6.16}$$

To the extent that $\hbar$ is non-zero, it is a measure of the extent of 'natural' fluctuations in the system. As path integrals can be evaluated at finite times the thermal properties of a quantum system can be calculated, it is by these methods that the possibility of a phase transition in a quark gluon plasma is studied in QCD.

For sufficiently large values of τ, specifically when $\tau \gg \hbar/(E_1 - E_0)$ where E_0 and E_1 are the two lowest energies of the system, all higher terms in the sum in equation 6.15 will become negligible compared with the first term. Therefore, if the propagator is evaluated at $x_f = x_i = x$ we find,

$$|\, \psi_0(x) \,|^2 = \lim_{\tau \to \infty} e^{E_0 \tau/\hbar} G(x, -i\tau; x). \tag{6.17}$$

In other words, at sufficiently large values of τ the propagator is proportional to the square of the ground state wavefunction, with the ground state energy entering into the coefficient of proportionality. If the propagator can be found these two properties of the system can be found. Alternatively, if we impose normalisation,

$$\int |\, \psi_0(x) \,|^2 \, dx = 1,$$

we can write

$$| \psi_0(x) |^2 = \lim_{\tau \to \infty} G(x, -i\tau; x)/Z. \tag{6.18}$$

In principle the ground state energy can be found from equation 6.17 evaluated at different times, in practice we use the expectation value of the energy, kinetic plus potential, in the ground state, i.e.

$$E_0 =< 0 \mid H \mid 0 >= \lim_{\tau \to \infty} \int \mathcal{D}[X] E(X) e^{\; iS[X]/\hbar} . \tag{6.19}$$

To calculate these various quantities, e.g. equations 6.17 to 6.19, we make use of the convolution result equation 6.10. Letting $\epsilon \to -i\epsilon$ in equation 6.11 and inserting into equation 6.10 enables us to write

$$\begin{aligned} G(x_f, t; x_i) &= \int dx_1 ... dx_M A^{\frac{M+1}{2}} \exp -\frac{\epsilon}{\hbar} \sum_{k=0}^{M} \{ \frac{m}{2} [\frac{x_{k+1} - x_k}{\epsilon}]^2 - \overline{V}_k \} \\ &= \int dx_1 ... dx_M (\frac{m}{2\pi\hbar\epsilon})^{(M+1)/2} e^{\; -\frac{\epsilon}{\hbar} E(x_0, x_1, ..x_{M+1})} , \end{aligned} \tag{6.20}$$

where

$$E(X) = \sum_{k=0}^{M} \{ \frac{m}{2} [\frac{x_{k+1} - x_k}{\epsilon}]^2 - \overline{V}_k \} \tag{6.21}$$

is an energy associated with a given segmented path X. A more general way of looking at this is as an energy which is a function of variables, x_k, defined on lattice sites, involving coupling to variables on other sites (here just nearest neighbours), similar to the formalism for spin variables in the Ising model described in section 10.4, except that here the variables can assume a continuum of values. The construction of an analogous quantity, the so called 'lattice action' in gauge field theory is the starting point for these methods in studies of QCD. Different configurations of these variables (in our case distinct paths) contribute to the propagator, which can be calculated by judiciously choosing 'important' configurations.

Using equation 6.16 we have for the partition function at $\beta \equiv 1/k_B T = (M+1)\epsilon/\hbar$

$$Z = \int dx_0 \int dx_1 ... dx_M (\frac{m}{2\pi\hbar\epsilon})^{(M+1)/2} \exp[-\frac{1}{\hbar}\sum_{k=0}^{M}\{\frac{m}{2\epsilon}(x_{k+1}-x_k)^2 + \epsilon\overline{V_k}\}] \tag{6.22}$$

just the same as equation 5.14.

Before we look at the practical evaluation of quantities, we digress slightly to note that if we insert equation 6.20 into equation 6.3, and use the notation $\epsilon = \Delta\tau$, we find the wavefunction at a later time which has evolved from an initial function is, $x_{M+1} = x$,

$$\Psi(x,t) = \int e^{-\overline{\mu}_M \Delta\tau} W(x_M \rightarrow x) dx_M ...$$

$$... \int e^{-\overline{\mu}_1 \Delta\tau} W(x_1 \rightarrow x_2) dx_1 \int e^{-\overline{\mu}_0 \Delta\tau} W(x_0 \rightarrow x_1) dx_0 \Psi(x_0), \tag{6.23}$$

where $\overline{\mu}_k = \overline{V}_k/\hbar = [V(x_k) + V(x_{k+1})]/2\hbar$, and $W(x_k \rightarrow x_{k+1})dx_k$ is the Gaussian probability distribution with variance $\hbar\Delta\tau/m$,

$$W(x_k \rightarrow x_{k+1})dx_k = \sqrt{m/2\pi\hbar\Delta\tau} \exp -\{\frac{(x_{k+1}-x_k)^2}{2\hbar\Delta\tau/m}\}.$$

Reading equation 6.23 from the right to left, we see that it is exactly the procedure we described as Diffusion Monte Carlo earlier in the chapter; starting from x_0 selected from the initial wavefunction, successive steps are made from x_k to $x_{k+1} = x_k + \Delta x$, where Δx is selected from a Gaussian with mean zero and $< \Delta x^2 >= \hbar\Delta\tau/m$, after which it survives (reproduces) with probability $e^{-\overline{\mu}_k \Delta\tau}$. DMC is an alternative description of the path integral method in this approximation.

11.6.4 Numerical Evaluation

Setting $x_i = x_f = x$, from equation 6.18 we can write

$$| \psi_0(x) |^2 = \lim_{T \to \infty} \int dx_1 ... dx_M \frac{1}{Z} e^{-\frac{\epsilon}{\hbar} E(x, x_1, ..x_M, x)},$$

or, in a redundant, but more helpful, way

$$| \psi_0(x) |^2 = \lim_{T \to \infty} \int dx_0 dx_1 ... dx_M \delta(x_0 - x) \frac{1}{Z} e^{-\frac{\epsilon}{\hbar} E(x_0, x_1, \ldots, x_M, x_0)} \tag{6.24}$$

This integral is of the familiar form

$$\int d\mathbf{x} f(\mathbf{x}) \pi(\mathbf{x}),$$

where $\pi(\mathbf{x}) = \frac{1}{Z} e^{-\beta E(\mathbf{x})}$, $\beta = \epsilon/\hbar$, is the normalised Boltzmann probability distribution, with a path $\mathbf{x} = \{x_0, x_1, \ldots, x_M, x_0\}$ defining a configuration, and $f(x)$ here is the rather trivial function $\delta(x_0 - x)$. In the same way, using equation 6.19 we can write

$$E_0 = \lim_{T \to \infty} \int d\mathbf{x} E(\mathbf{x}) \pi(\mathbf{x}). \tag{6.25}$$

A difficulty arises here if we try to use equation 6.21 for $E(\mathbf{x})$; because of the nature of quantum paths the kinetic energy diverges as $\epsilon \to 0$. We have

$$T_{\text{kin}} = \frac{m}{\Delta t} \frac{< \Delta x^2 >}{2 \Delta t},$$

but for quantum paths $\frac{<\Delta x^2>}{2\Delta t} = \frac{\hbar}{m}$, so $T_{\text{kin}} \to \hbar/2\epsilon$. There are different ways of overcoming this problem, here we use the Virial theorem to write an alternative form $E_V(\mathbf{x})$. This states that the expected energy in any state is the expectation value of the quantity $(1 + \frac{1}{2}\ \mathbf{r}.\text{grad})\, V$, so we take

$$E_V(\mathbf{x}) = \sum_{k=0}^{M} \left\{ -x_k \frac{\partial V}{\partial x}\bigg|_{x_k} + V(x_k) \right\}. \tag{6.26}$$

The integrals, equations 6.24 and 6.25, which we remember are for a one-dimensional problem, in almost all cases will require numerical methods to evaluate. From our earlier discussions we can expect to evaluate them using importance sampling, e.g. using the Metropolis algorithm for selecting paths, in a manner entirely analogous to the selection of configurations described in chapter 10 — we shall shortly look at a concrete example. We will also see how the wavefunction can be determined at many distinct values of x using the same set of configurations.[11] Clearly, a method as complicated as described is not one to be used for determining the properties of a simple system, its merits only become obvious when advanced quantum mechanical calculations are considered. Already we have mentioned its advantage as a possible approach when perturbation methods cannot be applied to the problem, another advantage arises from the ease with which the calculation can be formulated in discrete terms on a lattice. This enables a concise specification of the fields to be given in terms of arrays, and, in addition, it enables a cut- off to be placed on integrals, corresponding to renormalisation in perturbation theory, before extrapolating from the lattice to the continuum.

EXAMPLE 11.3: Motion in the Morse potential

In this example we look at the motion of a particle of mass m in a Morse potential,

$$V(x) = V_0(e^{-2a(x-x_0)} - 2e^{-a(x-x_0)}).$$

as we did in example 11.1, and use the same scaled variables, i.e. time in units of ω^{-1} with $\omega \equiv a\sqrt{2V_0/m}$, $\epsilon \to \epsilon\omega^{-1}$ and $z \equiv (x - x_0)/\sqrt{\hbar/m\omega}$, and we define $\zeta^2 = \hbar\omega/2V_0$. Our aim is to calculate the integrals, equations 6.24 and 6.25. The time τ is divided into M+1 segments (M odd) of length $\epsilon = \tau/M + 1$. We use the Metropolis algorithm starting from an arbitrary initial path to select sample paths from the Boltzmann distribution. When the influence of the initial path has died away, averages of the quantities over the remaining samples are calculated. The exponential term in equation 6.20 takes the form $e^{-\epsilon E(\mathbf{x})}$, with

$$E(\mathbf{x})/\hbar\omega = \sum_{k=0}^{M} \left\{ \frac{1}{2}\left(\frac{z_{k+1} - z_k}{\epsilon}\right)^2 + \frac{1}{2\zeta^2}\left(e^{-2\zeta z_k} - 2e^{-\zeta z_k}\right) \right\}. \quad (6.27)$$

The virial expression for the energy, equation 6.26, takes the form

$$E_V(\mathbf{x})/\hbar\omega = \frac{1}{M+1}\frac{1}{2\zeta^2}\sum_{k=0}^{M}\left\{(1 - \zeta z_k)e^{-2\zeta z_k} - (2 - \zeta z_k)e^{-\zeta z_k}\right\}.$$

Starting with an arbitrary path — we took the simple path $z_k = 0$, shown by the dashed line in Fig. 11.7 — the energy, equation 6.27, is calculated. Sweeping systematically through the sites from $k = 1$ to M paths are generated, which differ by at most the segments connected to the coordinate z_k ($\to z'_k = z_k + \Delta z$, with $\Delta z = (2\xi - 1)\alpha$) at a time level τ_k, and accepted with probability $W_{kk'} = \min[1, \exp -\epsilon\Delta E]$, where $\epsilon\Delta E$ is the difference in the action on the paths incorporating z' and z, as determined by equation 6.27. Successive paths are chosen in this way, and initially ignored until the influence of the initial path has died away — an example of such a 'thermalised path' with $\tau = 10$ and $\epsilon = 5/16$ is shown by the solid line in Fig. 11.7. Thereafter, when a new path is chosen (it may be identical to the previous one), the function being integrated is evaluated and added into a running sum; this sum will eventually be divided by the total number of terms accumulated to find its average.

A word needs to be said of the case where the function being integrated is the delta function, $\delta(x - x_0)$ in equation 6.24. A typical step in the algorithm will move z_k to z'_k (as shown by the dotted line in Fig. 11.8). Each coordinate so chosen, whether it be 'new' or 'old', can be thought of as the starting, or ending, point for a path CBD in the

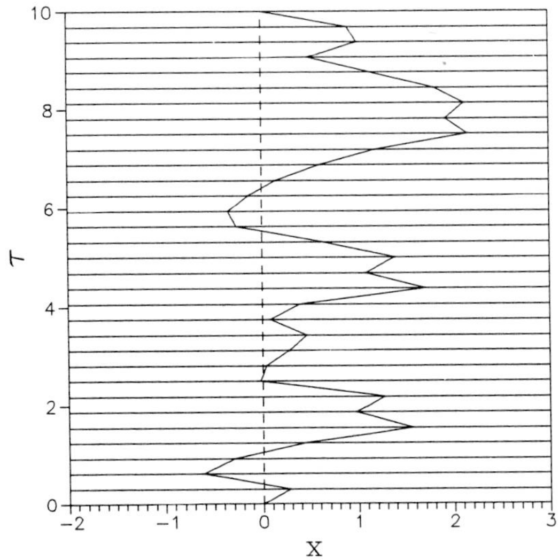

Figure 11.7: The dashed line shows the initial path chosen to link (0, 0) to (0, 10) in example 11.3. The solid path is an example of contributing paths when selection has become 'thermalised'.

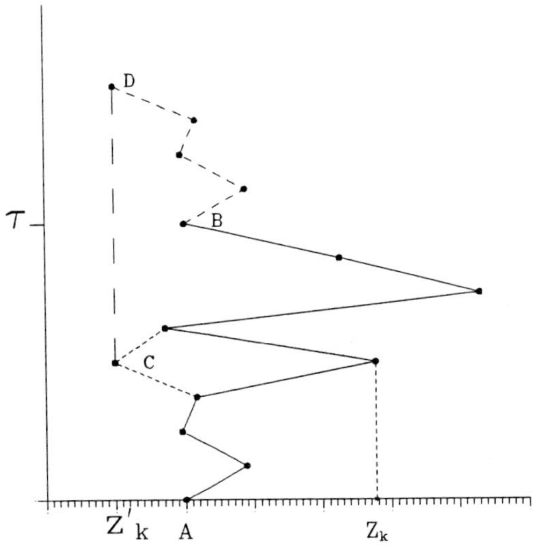

Figure 11.8: An example of a new path linking A to B generated by the move z_k to $z_{k'}$. The extension B to D is a duplicate of the segment A to C, showing that C can be viewed as the starting point for a legitimate path C to B to D.

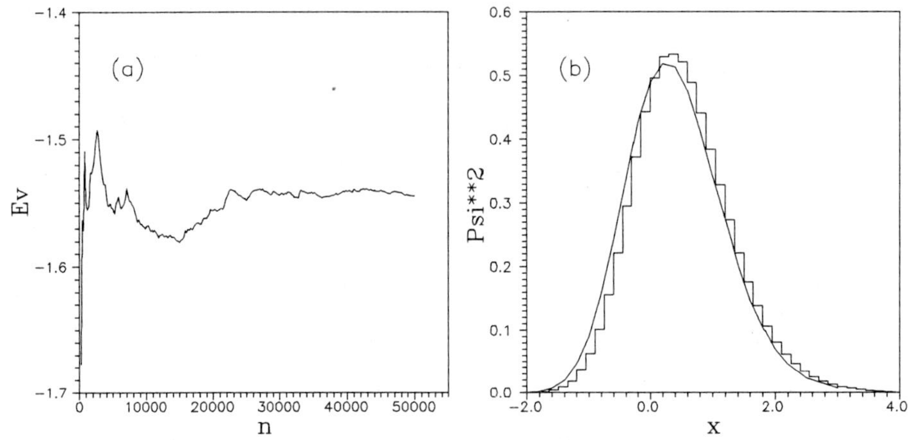

Figure 11.9: (a) The virial energy $E_V/\hbar\omega$ as a function of iteration number for the potential in example 11.3, and (b) an accumulated histogram approximating the square of the ground state wavefunction, compared with the square of the exactly known wavefunction.

figure, whose extension BD, shown by the dashed line in the figure is just a replication of the earlier segment AC. This new path makes a contribution unity to the integral of $\delta(z-z_0)$ at $z=z'_k$. In practice, a histogram is built up with unity only added to the bin in which z'_k is located at every probe, rather than to all the bins corresponding to the other locations on the path. Because we are systematically sweeping through the points, this will build up the distribution of contributing points in an unbiased way. When divided by the number of probes executed, this histogram will tend to $\mid \psi_0(z) \mid^2$.

In the figures we show some results obtained for $\tau = 20, M = 127$, $(\epsilon = 5/32)$, with $\alpha = 1$, and taking 2000 probes per site (iterations) up to thermalisation, followed by a further 510^4 iterations, with sampling done every 200 iterations. In Fig. 11.9(a) we see the virial energy, $E_V/\hbar\omega$, as a function of iteration number, its mean value over the last half of the samples is 1.54 ± 0.01, to be compared with the known energy level in the $\epsilon \to 0$ limit, $E_n = -\frac{1}{2}[\frac{1}{\zeta} - (n+\frac{1}{2})\zeta]^2$, where n is an integer satisfying $0 \le n \le (1/\zeta^2 - 1/2)$, which yields $E_0 = -\frac{49}{32} = -1.53$. In Fig. 11.9(b) we show the histogram of the square of the accumulated wavefunction, and compare it with the $\epsilon \to 0$ expression

$$\psi_0^2 = \frac{(8/e)^7}{2\Gamma(7)}(\frac{z}{2} + \frac{8}{7}e^{-z/2}).$$

The example 11.3 has illustrated the basic components of Monte Carlo applications in path integral evaluation. As in earlier examples in the book, more efficient methods — variance reduction — which speed up the calculation can be implemented, important examples of which are Over-relaxation and Multigrid methods. More information can also be generated in the simulation, for example the extraction of the gap to excited states by studying correlations between sites along the quantum paths. On the other hand, many problems do not lend themselves to solution using the imaginary time formalism, resulting in path integrals of a highly oscillatory kind. Although much more difficult to calculate, Monte Carlo methods can be devised for these cases also. All of these topics, however, are more complex than can be conveniently covered in our introductory discussion.

Notes

1. Feynman, R. P. (1982), *Simulating Physics with Computers*, Int. Jour. Theor. Phys., **21**, 467-88.

2. Schrodinger, E. (1932), *Une analogie entre la mécanique ondulatoire et quelques problèmes de probabilités en physique classique*, Ann. Inst. H. Poincare, **2**, 296-310.

3. Fermi, E. quoted in Metropolis N. and S. Ulam (1949), *The Monte Carlo Method*, J. Am. Stat. Assoc., **44**, 335-41.

4. The relation 11.2.2 is derived in any introductory Quantum Mechanics texts, e.g. in section 3.8 of B. H. Bransden and C. J. Joachain, *Introduction to Quantum Mechanics* (London: Longman, 1989).

5. Anderson, J. B. (1975), *A random-walk simulation of the Schrodinger equation:* H_3^+, J. Chem. Phys., **63**, 1499-1503.

6. Kalos, M. H., D. Levesque and L. Verlet (1974), *Helium at zero temperature with hard-sphere and other forces*, Phys. Rev. A **9**, 2178-95.

7. Schrader, D. M., T. Yoshida and K. Iguchi (1992), *The Binding Energy of Positronium Chloride, PsCl: A Quantum Monte Carlo Calculation*, Phys. Rev. Lett., **68**, 3281-83.

8. For an account of this topic see Rafelski, J. and S. E. Jones (1987), *Cold Nuclear Fusion*, Sci. American, **257**, 66-71. A calculation along the lines described can be found in Ceperley D. and B. J. Alder (1985), *Muon-alpha particle sticking probability in muon catalyzed fusion*, Phys. Rev., *A*, **31**, 1999-2004.

9. The more logical definition, including the normalisation constant in the denominator is used by a minority of authors.

10. Feynman, R. P. and A. R. Hibbs, *Quantum Mechanics and Path Integrals* (New York: McGraw-Hill, 1965).

11. Lawande, S. V., C. A. Jensen and H. L. Sahlin (1969), *Monte Carlo Integration of the Feynman Propagator in Imaginary Time*, J. Comput. Phys., **3**, 416-43; M. Creutz and B. Freedman (1981), *A Statistical Approach to Quantum Mechanics*, Annals of Physics, **132**, 427-62; P. K. MacKeown (1985), *Evaluation of Feynman path integrals by Monte Carlo methods*, Am. J. Phys., **53**, 880-85.

Appendix

Programmes

A.1 Subroutine CORREL

```
      SUBROUTINE CORREL(K, XK, S1, S2,RHO,M)
C     Algorithm of Chan Golub and LeVeque for mean and variance of a
C     sequence,
C     combined with calculation of correlation coefficients.
C     Returns S1, S2 at every call, and also the M values of RHO at call
C     SUMARY.
C     K : (1.LE.K.LE.N) is the order of variate, XK, in the sequence.
C     S1: tends to the mean m.
C     S2/(N - 1) tends to the variance s**2, where N, the total number of
C     events is final value of K.
C     M: order of highest correlation coefficient required.
C     Here order of correlations limited to < 500
C     If correlation coefficients not required set M=0
      DIMENSION RHO(*), X(500)
      SAVE X
      IF(K.EQ.1) THEN
      S1 = XK
      S2 = 0.
      X(1) = XK
      RETURN
      ELSE
```

```
      RK=FLOAT(K)
      S1 = S1 + (XK - S1)/RK
      S2 = S2 + RK * (XK - S1) * (XK - S1)/FLOAT(K-1)
      X(K) = XK
      DO 1J = 1, K - 1
      IF(J.GT.M) THEN
      RETURN
      ELSE
      RHO(J)=RHO(J) + X(K - J) * XK
      ENDIF
    1 CONTINUE
      ENDIF
      RETURN
      ENTRY SUMARY(N,AVE,VAR,RHO,M)
      AVE=S1
      VAR= S2/FLOAT(N-1)
      DO 2J = 1, M
    2 RHO(J) = 2.*FLOAT(N-J)*(RHO(J)/FLOAT(N-J)-
    AVE*AVE)/FLOAT(N)**2
C   This is rhodash, equation (3.4) in section 2.3.3
      RETURN
      END
C.................................................................
```

A.2 Subroutine CORVAR

```
      SUBROUTINE CORVAR(N,VAR,RHO,M,STDV,TAU,IWARN)
C   STDV, output is corrected std. devn. on AVE, corrected for
C   correlations, after Daniel et al. Call should have been preceded
C   by calls to CORREL and SUMARY (via CORREL).
C   IWARN returned greater than zero indicates correlations are
C   significant at least up to K = N/4; returned less than zero
C   indicates
C   the procedure was aborted.
C   At call, M= dimension of RHO,
C   At return, M = KK, the highest order of significant correlations (J)
C   VAR is corrected variance s**2, STDV is sm, and TAU the
```

```
C     correln.length
      DIMENSION RHO(*)
      IWARN=0
      FN=FLOAT(N)
      TERM=4.*VAR**2/FN**3
      SUMRJ2 = 0.
      KK = -1
      DO 1 K = 1, M
      SUMRJ2=SUMRJ2+RHO(K)**2
      RMAX2=TERM+2.*SUMRJ2/FN
      IF(RHO(K)**2.LT.RMAX2) THEN
      KK = KK + K
      GOTO 11
      ENDIF
    1 CONTINUE
   11 M=KK
      FKK=FLOAT(KK)
      IF(KK.LT.0.OR.2*KK.GE.N) THEN
      IWARN=-1
      WRITE(*,4)
    4 FORMAT(' Condition never satisfied, something wrong?',/,. ' or
    # correlns significant up to N/2, procedure Abandoned'/)
      STDV=SQRT(VAR/FN)
      RETURN
      ELSE IF(4*KK.GT.N) THEN
      IWARN=1
      ENDIF
      IF(KK.EQ.0) THEN
      STDV=SQRT(VAR/FN)
      TAU=0.
      RETURN
      ELSE
      RKK=0.
      DO 2 K = 1, KK
    2 RKK=RKK+RHO(K)
C     RKK here is RhoJhat
      ENDIF
      STDV=SQRT((VAR+FN*RKK)/(FN - 2.*FKK-1.))
```

```
C    STDV is sm
        RJ = FN * (RKK+2.*FKK*VAR/FN * *2)/(FN - 2.*FKK)
C    RJ is rJ
        SSQ=FN*VAR/(FN - 1.) + FN*RKK/(FN - 2.*FKK)
C    SSQ is s**2
        VAR=SSQ
        TAU= FN * RJ/(2.*SSQ)
        RETURN
        END
C.....................................................................
```

A.3 The Random Number Generator RANECU

Here we give the FORTRAN code for a generator of uniform random numbers on [0,1]. RANECU is a multiplicative linear congruential generator suitable for a 16-bit machine, due to l'Ecuyer[8]. It combines three simple generators, and has a period exceeding 810^{12}. It is constructed for more efficient use by providing for a sequence of such numbers, LEN in total, to be returned in a single call, (following James; Comp. Phys. Comm., **60**, 329 – 344, (1990)). A set of three non-zero integer seeds can be supplied, failing which a default set is employed. If supplied, these three seeds, in order, should lie in the ranges [1,32362], [1,31726] and [1,31656] respectively.

```
        SUBROUTINE RANECU(RVEC,LEN)
C    Portable random number generator for 16 bit computer, after
C    l'Ecuyer, Comm. ACM, 31, 742(1988), based on Fortran version
C    of James in Comput. Phys. Comm., 60, 329 - 344(1990).
C    Generates a sequence of LEN pseudo-random numbers, returned in
C    RVEC.
        DIMENSION RVEC(*)
        SAVE ISEED1,ISEED2,ISEED3
        DATA ISEED1,ISEED2,ISEED3/1234, 5678, 9876/
C    Default values, used if none supplied via an ENTRY call at RECUIN
        DO 100 I = 1,LEN
```

```
      K=ISEED1/206
      ISEED1 = 157 * (ISEED1 - K * 206) - K * 21
      IF(ISEED1.LT.0) ISEED1=ISEED1+32363
      K=ISEED2/217
      ISEED2 = 146 * (ISEED2 - K * 217) - K * 45
      IF(ISEED2.LT.0) ISEED2=ISEED2+31727
      K=ISEED3/222
      ISEED3 = 142 * (ISEED3 - K * 222) - K * 133
      IF(ISEED3.LT.0) ISEED3=ISEED3+31657
      IZ=ISEED1-ISEED2
      IF(IZ.GT.706)IZ = IZ - 32362
      IZ = IZ+ISEED3
      IF(IZ.LT.1)IZ = IZ + 32362
      RVEC(I)=REAL(IZ) * 3.0899E - 5
  100 CONTINUE
      RETURN
      ENTRY RECUIN(IS1,IS2,IS3)
      ISEED1=IS1
      ISEED2=IS2
      ISEED3=IS3
      RETURN
      ENTRY RECUUT(IS1,IS2,IS3)
      IS1=ISEED1
      IS2=ISEED2
      IS3=ISEED3
      RETURN
      END
```

Bibliography

Abramowitz, M. and Stegun I. A. (eds), *Handbook of Mathematical Functions*, (New York: Dover, 1965).

Binder, K. (ed.), *Monte Carlo Methods in Statistical Physics*, 2nd edn, (Berlin: Springer-Verlag, 1986).

Binder, K. (ed.), *The Monte Carlo Method in Condensed Matter Physics*, (New York: Springer-Verlag, 1992).

Bratley, P., B. L. Fox and L. E. Schrage, *A Guide to Simulation*, (New York: Springer, 1983).

Cooper, Necia Grant (ed.), *From Cardinals to Chaos*, (Cambridge: C.U.P., 1989).

Dubewicz, E. J., *The Handbook of Random Number Generation and Testing*, (Columbus, OH.: American Science Press, 1988).

Eadie, W. T., J. Drijard, F. E. James, M. Roos and B. Sadoulet, *Statistical Methods in Experimental Physics*, (Amsterdam: North-Holland, 1970).

Fishman, G. S., *Principles of Discrete Event Simulation*, (New York: Wiley-Interscience, 1978).

Gardiner, C. W., *Handbook of Stochastic Methods for Physics, Chemistry and the Natural Sciences*, 2nd edn, (Berlin: Springer, 1985).

Gould, Harvey and Jan Tobochnik, *An Introduction to Computer Simulation Methods*, (Reading, MA: Addison Wesley, 1988), 2 volumes.

Hammersley, J. M. and D. C. Handscomb, *Monte Carlo Methods*, (London: Methuen, 1964). Apart from the sound mathematical basis for the subject provided the book contains a valuable bibliography on early contributions to the development of Monte Carlo methods.

Jenkins, T. M., W. R. Nelson and A. Rindi (eds), *Monte Carlo Transport of Electrons and Photons*, (New York: Plenum, 1988).

Kalos, M. H. and P. A. Whitlock, *Monte Carlo Methods Vol 1*; *Basics*, (New York: Wiley-Interscience, 1986).

Knuth, D. E., *The Art of Computer Programming*, Vol 2/*Seminumerical Algorithms*, (Menlo Park, CA: Addison-Wesley, 1981).

MacKeown, P. K. and D. J. Newman, *Computational Techniques in Physics*, (Bristol: Adam Hilger, 1987).

Morgan, B. T., *Elements of Simulation*, (London: Chapman & Hall, 1984).

Press, W. H., B. P. Flannery, S. A. Teukolsky and W. T. Vetterling, *Numerical Recipes, The Art of Scientific Computing*, (Cambridge: Cambridge University Press, 1986).

Rubinstein, R. Y., *Simulation and the Monte Carlo Method*, (New York: Wiley, 1981).

Shreider Yu. A. (ed.), *Method of Statistical Testing - Monte Carlo Method*, (Amsterdam: Elsevier, 1964).

Wax N. (ed.), *Selected Papers on Noise and Stochastic Processes*, (New York: Dover, 1954).

Wong, S. S. M., *Computational Methods in Physics and Engineering*, (Englewood Cliffs, NJ: Prentice Hall, 1992).

Index